BITTER HARVEST

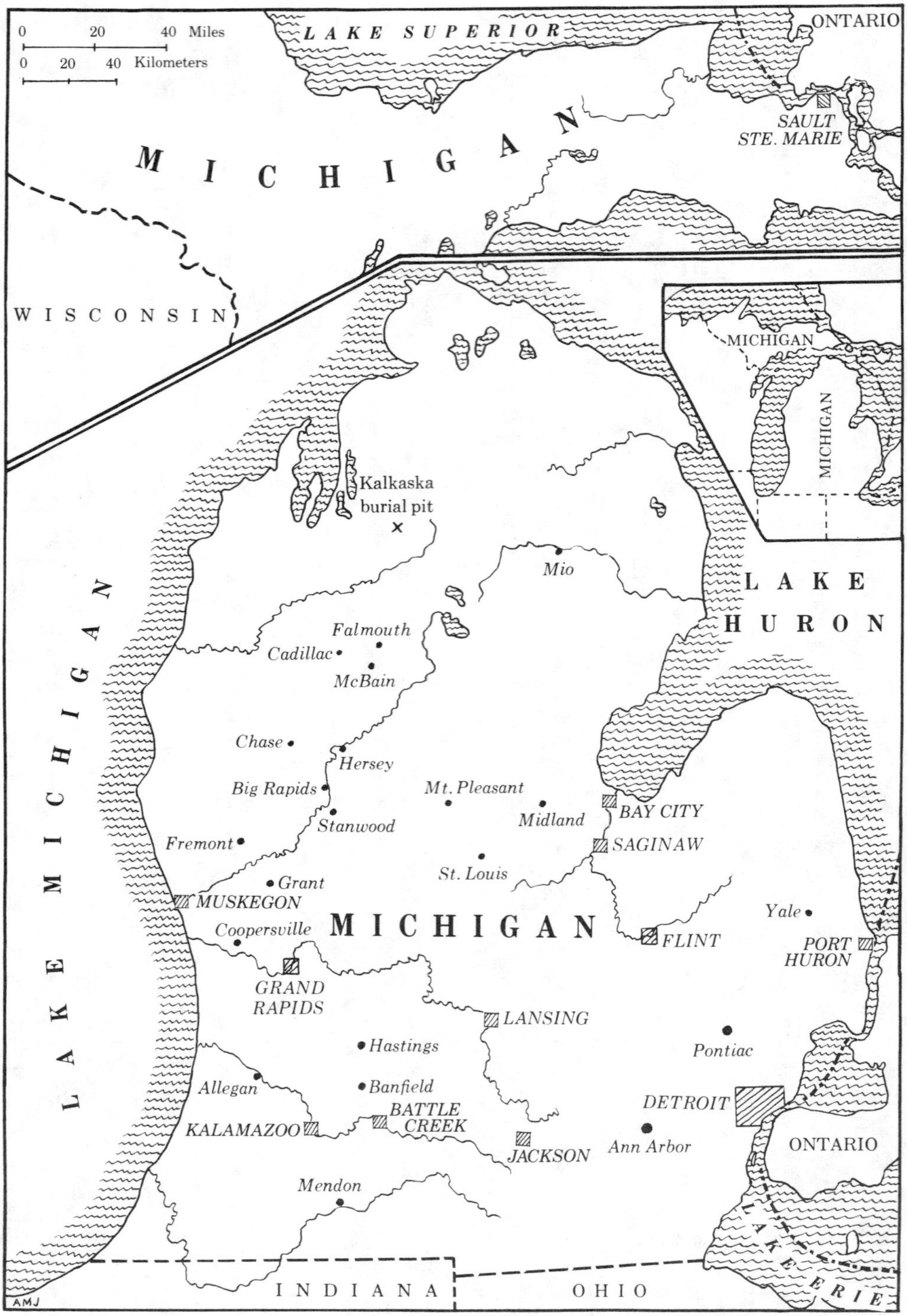
0 20 40 Miles
0 20 40 Kilometers
LAKE SUPERIOR
ONTARIO
SAULT STE. MARIE
MICHIGAN
WISCONSIN
MICHIGAN
MICHIGAN
Kalkaska burial pit
LAKE HURON
Mio
LAKE MICHIGAN
Falmouth
Cadillac
McBain
Chase
Hersey
Big Rapids
Mt. Pleasant
Midland
Stanwood
BAY CITY
SAGINAW
Fremont
St. Louis
Grant
MUSKEGON
Yale
MICHIGAN
Coopersville
FLINT
PORT HURON
GRAND RAPIDS
LANSING
Hastings
Pontiac
Allegan
Banfield
DETROIT
BATTLE CREEK
KALAMAZOO
JACKSON
Ann Arbor
ONTARIO
Mendon
LAKE ERIE
INDIANA
OHIO
AMJ

Joyce Egginton

BITTER HARVEST

SECKER & WARBURG
London

First published in England 1980 by
Martin Secker & Warburg Limited
54 Poland Street, London W1V 3DF

SBN: 436 14150 7

Book deisgn by A. C. Goodman
Typefaces are Video Gael and Foundry Baskerville
Manufactured by The Haddon Craftsmen

Printed in the United States of America

For John and Stephen

May their generation
find a better way

CONTENTS

ACKNOWLEDGMENTS

THE idea for this book grew out of two conversations—one in New York with Irving Selikoff, the other in Lansing with Edith Clark and Garry Zuiderveen. These three gave me my first insight into the Michigan disaster. My gratitude to them is extended to the many farmers, farmers' wives, scientists, doctors, veterinarians, attorneys, politicians, and government officials who gave me the benefit of their experience, many times over. There are too many to be named, although special mention should be made of Frederic Halbert, Alpha Clark, Patricia Miller, Gary Schenk, Paul Greer, Frederick Boncher, Richard Shuster, and Larry Robertson. More than 150 interviews were conducted, most of them lasting for several hours, and some people were seen a number of times as events moved and their perception of the situation deepened. No matter how often I went back to them, these witnesses were always patient and helpful. The warmth and trust shown to me by many farm families is a particularly pleasant memory.

My research would have been incomplete without the thousands of pages of testimony and documents which became available through various public hearings, at federal and state level, and through the suit of one farmer and his wife against the chemical and feed companies. The diligent efforts of a few attorneys and politicians to bring out all the details of the Michigan disaster may be without precedent. They certainly revealed aspects of a chemical contamination which are not usually made public.

My personal thanks go to my editor at W. W. Norton and Company, Starling Lawrence; to my agent, John Cushman; and to the Fund for

Investigative Journalism, for their belief in this project from the beginning and their encouragement along the way. I am also indebted to Southampton College Library of Long Island University for use of the John Steinbeck writers' room. My husband, Carey Longmire, gave me his thoughtful criticism of the manuscript, and, for almost three years, our children tolerated my preoccupation and my absences with an affectionate understanding which will always be remembered.

J.E.

LIST OF ABBREVIATIONS

APHIS	Animal and Plant Health Inspection Service (of the U. S. Department of Agriculture)
DNR	Department of Natural Resources (of Michigan)
EPA	Environmental Protection Agency (U. S. Government)
FDA	Food and Drug Administration (U. S. Government)
MDA	Michigan Department of Agriculture
NIOSH	National Institute for Occupational Safety and Health (of the U. S. Department of Health, Education and Welfare).
USDA	United States Department of Agriculture
WARF	Wisconsin Alumni Research Foundation

Designations of chemicals:

BP-6	(or Firemaster BP-6) Michigan Chemical Corporation's trade name for hexabrominated biphenyl, which is a version of PBB.
FF-1	(or Firemaster FF-1) Michigan Chemical Corporation's trade name for BP-6 after it had been ground to crystals.
PBB	Polybrominated biphenyl
PCB	Polychlorinated biphenyl
ppm	parts per million

PREFACE

The Courageous Few

In the late spring of 1973 a truck driver known as Shorty made a routine delivery from Michigan Chemical Corporation to Farm Bureau Services, which operated the largest agricultural feed plant in Michigan. Shorty's truck carried about a ton of a crumbly whitish substance which was packed in fifty-pound brown paper sacks and described on the inventory as Nutrimaster, a new trade name for magnesium oxide. An innocuous alkaline, magnesium oxide was a recent discovery of the dairy industry; mixed into feed, it helped a cow's digestion and thereby increased her milk supply.

Most of the company's other products were chemicals for industrial use, including a range of fire-retardants which had been developed to serve another expanding market—this one created by the federal government's insistence upon higher standards of inflammability in a variety of manufactured goods. Some of these chemicals were given the trade name of Firemaster.

One version of Firemaster, a crystallized polybrominated biphenyl (PBB), had a close physical resemblance to magnesium oxide. But while Nutrimaster was harmless, Firemaster was highly toxic. Someone at the chemical company confused the two; Shorty took the wrong bags to the feed plant, and no one there noticed the mistake. PBB was mixed into several large batches of cattle feed which was sold to farmers throughout Michigan. The results were devastating. Tens of thousands of farm animals became deathly ill. Milk production fell. Calves died in their barns. Cows aborted. Lambs were born with gross deformities. Chickens developed strange tremors . . . and no one could understand why.

The plague raged for nine months before a farmer with an unusual knowledge of chemistry and a great deal of perseverance tracked down its cause. By then virtually all the nine million people in the state of Michigan had eaten contaminated meat and dairy supplies. Almost nothing was known about the effect of PBB on the human body, except that it accumulated in the tissues and belonged to a particularly toxic and persistent family of chemicals. It was suspected of causing cancer and genetic damage, and there was no known antidote.

When the state authorities realized the extent of the contamination, it was already too widespread for them to handle. There were not enough public funds to destroy all the contaminated animals, indemnify the farmers, seize all the contaminated food, and care for the people whose health had been damaged. Consequently, for many months very little was done in the bureaucratic hope that if the problem was ignored it might disappear. Instead, it spread and worsened. "They tried to sweep it under the rug," said one Michigan legislator, "and they swept so hard it came out the other side."

All the relevant institutions of democracy failed. Rather than working together for the public good they operated, individually and collectively, against it—not out of malice, but simply because this was an event outside their experience and beyond their budgetary means to resolve. The state agencies underestimated the seriousness of the situation and, when they finally acknowledged it, tried to cover their earlier inadequacies. The governor of Michigan took his advice from the state agencies, and more than two years passed before he realized he had been misled. Various departments of the federal government decided that, since the contamination was confined to Michigan, it was not their responsibility. Although an entire state was affected, it could not be designated a federal disaster area, entitled to relief funds, because the disaster was not one of nature's excesses but the result of an industrial accident for which industry should compensate. The aggrieved were left to seek redress through the legal system, where they were outmatched by the superior resources of those corporate giants whose employees—through carelessness and human error—were responsible for poisoning their bodies and their land.

Scientific research did not serve the people as it should have done because, in the early days after the contamination, the only authorities prepared to fund it had a vested interest in demonstrating that PBB did little harm. When farmers turned in desperation to the courts, they found that it was beyond the expertise of the judiciary to make the scientific judgment they sought. The press, which should have been a public watchdog, failed in its function. Not until the results of the contamination were glaringly obvious did it stop accepting the explanations of the bureaucrats and start investigating farmers' complaints. By then it was almost too late to prod government and industry into action.

The contamination of Michigan was probably the most widespread, and the least reported, chemical disaster ever to happen in the western world, but the handling of it was far from unique. The bureaucratic denials, the indifference of industry, and the isolation of the afflicted have had their parallels in the incidents of Minamata, *yusho,* Seveso, and Love Canal. In many such instances the cause of the tragedy has remained virtually hidden, but in Michigan one man's resourcefulness led to its discovery. The stubborn courage of several other people helped to prevent its consequences from being ignored.

These people—mostly farmers, veterinarians, and attorneys—were unlikely to have assumed roles of leadership in normal times. Before PBB they went quietly about their jobs, disinterested in public recognition. Their common qualities were an extraordinary independence and an innate sense of justice which led them to risk their livelihoods, their reputations, and in some cases perhaps, their lives.

Edmund Burke has been quoted as stating, almost two centuries ago, that the only thing necessary for the triumph of evil is for good men to do nothing. In Michigan, a few stood up and spoke out; they are the heroes of this story.

PART ONE

AN UNKNOWN TOXIN

Abraham Lincoln once said that the greatest danger to our country lies not from the outside, because as freedom-loving people we would band together to protect ourselves from invasion. Rather, our greatest threat lies from within. I believe that he was right. We must protect ourselves from those among us who, through ignorance, stupidity, or criminal neglect, would destroy our civilization.

If we allow these environmental disasters to continue, sooner or later we will kill or seriously impair the health of millions of our fellow countrymen.

—THOMAS H. CORBETT, physician
(referring to Michigan's PBB disaster)

1

Prelude to Disaster

THE story has a different beginning for everyone involved in it, and Rick Halbert can remember exactly when it started for him. It was September 20, 1973, when he watched his cows amble into the milking parlor and had an uneasy feeling that something was wrong. His was not one of those positive diagnoses which dairy farmers are used to making when some recognizable condition, charted in the veterinary manuals or common to everyday experience, afflicts their herds. Dairy cows are sensitive animals with complex and delicate digestive systems, easily upset by changes of diet, surroundings, or routine, and their owners—governed by the commercial considerations of milk yield—are ever watchful for symptoms which might affect production. On that September day the Halbert herd produced the usual daily yield of 13,000 pounds of milk. But the cows seemed strangely lethargic and for some reason did not want to eat.

Rick Halbert was worried. He was a conscientious and knowledgeable farmer whose family had farmed in Bedford township, Michigan, for more than a century. He had grown up on his parents' farm near the village of Banfield, twenty miles north of Battle Creek, responsive, like all farm children, to the pattern of seasons, the vagaries of climate, and the breeding cycles of animals. Yet he had been more privileged than most. His father, Frederick P. Halbert (generally known as Ted Halbert), was a successful and respected member of the farming community, active in agricultural organizations and well known throughout the area; he had expected his sons to work on the farm during school vacations—but not to work too hard. When it came to the choice of a career he did not assume that Rick (who had also been named Frederic, but without the final consonant) would follow family tradition. Rick's grandfather, the first Frederick Halbert, had farmed modestly on the margins of Battle Creek. As a young man, Ted started with two hundred acres of his own, which he bought in the early 1940s and expanded upon as neighboring land became available.

Rick's mother, who had reared her children through those struggling years, understood too well the uncertainty and tyranny of a dairy farmer's life, how his entire existence is defined and governed by market conditions and milking times, and she wanted a broader, better future for her sons. Her feeling that agriculture was beneath her dignity influenced Rick. Thus it was that he did not come to farming until 1971, when he was twenty-eight years old, and then by a curious and circuitous route which, at the time, seemed irrelevant. Had he not taken it, however, this story would have been quite different. Indeed, it might never have been pieced together. The mysterious disease which began to afflict the Halbert herd in the fall of 1973 and which spread, with a wide range of seemingly unrelated symptoms, throughout the state of Michigan might have been written off as a veterinary curiosity, never fully diagnosed or understood. And the equally wide range of medical symptoms—neurological, gastric, orthopedic—which later began to appear in people, many of whom had nothing to do with farming, might never have been connected with the cattle scourge but would probably have been ascribed by various doctors to a variety of causes, some physical, some psychosomatic.

Before becoming a dairy farmer Rick Halbert was a chemical engineer. From boyhood he had been fascinated by scientific phenomena such as the chemistry of fire and the composition of electricity. His interest blossomed under a chemistry instructor at Kellogg Community College, Battle Creek, which Rick attended for two years after graduating from high school. There he won an Outstanding Student award and transferred to Michigan State University at East Lansing, where his training as a chemical engineer began in 1965. Two years later he graduated and took a summer job with Eastman Kodak in Rochester, New York. The Vietnam war was then at its height, but he was exempted from military service, Eastman Kodak having attested that he was doing work of significant importance to the national economy. He was relieved about that: "I was not enthusiastic about being drafted."

Later that year, 1967, he returned to Michigan State where he was accepted into "a sort of work and learn program" while studying for his master's degree, which he gained in the summer of 1968 when chemical engineers were in such demand that he was virtually able to choose his own employer. He applied to the corporate headquarters of Dow Chemical in Midland, Michigan, where he worked for more than two years in a program called "corporate special assignment."

In his description, "It meant that I had no permanent job, but that I could take an assignment in planning engineering, or in a technical development center, or working for a business manager. It was a great program whereby a person could test the various waters of the company." There Rick Halbert worked with a variety of modern industrial chemicals and gained an experience which was to prove of enormous consequence to himself and, eventually, to all the nine million people then living in the state of Michigan.

He had married in the meantime and his wife, Sandra, a lovely clear-skinned blonde of Scandinavian parentage, had borne him two children and was expecting a third. They had a pleasant house in Midland on a sizeable lot but as time went on it seemed to Rick distressingly small. He missed the limitless horizons of the farm country and started worrying about the vulnerability of a family man solely dependent upon a salary from a large corporation. He remembered hearing how, during the recession of a decade earlier, his own company had suddenly withdrawn job offers made to a number of college graduates and it bothered him that, well as he was doing at Dow, he might also become a victim of economic cuts or of some internal restaffing over which he had no control.

About this time Ted Halbert was rebuilding a barn, and Rick would return to his old home on weekends to help with the wiring. He felt happy and comfortable in the familiar farm surroundings and, just as he had convinced himself of his vulnerability in industry, so he now reasoned that a farmer was one of the few workers who could control his own destiny. "This is a total fallacy," he said later. "You cannot control the weather. After I made the decision I often thought of the story of the two knights jousting. One of them said, 'Let's give this up and turn to farming,' and the other said, 'No, not farming, it's too risky.' "

Nevertheless, working on the barn, Rick realized that farming was what he really wanted to do, more than chemistry, more than industrial advancement, more than anything. He resigned from Dow, borrowed sixty thousand dollars, and bought 260 acres only four miles from his father's farm. The deal included a Victorian farmhouse which the previous owners had modernized at the sacrifice of its character, a fact which troubled Sandra, who had considerable artistic ability and an innate sense of the fitness of design. Years later she was still regretting the gingerbread molding which had been stripped from the house and thrown away as extraneous, still saving the broken pieces she had salvaged, hoping to use them some day. Meantime she worked hard to invest this house with a new personality. She replaced one of the plain glass window panes with stained glass which she had created herself, and over the livingroom fireplace hung a life-size reproduction of a halbert—a long, medieval weapon, like a combined spear and battle-axe. So the family—which by this time had grown from two to three small daughters, Stephanie, Kristen, and Lisa—had settled, and Rick made the mental reservation that if farming did not work out for him he could always turn to teaching. His father expressed no great surprise at Rick's change of career. His mother was appalled. She felt her son was wasting his education.

In fact, modern dairy farmers need to be experts in a number of disparate fields if they are to survive the intense competition. They must understand animal nutrition: what blends of feeds and additives to use under varying conditions, how to compensate the cattle for deficiencies in locally grown grain, what proteins and minerals to use as supplements—and when to use them. They must be engineers, so that if a valuable tractor breaks down, the harvesting does not have to be held up for a repair. They must be knowledgeable about the breeding and bloodlines of cattle so that by culling and adding animals of the appropriate strain they can build up a herd which has the same qualities of individual excellence and interdependence as a championship baseball team. They must know a great deal about veterinary work, since highly bred dairy cattle are more delicate and vulnerable to disease than old-fashioned pastured cows, and in rural areas there are rarely enough veterinarians to go round. They must be efficient business managers, attuned to world financial markets. It costs hundreds of thousands of dollars to equip a modern farm with machinery alone, and, at a moment when heavy financing of equipment is needed, a sudden increase in interest rates or a fluctuation in the price of farm products could mean disaster. Even on the best-run farms the debts are huge and the profit margins hazardous.

In short, farmers must develop a battery of skills which would have been extraneous in agriculture back in the days when a family could make a decent living with a few acres and an assortment of domestic animals. That older generation learned from the conventional experience of growing up on farms and by developing from their own and others' wisdom an unerring feel for the land. The new generation of dairy farmers must go to agricultural college, essentially to master the complexities of running an efficient milk factory of which every cow, every piece of machinery, and every silo full of feed is a vulnerable component which must not be allowed to fail.

When Rick Halbert was in high school in the early 1960s, the extent of this developing mechanization was barely recognized by farm families. Raising cattle did indeed seem to many the waste of a higher education, just as academics seemed superior to agriculture. But when Rick eventually turned to farming his qualifications as a chemical engineer were not at all irrelevant. He was just a little, only a very little, ahead of the times. And when mysterious symptoms appeared in his cattle, his scientific knowledge was crucial.

The same revolution which transformed dairy farming into an industry has made it almost impossible for a young man to start on his own. The necessary cash investment is much too great. If he were to start from scratch by renting both the land and the machinery, a farmer could be destroyed by the first crop failure. So the pattern of father and son partnerships has grown. The older man provides the equity, the younger man the zeal, and jointly they enlarge and modernize the original family farm. It is not always a satisfactory arrangement. Some fathers treat their sons like hired hands and insist that the proven methods cannot be bettered. Some sons are unwilling to temper their knowledge of the new technology to their parents' experience. But in the Halberts' case it was a happy combination. By this time Rick's father had become a prosperous farmer, well-organized and well-entrenched; he served on the boards of directors of the Michigan Milk Producers' Association and of the Battle Creek Farm Bureau and was happy to hand over most of the farm management to his sons. He was in a position to advance the money for Rick to buy, at a modest price, half his cattle and machinery, and thus to become a junior partner in the family enterprise. Later Rick's brother Mark joined the business. Their father's original farm had by now been expanded to more than one thousand acres—some of it wooded and some swamp, but a great deal of it fertile and cultivated cropland—and the Halberts had become the largest milk producers in the Battle Creek area.

When Rick went into business with his father there were about 150 milking cows on the farm, at a time when the average Michigan dairy herd had only 50. Together they built up the herd so that by the fall of 1973 they had a total of 750 animals of which 300 were milking cows. These were healthy, well cared for animals, scientifically managed and producing well. But like all other dairy farmers in the business, the Halberts were constantly looking for ways to increase their milk yield, and the most obvious and accepted way was to bolster their own home-grown cattle feed with a balance of nutrients appropriate to the herd. This process has to be evaluated often, depending upon seasonal changes in diet and weather, the condition of the cows, and the quality of the soil.

For years the Halberts had bought most of their supplements from Farm Bureau Services Inc., a statewide farmers' cooperative and an affiliate of Michigan Farm Bureau, which has a central mixing plant just outside Battle Creek. There, tons of feed were mixed for all kinds and conditions of animals and trucked out to farms and to farm cooperatives in different parts of the state. Being relatively close to the Battle Creek plant and one of its best customers, Rick Halbert was able to deal directly and personally with the staff, and in the autumn of 1972 he began urging them to change the formula of one of the plant's best quality feed supplements for milking cows, which was generally known by its tag number, Dairy Ration 402.

The 402 mixture was a high-moisture corn and silage balancer with 24 percent protein, an enriched formula which came in greenish pellets about half an inch long and a quarter of an inch in diameter. The pellets would be spread upon the grain like jam on bread (a system which farmers call top dressing), and normally the cows would gobble them up greedily. This was, therefore, a sure way to get nutrients into their systems.

Because the local soil was low in magnesium, without which cows are prone to a potentially fatal disease known as wheat pasture fever, Rick Halbert urged Farm Bureau to increase the magnesium oxide content of 402. Another good reason for doing so was that magnesium oxide works as an antacid in a cow's digestive system, compensating for the natural acidity in high-moisture corn, which many modern farmers find more convenient to grow and store than the old-fashioned, less acid, variety. This antacid also causes a cow to increase her output of butterfat, and a high milk yield with a high butterfat content means more money for the farmer. This relationship between magnesium oxide intake and butterfat output was a fairly recent finding in agricultural research, and more and more dairy farmers were enriching their feeding programs to take advantage of it.

At first the Halberts bought hundred-pound sacks of magnesium oxide directly from Farm Bureau Services and dribbled it on top of the cattle feed. They soon felt this method to be crude and unscientific so Rick asked the Battle Creek plant to increase the magnesium oxide content of 402 so there would be eight pounds of it in every ton of feed. This "improved" formula was first made up for the Halberts on October 12, 1972, according to the plant's records, with a second delivery going to the farm late that winter. By spring 1973 the cows' diet had changed, as it always does at that time of year, to homegrown alfalfa roughage —the natural "green chop" of the summer months, rich in magnesium and needing no additive. At the end of August 1973, when the alfalfa supply was exhausted, the cows were put back on 402, which Rick re-ordered from Battle Creek. He still thought that the magnesium-enriched blend was being mixed for him alone. In fact, Farm Bureau Services had accepted his suggestion for increased magnesium oxide as an improvement over their earlier mixture, and in the spring of 1973 adopted the Halbert formula for all customers. The list of ingredients on the tag which came with the August delivery of 402 to the Halbert farm read:

Soybean Meal, Ground Corn, Dried Apple Pomace, Cane Molasses, Wheat Middlings, Dehydrated Alfalfa Meal, Calcium Carbonate, DiCalcium Phosphate, Salt, Vitamin A Palmitate, Calcium Iodate, Manganese Sulfate, Magnesium Oxide, Iron Carbonate, Vitamin D-3 Supplement, Vitamin E Supplement, and Bentonite.

Between the end of August and the first week in October 1973 the Halberts bought sixty-five tons of this feed, with the Battle Creek plant making deliveries to the farm twice a week. The first delivery came on September 4. No one suspected it until Rick Halbert made the discovery himself some eight months later, but in these latest batches of the "improved" 402 the crucial ingredient was not magnesium oxide. It was a highly toxic industrial chemical, polybrominated biphenyl.

Large quantities of this poisoned feed went to farmers and farm suppliers in different parts of the state; later, there was no precise way of telling where or how much. After the extent of the catastrophe became known, however, Rick Halbert was moved to remark that it was as if he had unwittingly been responsible for opening a Pandora's box, letting loose a thin dust of deadly toxin that was to settle all over Michigan. But neither the cause nor the effect was remotely foreseen that morning of September 20 when Rick Halbert, noting that his cows were losing their appetites, began to worry.

The Farm Bureau Services mixing plant near the village of Climax, five miles south of Battle Creek, stands against the skyline just above Interstate Highway 94 roughly halfway between Detroit and Chicago. Surrounded by cornfields and approached by a country lane, the plant's main building consists of a long blue warehouse topped by a big cream-colored square tower with the words "Farm Bureau" painted in large brown letters. A single-track railway siding curves across the meadows to the plant. There is an isolated and peaceful atmosphere about the place, despite the hustle as workers in their yellow hard hats go about the business of running the huge machinery that blends all kinds of animal feed. Feed and ingredients for feed are stacked in large brown sacks, one upon another, forming a grid of parallel walls inside the warehouse.

In 1973 Farm Bureau Services bought its ingredients from a variety of sources, one of them being Michigan Chemical Corporation in the small town of St. Louis, Michigan, about sixty miles to the north. This company was Farm Bureau's regular supplier of magnesium oxide and salt, both important items for dairy farmers. The magnesium oxide was mixed into a variety of animal feeds as well as being supplied undiluted to farmers and farm cooperatives who preferred to do their own mixing. Some of the salt went into feed mixtures; much of it was sold loose. Most dairy farmers keep a salt box in the exercise lot for cows to lick, and this has a double benefit. It not only supplies essential iodine but it also induces thirst, causing the cows to drink more water and thus increase their milk production.

Paul Mullineaux, production manager of Farm Bureau Services' Battle Creek plant, did not know whether Michigan Chemical Corporation made anything other than feed ingredients nor did it occur to him to ask. In fact, the magnesium oxide and salt were mere by-products of Michigan Chemical Corporation's St. Louis plant—and this sprawling, shabby factory was not the small-town industry which it seemed to the casual observer, but a subsidiary of Velsicol Chemical Corporation, which in turn was a subsidiary of Northwest Industries, one of the largest industrial giants in the United States. Its main output was a range of fire retardant chemicals manufactured for industrial use, all of them highly toxic.

A brochure put out by the company at this time described Michigan Chemical as "the masters of flame retardant chemistry." This was not an idle claim. The ready supplies of bromine in Michigan's huge underground salt deposits formed the basis of a new breed of chemicals, of which polybrominated biphenyl (PBB) was one of the newest. A close —though much more toxic—relative of polychlorinated biphenyl (PCB), it was developed by Michigan Chemical as a fire retardant which could be incorporated in hard plastics likely to be subject to heat, such as the dashboards of automobiles and the casings for telephones, hair dryers, radios, and television sets. For market appeal, the product was named Firemaster.

The company was fond of the suffix "master" and used it often. There was a substance for controlling dust on highways named Dustmaster. There was a form of magnesite, used to help remove deposits in high-temperature boilers, known as Boilermaster. And there was magnesium oxide, of the kind that served as a nutrient in cattle feed, called Nutrimaster. Both the Nutrimaster and the Firemaster were packaged in fifty-pound bags of heavy brown paper with the name of the product printed in large letters. The bags were identical in size and shape, but the Nutrimaster bag had a broad blue stripe, running diagonally from corner to corner, and some Firemaster bags had a similar stripe of red. The PBB was marketed in two forms: a thick, sticky, yellowish-brown substance which looked like chunks of toffee, or a crumbly, grayish-white powdery substance which tended to clump together on exposure to dampness—and which looked remarkably like magnesium oxide.

Except for the color coding on the bags, Nutrimaster and the powdery form of Firemaster could easily have been mistaken one for the other—which is exactly what happened when, during a temporary nationwide paper shortage in the winter of 1972–73, Michigan Chemical

Corporation ran out of preprinted bags and made do by simply hand-stencilling the trade names in black. Sometimes the lettering was smudged. Even when it wasn't, the names Nutrimaster and Firemaster were so similar that, at a quick glance, one might be read for the other. Yet it did not seem to occur to anyone at the St. Louis plant that here were all the ingredients of a potential mix-up, and that the precautionary measure of labelling one chemical "for animal consumption" and the other "poison" would be wise. They reasoned that there could be no confusion because PBB was produced in a different part of the plant from magnesium oxide and dispatched to customers from a different loading dock.

Months, even years, later, after many searching inquiries into the causes of the chemical disaster that poisoned Michigan, it was—and will remain—impossible to document how many bags of Firemaster were delivered to the Battle Creek feed plant in place of Nutrimaster, by whom, and when. All that could be pieced together was that at some time during the spring of 1973 somebody at Michigan Chemical Corporation loaded a quantity of fifty-pound bags of Firemaster on to a truck, thinking it to be Nutrimaster, and dispatched it to Farm Bureau Services at Battle Creek, where somebody unloaded it, failed to notice anything wrong, and stacked it with the magnesium oxide. The rest was inevitable.

2

A Mountain of Trouble

IF there was anything fortuitous about the ensuing disaster, it was the blind act of fate which caused Rick Halbert to be one of its earliest victims. Of all the dairy farmers in Michigan, he was probably the only one with a refined training in industrial chemistry and recent, firsthand experience with man-made toxins. He also had a fine mind, the limitless energy of youth, and the tenacity of a skilled researcher who will neither grasp at any possibility nor be fobbed off with an explanation that cannot be proved beyond all doubt. In spite of all the technological training of the new agriculture, a man like Rick Halbert in the farming business was still unique. He didn't even look like a farmer. He was slim and boyish with a quiet voice and studious manner which made him seem more fitted to the atmosphere of a university library than to the manure smells of a dairy barn. Even his adversaries (and there were to be many as time went on) tempered their scorn with admiration; some of them said he reminded them of a schoolboy determined to know all the answers.

When the cows first seemed sick to him, he thought of all the conventional veterinary explanations. Along with the Dairy Ration 402 the Halberts had been feeding their cows some high-moisture corn which they had bought from a neighbor. It had been a poor year for corn in Michigan; because of cold, wet weather some of it had molded before harvesting. Perhaps this was upsetting the digestive systems of the Halbert herd. The listlessness of the milking cows, however, plus their dramatic and continued drop in milk production presaged something much more serious. The animals were barely eating, almost all of them had stopped chewing their cuds (a symptom of serious trouble in a ruminant), and it seemed to Rick "as if their mammary tissue was drying up before our eyes." Up to September 21 the herd was consistently yielding 13,000 pounds of milk a day; by October 8 it had dropped to 8,000 pounds. This was a desperate situation for a dairy farmer. It meant the difference between cows that were providing income and cows that not only wiped out the farm's profit but were a drain upon capital.

In the last days of September, Rick made several anxious telephone calls. He talked to Paul Mullineaux, production manager of the Battle Creek feed plant, to Anthony Gruscynski, the local manager for Farm Bureau Services at Battle Creek, and to Dr. James McKean, Farm Bureau veterinarian. He told them all that he believed his cattle were suffering from a feed problem, and he asked a lot of searching questions. He immediately suspected the magnesium oxide content of 402, partly because it was his impression that this was his first batch of Farm Bureau feed which was made up to the "improved" formula. He asked if someone at the Battle Creek plant could have made an error in reading the chemical formula, mistaking manganese (Mn) or molybdenum (Mo) for magnesium (Mg). He remembered hearing of another farmer in the county who had once received a high concentrate of manganese in the feed he bought "and it didn't do the cows much good." Rick Halbert's question was tantalizingly close to uncovering the truth, if only someone had stopped to do some investigating. But the answer came, No, there could not possibly have been such a confusion.

Rick pursued other possibilities with the Farm Bureau staff. It seemed to him that some of the symptoms in his cows resembled those of the mysterious "X-disease"—or bovine hyperkeratosis, as it was later called—which plagued cattle in many areas of the United States between 1949 and 1952. This disease, rarely heard of any more, suddenly appeared and became widespread until researchers discovered its cause: a slow poisoning of the animals by chlorinated naphthalenes. At that time chlorinated naphthalenes were used in certain lubricating oils, which in turn were used on farm machinery. Some cattle developed a taste for this grease and enjoyed licking it from tractors and the hinges of barn doors. Others were poisoned less directly because the oil was used on mixing machines which made their feed pellets. All this was history by 1973, by which time it was unlikely that any of Rick Halbert's generation of farmers had ever seen a case of X-disease. But, searching through his veterinary manual, Rick was impressed by a similarity between the description he read and the signs he saw in his cows—a similarity which was to become even more startling as months went by and the toxic symptoms in his herd progressed, producing the mangey appearance, matted hair, thickened skin, diarrhea, emaciation, and high calf loss which had been the classic manifestations of X-disease.

He asked at least one Farm Bureau official whether there could be chlorinated naphthalenes in the lubricants used on the pellet machines. No, came the answer, all the oil used at the Battle Creek plant was approved as non-toxic by the Federal Department of Agriculture. Again, although the answer was right, the question came poignantly close to uncovering the facts. It was to take Rick Halbert another seven months before his research would lead to the discovery of polybrominated biphenyl in his animal feed, and almost another two years after that before another scientist* would find that, because it is almost impossible to produce a *pure* man-made chemical, this PBB was in turn contaminated by the brominated naphthalenes which were by-products of its manufacture. Brominated and chlorinated naphthalenes are closely related to one another, and their toxic effects are most probably similar. So, far into the future, the enigma would remain. Was it the PBB in its entirety which poisoned the Halbert herd, and later hundreds and thousands of animals throughout Michigan? Or was the primary culprit some small component of PBB, such as naphthalene? Was it several components, interacting with one another, or, interacting with other chemicals which cattle—and people—inevitably ingested from other sources, such as polychlorinated biphenyl (PCB), the widely used industrial chemical which by this time was almost everywhere, from the succulent flesh of the thick whitefish in Michigan's lakes to the plastic sealants on silos in which cattle feed was stored on many Michigan farms?

Some of the questions may never be answered, but in his tenacious research Rick Halbert was to settle some and raise others. He would chase down blind alleys, turn off at tangents, come within a hair's breadth of the answer and miss it, pile frustration on frustration, until by digging and sifting and backtracking and eliminating he would eventually arrive at PBB. Such is the nature of most scientific research, but usually it is pursued in laboratories, systematically, unhurriedly, and by men who have the security of a monthly pay check. Usually these men have some idea of what they are looking for. Rick Halbert had none of these advantages. He had no hard clues as to what caused his cattle to sicken (although he suspected the feed), no research equipment, no encouragement from any but the members of his family, who shared, increasingly, his sense of desperation. Unless he could find an answer soon, the Halberts' possibility of financial ruin was very real.

On October 2, 1973, he called in his veterinarian, Dr. Ted Jackson. He told Jackson that the milk production was down, the cows had runny

* Dr. Patrick O'Keefe of the Department of Chemistry, Harvard University, who isolated brominated naphthalene from polybrominated biphenyl in February 1976.

eyes and they were not chewing their cuds. Jackson also noticed that the udders of cows which had recently calved were beginning to shrink, and he was concerned because their obvious distaste for the feed had already led to weight loss. The two men discussed the possibility of a contagious disease, although Jackson leaned to the view that there was some kind of mold problem from the corn. It was an understandable, even obvious, conclusion in view of the heavy rains during the previous year's harvest. Most of the stored corn in Michigan at this time was damp and unappetizing. Moreover, the Halberts had been putting urea on their corn sillage to increase the protein content. Rick wondered if this could have been contaminated by heavy metals. Urea, if introduced too quickly into cattle feed, can produce digestive disturbances, even toxicity. It even passed through Rick's mind that a disgruntled employee of the farm, recently fired, may have put something disagreeable in the water tanks out of spite, or, that a neighbor who resented the Halberts' farming success may somehow have poisoned the feed. "I had to be paranoiac," he said later. But none of the suggested explanations really fitted the symptoms, and Jackson was as baffled as he.

Like most veterinarians in the state, Ted Jackson had graduated from Michigan State University, which was founded in 1857 as the nation's first agricultural land grant college and which had developed one of the finest veterinary departments in the country. Traditionally, if Michigan veterinarians run across a problem in their practice, they are encouraged to take advantage of the expertise of their alma mater. They may also seek help from specialists at the Michigan Department of Agriculture, which has a diagnostic laboratory on the university's campus at East Lansing. This close link between the two state-financed bodies had always been an advantage to farmers and their practitioners, making available to them the enlarging fund of knowledge and research facilities which had helped to build up Michigan agriculture. The MDA's diagnostic laboratory at the university was the obvious place for Ted Jackson to take the Halberts' problems. Many months later, as many more of the state's veterinarians sought similar help for the owners of contaminated dairy farms, their confidence in the impartiality of the Michigan Department of Agriculture and even of the state university would be eroded; as it turned out, in such an unprecedented crisis the interests of the institutions conflicted with those of the farmers and veterinarians they were meant to serve.

Ted Jackson, however, received the prompt, unprejudiced help he expected. Two of MDA's veterinary diagnosticians—Dr. Donald Grover and Dr. Frank Carter—visited the Halbert farm within forty-eight hours. They found cows which were obviously sick but not with a disease they could classify. Some of the symptoms indicated a viral infection, yet the cows did not have the elevated body temperatures normally associated with a virus. Nevertheless, the two men took routine eye and nasal swabs to test for the upper respiratory infection known to dairy farmers as IBR (infectious bovine rhinotrachitis)—a test which would also reveal familiar types of bovine influenza or viral diarrhea. Dr. Grover was suspicious of the high-moisture corn, and took a sample of it back to his laboratory.

As soon as they learned that the swab tests were negative, Ted Jackson and Rick Halbert tried another expert on the vast East Lansing campus. They sought the advice of Dr. Donald Hillman, dairy nutrition expert in the university's Department of Dairy Sciences, who also visited the farm and inspected the herd. His suggestion, with which Jackson concurred, was that since this seemed to be a feed problem of unknown origin, every single item in the cows' diet should be changed. This was not easy. There was no difficulty in taking away the Dairy Ration 402, which was an addition to the basic grain, but replacing both the high-moisture corn and the corn silage which the animals had been getting would be harder, as there were few alternatives in October. However, the Halberts scrounged all the baled hay and cut all the fresh alfalfa they could find in their fields, managing to gather just enough to last until the winter feed period began. They also removed all the 402 pellets from the feed bin. By now there were about five tons of pellets left. Rick said, "We dumped them on the floor of my Dad's barn." (This was at some distance from the dairy operation.) "For a long time they stayed there, like a small green mountain." Several times he was tempted to throw them all out, but Jackson cautioned him to keep them as evidence.

There was still no proof that the pellets had caused the mysterious cattle disease. The cows had virtually stopped eating everything, silage included. Yet on a new diet of the hay and alfalfa, with molasses blocks hung in their feed stalls, their appetites slowly improved. But their milk production continued to be devastatingly low. In farming language, the cows were downright unthrifty.

The laboratory tests failed to find any of the common contaminants in the high-moisture corn, including mold toxins. Dr. Grover then started analyzing the 402 feed—testing specifically, at Rick's suggestion, for chlorinated naphthalenes, cadmium, pentachlorophenol (a wood preservative, sometimes used in farm buildings, which can poison animals), and heavy metals, such as lead, arsenic, and mercury. All the tests were negative. At their own expense, the Halberts had a laboratory in nearby Kalamazoo test the feed for excessive urea and nitrates, but both were found to be in acceptable amounts.

There was a limit to how much of this kind of work could be done. Laboratories are not research institutions. They do not have the same technical capabilities. One could not give a laboratory a batch of cattle feed with a broad directive to find out whether it contained a foreign substance, and if so, what. This would be a job for skilled scientific researchers with unlimited time and money. It might mean analyzing the feed hundreds of thousands of times over, searching, one by one, for every toxin on record, always in the knowledge that if no known toxin was found, the feed might still be contaminated by some substance or combination of substances which had never been recognized before. It would be more difficult than looking for a needle in a haystack. It would be like taking a haystack apart in a search for something which one didn't even know was a needle.

Rick perused every chemistry textbook and veterinary manual in his library, suggesting, whenever he came upon a clue, some new test which might be made, some detail which should be checked. At one point his research led him to suspect the soybean content of 402, having read that one method of extracting the oil from the beans used trichloroethylene and that trichloroethylene contained a toxic substance which could poison cows. It seemed a possible explanation, in view of the current high price of soybeans. Had Farm Bureau Services switched to a different soybean supplier in order to save a little money—a supplier who, unknown to them, used this potentially poisonous chemical? Assured that this couldn't have happened, Rick continued to pester Farm Bureau Services' officials with his earlier questions about the possibility of a confusion over the magnesium oxide, or of the presence of chlorinated naphthalenes. The answers were always the same, firmly insisting there could not have been a mistake, that every precaution had been taken. It did not yet occur to him to ask whether any other farmers had similar complaints. Rick Halbert had had Dairy Ration 402 blended to his own specifications, and he did not know that it was now being widely distributed. If these green pellets had caused his cattle to sicken, and he still had no proof that they had, then he had every reason to believe that his herd alone would be affected. One October day, in desperation, Rick stuffed a handful of 402 into his own mouth in the hope that he could discern why his cows found it so unpalatable. So far as he could tell, it tasted like normal cattle feed.

His observation of a number of young heifers on the farm added to his conviction that the silage had not been responsible for the symptoms in his milking herd. Before the mysterious disease struck, these heifers had been eating the same silage as the cows, but not the high-moisture corn or the 402. They still seemed healthy. So he reintroduced the silage to the cows and, although some were still losing weight and all of them looked ill, they seemed content with it and there were signs that their conditions were stabilizing.

Rick's brother Mark, who had been looking after the calves on the farm, suggested a follow-up experiment. He felt they should isolate a group of young calves, feed them nothing but 402 pellets, and observe the results. It seemed a good idea, so Rick and Mark took a dozen three-month-old heifers and put them together in a bullpen. Offered the pellets, the calves almost refused to eat. After a week of this the brothers had no choice but to let the animals starve or boost their feed with silage, which they did. The calves cautiously nibbled the silage while the green pellets lay untouched in the manger. This seemed to be an inconclusive end to the experiment, and being much more concerned about the unproductive cows, Rick pushed it to the back of his mind.

During October one of the milking cows died. The death was unusual in that it was not associated with calving, so Ted Jackson did a post-mortem. He found necrosis of the liver and sent a sample of this tissue to the university laboratory for analysis. The laboratory confirmed his diagnosis: acute necrotic lesions of the liver, with no apparent explanation. It was baffling.

About the end of October the cows were put on their winter feed of oats and soybean meal. They seemed to feel somewhat better but their milk production was still very low. Other curious symptoms, related to the reproductive process, were beginning to appear. The cows seemed to come into heat with much greater frequency than the calendar showed they should, climbing on one another's backs in the homosexual way that cows do when the urge is upon them and there is no bull around. It is a common enough sight on dairy farms in this age of artificial insemination, where a cow's existence is one of repeated pregnancies in surroundings as sexually restrictive as a convent. Some of the cows which seemed to be in heat had been recently inseminated and, under normal circumstances, ought to have been pregnant. Rick Halbert and Ted Jackson wondered whether they had suffered fetal re-absorbtions, so they re-bred a number of them, this time with more success.

Early in November Rick noticed another strange thing. All the rodents on the farm had disappeared. Wherever there is grain, rats breed with such tenacity that it is almost impossible to get rid of them, no matter how hard a farmer tries. Rick asked the various farm employees if they had seen any rodents lately, and all agreed they had not. However, one man remembered that some distance from the grain supply there was a nest of rats in the crawl space under the farmhouse kitchen. This could be reached from inside the house by a small trap-door in the kitchen floor. Rick gathered between ten and fifteen pounds of the pellets from his father's barn, opened the trap-door and dumped them into the darkness below. He checked the area a week later and there was no sign of a rat.

"It was as if the Pied Piper had come," he said.

All these things were strong indications of some toxin in the 402 feed, but still there was no proof. And proof was essential for the Halberts to have any hope of recovering their financial losses from Farm Bureau Services. The losses in milk production plus the cost of keeping unthrifty cattle alive were already reaching the point where—as Rick in his desperation told Dr. McKean of Farm Bureau Services some weeks later—it was like drawing eight hundred dollars a day out of the business's bank account and throwing the money away. In order to succeed with an insurance claim he had to demonstrate that it was Farm Bureau Services' feed, not any fault of farm management, which caused the mysterious cattle disease. Throughout his research this was Rick's motivation. He had no reason to suspect that the plague had spread beyond his herd. He was concerned only with restoring this herd to health and the farm to financial stability.

When in November he was due for another feed delivery to bolster the oats and soybean meal, Rick told Farm Bureau Services that he wanted no more of the 402. Instead he ordered a cheaper mixture, formula 412, which contained less protein and was more widely used by local farmers. Rick reasoned that if there was anything wrong with the 412 he would have heard about it from his neighbors. What neither he nor the people at Farm Bureau Services realized until much later was that almost all feeds produced at Battle Creek plant for many months were certain to be contaminated since all the ingredients were churned together in the mixer into which PBB had been inadvertently tipped.

This machine operated like a cement mixer with a huge revolving drum which amalgamated feed in three-ton batches. At the end of every mixing process a few pounds of feed would inevitably be left in the bottom of the drum where it would stay until it was mixed in with the next load. The drum was thoroughly emptied only on those rare occasions when the preceding mixture contained antibiotics. Otherwise, even though the next load of feed might be produced to a formula different from the last, the residue was considered to be of little consequence—as indeed it was until the residue contained PBB. After that, whatever came out of the machine was ineluctably poisoned, every load a little less than the last, but with a poison so pervasive that probably no feed mixed at the Battle Creek plant for many months after Rick Halbert received his powerful dose was safe for animal consumption. Yet from the late summer of 1973—and perhaps earlier—until the spring of 1974, contaminated feed, tons and tons of it, was sent out from the Battle Creek plant to farmers and farm cooperatives throughout Michigan.

Early in December the calves in Mark Halbert's experiment began to die. It had been about six weeks since they were first offered the green pellets, and in the last days of their lives they refused all food. Rick observed in a tone of horror that "it was as if someone had sealed off their mouths." In their lifetimes of farm experience neither he nor

Mark had seen animals react in this fashion. Nor had Ted Jackson, who arranged for two of the bodies to be sent to the diagnostic laboratory at Michigan State University for thorough examination.

The autopsy report was inconclusive and—to Rick Halbert—infuriating. It stated that the calves died of malnutrition. "This did not sit well with me," he said. "Since they refused to eat, obviously they were under-nourished. I thought the laboratory report was an insult to me because it seemed to presume that I didn't know what I was doing." What Rick wanted to know was *why* the calves had not eaten. He suspected poisoning because they had ground their teeth a lot, a sign of severe abdominal pain and, often, of heavy metal toxicity.

For days he pestered Donald Grover to address himself to this question. Rick suggested that his farm provide Michigan Department of Agriculture with some calves and a quantity of 402 pellets so that the experiment could be repeated under controlled laboratory conditions. Grover replied that he did not have the facilities to house large animals or the capability to do this kind of testing. "This is a regulatory agency, not a research organization," he explained. Rick went back to Michigan State University's Department of Dairy Sciences and repeated the request, but there too he was told it would not be feasible. "I was told at the university that, as a taxpayer supported institution, MSU had no business getting involved in a squabble between a farmer and a feed company. At least that is my perception of what was said."

After that discouragement he tried a veterinarian in Indiana whose name he had heard in connection with another experiment on poisoned feed. This man, apparently fearful that he might become involved in a legal action, also refused. All this made Rick the more determined to get his research done, so back he went to Grover, who finally agreed to run the feed experiment on some mice, using the facilities of Michigan Department of Agriculture's Geagley Laboratory in East Lansing.

Grover took ten mice, feeding to five of them nothing but the ground-up 402 and to the other five, normal grain. Those offered the 402 ate cautiously and without appetite, perceptibly losing weight. Within two weeks all five were dead. The others continued to flourish. Grover was convinced that whatever had been concluded about the calves, the first group of mice had not died of starvation. Although they clearly disliked the feed, they continued to eat sufficient to maintain bodily function. Grover repeated the experiment on another group of mice, this time adding proteins to the 402 in case the problem was a protein deficiency in the feed. But these mice died as quickly as the last. Grover was troubled and said he would like to have pursued the research. "We tried to do everything that Halbert requested," he said. "But we had exhausted all that we were capable of doing in our laboratory."

Rick, meantime, had been making almost daily telephone calls to Geagley Laboratory to inquire after the health of the mice. He had no idea, nor was he given any hint of it in those telephone conversations, that Grover was also investigating a complaint from another dairy farmer, Peter Crum of Coopersville, whose cows' milk production had been falling drastically for months. He, too, had used Farm Bureau's Dairy Ration 402. It would have been a breach of professional confidence for Grover to tell Rick of Crum's problems, despite the fact that Grover now had the strong suspicion that something was very wrong with at least two batches of 402. But if a sophisticated analysis was to be made of this feed it would need equipment and expertise unavailable to him. So Grover went through the routine of channeling a request through the Federal Department of Agriculture's office in Lansing for some of the Halberts' feed to be analyzed at the department's research station in Ames, Iowa.

All this would take time and was of no immediate comfort to Rick. His thoughts were concentrated on the single issue of proving that the feed was poisoned so that Farm Bureau Services could be held liable. By now it was almost the end of December 1973 and the Halberts' milk production losses had mounted to forty thousand dollars. The death of the mice was the best news Rick had heard for weeks. When Grover told him that the first group died, "my spirits shot up," Rick said. "We now had a lot of circumstantial evidence and some direct evidence." He also noted with interest that the mice had been affected in a manner similar to the cows, suffering weight loss which was disproportionately large compared with their loss in appetite. It was as though it—whatever "it" was—was short-circuiting the energy process, Rick remarked.

On the evening of December 28 he telephoned Donald R. Armstrong, executive vice president of Michigan's Farm Bureau Services, at his home. He told Armstrong about his experiences with his herd, the various tests that had taken place, the deaths of the calves, and this latest experiment with mice.

"From now on this is no longer just our problem," Rick told him. "It's your problem too."

On January 9, 1974, Donald Armstrong met with Rick and Ted Halbert in the offices of Farm Bureau's Battle Creek plant. Dr. James McKean and the local Farm Bureau manager, Anthony Gruscynski, were also at the conference. Armstrong did most of the talking. He was concerned about the Halberts' complaint, not only because they were reputable farmers and good customers, but also because he had become aware of similar charges of bad feed from at least one other farmer, Jerome Petroshus of Allegan. Word of Peter Crum's problems had apparently not reached Armstrong's level. Armstrong did not tell the

Halberts about Petroshus. The Halberts were also unaware that as a result of these complaints, all involving Dairy Ration 402, Farm Bureau Services had ceased production of this feed in December.

Armstrong did his best to reassure Ted and Rick Halbert. He had known them both for some time and it was a friendly conversation. He told them that his organization would do everything possible to discover the cause of their cattle's disease. He promised additional tests, this time at the WARF (Wisconsin Alumni Research Foundation) Laboratory in Madison, Wisconsin, and he undertook to finance a fairly elaborate experiment on calves at the Agway feed research station near Syracuse, New York. Both organizations had impeccable reputations in their fields and were far enough from Michigan to be beyond suspicion of local prejudice. It was Armstrong's idea that Agway would not only try to find out whether the feed made the calves ill, but also, if and when they sickened, what treatment could best restore them to health. He was probably suspecting, as Grover had done, some vitamin deficiency in the feed—something which could be remedied. It was agreed that the Halberts would supply one thousand pounds of the green pellets from their 402-pile for the experiment, that there should be a free exchange of information between them and Farm Bureau Services on the results of all tests, and that they would all meet again a few weeks hence to report progress.

One other item was discussed at this session: an offer from Farm Bureau Services to buy back from the Halberts their unused 402 feed. Having decided to hold on to it as evidence and for experiments, Rick and Ted refused. Farm Bureau Services did not pressure them to change their minds nor warn them not to give any more of the pellets to their animals, even though the company was planning (and perhaps had already begun) a duplication of the mouse experiment, this time using 402 feed from both the Halberts and Jerome Petroshus. Again the mice died. Yet two months would pass before Rick would make the chance discovery that, as the result of this last experiment with mice, Farm Bureau Services issued an order early in January for the recall of all 402 feed.

This instruction got as far as most of the farm cooperatives and small suppliers who sold directly to farmers. But it did not reach all the farmers who had barns and silos already stocked with 402 and who were innocently continuing to use it.

Farm Bureau Services also continued to produce and sell other formulas of cattle feed which contained essentially the same ingredients as 402—only the quantities were different. This was comparable with removing bacon from the market because it was suspect, while continuing to sell ham, cured with the same chemicals and cut from the same pig.

3

Working Against Time

TED JACKSON mistrusted Michigan Farm Bureau. He felt it had developed into a big business conglomerate whose original purpose, to help and advise farmers, had become obscured in the process. He felt that its political influence within the Michigan Department of Agriculture might actually work against the interests of farmers. He told Rick Halbert he could not understand (and yet he could) why, when the mice died at Geagley Laboratory, State Police weren't ordered to the Farm Bureau Services plant at Battle Creek to close it down. Later, when it became obvious that the feed contamination went well beyond the Halberts' farm and that Farm Bureau was making only minimal efforts to contain it, he remarked angrily to one of his associates that some Farm Bureau officials ought to be shot. The fact that his brother Andrew, a farmer, was on Michigan Farm Bureau's board of directors did not temper his outrage. Ted Jackson was not a man to mince his words.

His first glimpse (and it was barely a glimpse) of the extent of the feed contamination came early in December 1973, about the same time that Dr. Grover began his experiment with the mice. It was the week of the annual convention of the National Bovine Practitioners' Association, at Fort Worth, Texas. Ted took his wife, Lois, who had often accompanied him on his rounds and had even helped with animal surgery. Chatting with his colleagues at the convention, Jackson told of his frustration in being unable to solve what he thought was a feed problem affecting an entire herd in his area. One of the veterinarians to whom he told this story was Dr. Benjamin Hekhuis of Coopersville, some forty miles northwest of Battle Creek. Hekhuis said that he too had a farmer client whose cows had developed a variety of similarly puzzling symptoms, indicating protein deficiency, after being given some Farm Bureau Services feed which they seemed to find unpalatable. This farmer's main complaint was the same as Rick Halbert's: a severe drop in milk production coupled with a serious weight loss among the cows. Hekhuis told Jackson that his client had been so badly affected financially that he was considering selling his farm and sending the entire herd to the slaughterhouse.

The name of Hekhuis's client was Peter Crum—the farmer whose complaint Dr. Grover had been investigating, along with Rick Halbert's, on behalf of the Michigan Department of Agriculture—but professional ethics prevented Hekhuis from disclosing this. It was Ted Jackson's first intimation that the mysterious cattle disease had spread beyond the Halbert farm, yet he was not sure that he and Hekhuis were talking about the same thing. He reasoned that if Rick Halbert's feed was blended to his own formula at the Battle Creek plant, it was unlikely that another farmer buying another feed mixture from a small cooperative some distance away could have an identical problem. Ted Jackson had no more idea than did Rick Halbert that the 402 feed wasn't special to the Halbert farm; that it was being sold in various outlets to scores of Michigan farmers like Peter Crum. He still thought he was dealing with a singular problem—which brought him back to his suspicion of mycotoxins.

One of the main ingredients of 402 was corn, and since the Michigan corn of the previous year's harvest had been so rain-soaked and moldy, Jackson now wondered whether the corn in the 402 mixture—rather than the corn gathered from the Halbert farm—might have developed some fungus which was poisoning the cows. Rick had already raised this possibility with Farm Bureau Services and been assured that all corn used at the Battle Creek plant came out of government storage, which meant that it would have been harvested several seasons earlier. But the suspicion lingered.

Late in December Jackson was going through some medical literature when he came upon the newly published proceedings of a symposium on mycotoxins, which had taken place in New Orleans several months earlier. He read the document, fascinated. Mycotoxins—poisons which develop from fungi—were being widely discussed among agricultural experts at the time. Hundreds of varieties exist. Some had been classified; many were still unresearched; and, like medical diseases which suddenly become well known, they were suspected of causing a wide variety of ills. Jackson noted that the coordinator of the New Orleans symposium had been Dr. Allan C. Pier, chief of bacteriology and a senior researcher in mycotoxins at the National Animal Diseases Laboratory in Ames, Iowa. He read Pier's remarks with deepening interest, comparing the description of mycotoxicosis with symptoms he had noted in the Halbert herd. Not all of them fitted, but here at last was the name of a man highly experienced in an area which Jackson did not feel had been sufficiently explored by the Michigan researchers. Dr. Pier was also at the federal government level (the National Animal Diseases Laboratory comes under the United States Department of Agriculture) and thus removed from the influence of Farm Bureau, which Jackson so mistrusted.

On January 8, 1974, Ted Jackson wrote Dr. Pier a long letter, giving details of his observations of the Halbert herd and enclosing reports of the Michigan State University pathologists who had examined the carcasses of some of Rick's cattle which had recently died. Typically, one of the autopsy reports (concerning a dead bull) noted that the animal had been in very poor condition with serious atrophy of fat throughout the body, and ascribed the death to malnutrition. Jackson did not accept this as the real cause of death any more than Rick Halbert had with the calves—there were too many other strange symptoms.

By this time hooves of some of the cattle had grown grotesquely long, curling upward and inward, corkscrew fashion, so that walking was painful and difficult. Some cows seemed too weak to stand. Their rich glossy coats had become dull and matted, and the skin on their necks and shoulders had grown thick and wrinkled like an elephant's hide. Cows which had been in late pregnancy when they ate the poisoned feed were by now in a tragic state. Many had gone past delivery dates and when they finally went into labor the pelvic ligaments failed to relax, remaining taut and unyielding. Consequently half of the calves produced on the Halbert farm in January, 1974, were born dead.

Rick described the experience graphically: "We would find cows having calves with the head in the birth canal and the two front feet folded under. Time after time. Usually the feet and nose are born first. What we had was the nose first and then the feet. A cow cannot have a calf that way. There is not room. We would have to push the head back in, lassoo the feet with obstetrical chains, and then deliver. Sometimes we would pull the legs off a calf trying to extricate it. Most of these calves had died probably two or three days before labor, and it would tear the cows up trying to deliver them." Many of these injured cows gave up eating, their udders shrank, and they developed extreme cases of mastitis. Milk production was negligible.

Lois Jackson, who accompanied her husband on a visit to the farm when he photographed some of the diseased cows, recalled, "Those animals looked just awful. I remember wondering what on earth was wrong; whether they had anything contagious."

Her husband tried to convey some of this horror and bewilderment in his detailed report to Dr. Pier. His letter ended, "I'm frustrated,

Doctor, so any help you can give will be appreciated." Many months later when the pieces of the puzzle had been put together, Rick Halbert looked back over all the research that had been achieved or painstakingly attempted and came to the conclusion that one of the most valuable contributions towards the ultimate solution had been Ted Jackson's contact with Allan Pier.

Ted Jackson was no more of a conventional animal doctor than Rick Halbert was a conventional farmer. The archetypal country veterinarian is a big, brawny man, hard-muscled from heaving cows and horses, earthy, jocular, and unrefined. Jackson was slight in build, prematurely bald, quiet, thoughtful, often described by those who knew him well as a real gentleman.

If he had followed his doctor's advice he would never have become involved with Rick Halbert's cows. In 1961, when he was just forty, he had his first heart attack and was strongly urged to give up the large animal practice which he had built up in Battle Creek, and to do only the lighter work of caring for small pets. Jackson despised the idea. As a farmer's son, he saw cows and sheep and horses as part of the essential order of mankind, whereas pet dogs and cats were likeable enough but not necessary. Most rural veterinarians share these feelings, and most of them feel a special affinity for cows: gentle and intelligent creatures lumbering passively into the milking parlor, locked in a life cycle of maternity.

"It is the nicest thing in the world, doctoring cows," said Alpha Clark, a Michigan veterinarian who never knew Ted Jackson but who would later take up the cause of the chemically contaminated farms where Jackson had to leave it. "A cow is the second mother in the world. She feeds all our babies, and hers is the most perfect food on earth." To a dedicated country practitioner there is something spiritually satisfying about curing a cow. Jackson felt this as deeply as anyone. Besides, he was so involved in his job that he was incapable of changing his way of life.

The work took its toll and in the summer of 1968 he had to have major heart surgery. "That was when they told us how bad he was," said Lois. "They took an internal mammary artery and put it into the left ventricle. They told us he was a high risk for an operation, but that without it he would live only a year."

After the surgery, his specialist advised him to retire from work and apply for a Social Security disability pension. Knowing his patient's stubborn resistance, Jackson's personal physician tempered this advice a little. Give up the farm practice, he repeated, and limit yourself to small animals. Jackson translated this suggestion in his own obstinate fashion. He added to his clinic an examining room for small animals and took on a newly graduated young veterinarian as his assistant. And he continued to do most of the farm work himself. For a man in his fragile condition to heave cows and horses around was madness, but there was nothing anyone could do to stop him.

In the summer of 1969 he had three more heart attacks, one after another, spent a month in hospital, and suffered two more heart attacks after he returned home. Still he refused to consider retirement. He said his life would not be worth living if he had to vegetate. Certain that his working days were numbered, Lois did the only practical thing she could. She went back to college, gained her master's degree, and took a job as a librarian. She also tried to keep her husband's work load light, insisting to his secretary, Ruth Guernsey, that she make only a limited number of appointments for him. But on many days the appointment book would have more and more names written into it—by Ted Jackson himself.

"It was exasperating," said Mrs. Guernsey. "I would try to book a decent schedule for him so he wouldn't overwork, but then he would pick up the phone before I could get to it. And he couldn't tell anyone no."

He had never been a patient man and his irritation with himself grew more intense on days when, slowed by his heart condition, he was unable to work as hard as he wanted. It was also in his nature that when confronted by a problem he could not rest until he had unearthed its cause. "Professionally Ted always had to know," said Lois. "He could not cut off when his work on a case ceased to be profitable, so we did not come out economically sound on a lot of things."

For this reason few of Jackson's colleagues would have given the time he gave to the stubborn problem of the Halbert herd. "I wouldn't have done it," one of them admitted. "I would have looked too hard at the ledger sheet and so would most other vets. I might have given up at the first blind alley, but Ted was not like that."

At this point in his life, aged fifty-two and in a most precarious state of health, Ted Jackson was even more than a compulsive worker. He was a compulsive worker who knew his time was running out.

4

Clue from a Machine

DR. PIER was fascinated by Jackson's report. He tested the feed samples for several known mycotoxins but the results were negative. He then fed a little of the 402 to some laboratory mice, who, unlike the mice in Michigan Department of Agriculture's experiment at Geagley Laboratory, refused to eat it. Dr. Pier's mice also died, but rather more slowly, in fact at much the same rate that starvation would have killed them. So he was unconvinced that the feed was poisoned. He also tried feeding it to some calves, but since Jackson had sent him only about two pounds of the 402 he was able to give them only a little. The calves lost weight but developed no other noticeable symptoms.

It ran through Pier's mind that Halbert's farm management might be at fault. From Jackson's description there were almost too many symptoms, and nothing conclusive. The range of apparently unrelated complaints strained credulity. Lost milk production, mastitis, metritis, stillborn calves, uterine problems, deformities, loss of hair, teeth ground in pain, lack of resistance to disease. All these problems can happen in a dairy herd and none of them may have a clear-cut explanation. The cause could be a nutritional deficiency, lack of care in observing breeding cycles, barns in poor condition. Any single problem could have been due to a number of sources; all of the symptoms together, however, amounted to no recognizable disease. Any or all of them could have been caused by poor husbandry. These factors would plague Michigan farmers far into the future, long after PBB was identified in the feed. Contaminated herds looked and acted like herds which had been badly neglected. Since every farmer has his own level and quality of management, to a researcher, who has not seen the farm in question, a cow which had been inadvertently poisoned could easily be a cow which had been abused.

In several other respects Pier was at a disadvantage. There was no coordination between one piece of research and another. He was not informed, then or later, about Farm Bureau Service's calf experiment at Agway, which was running concurrently with his tests. Nor did he know that, a couple of weeks after he received his sample of 402, Jackson also had sent a much larger sample—about two hundred pounds—to the National Animal Diseases Laboratory at Ames. But it did not reach Pier's unit, the Agricultural Research Service. Following required procedure, Jackson used the U.S. Department of Agriculture's veterinarian in Lansing, Michigan, as intermediary, and this veterinarian, having already been advised of the Halberts' difficulties by Dr. Grover, directed it, not to Allan Pier but to a different unit in the National Animal Diseases Laboratory, known as the Animal and Plant Health Inspection Service (APHIS) for which he himself worked. Although both units were part of the same group of laboratories controlled by the same government department, they had little communication with one another. The Agricultural Research Service (which was responsible for development research in agricultural fields) and APHIS (which was responsible for controlling and eradicating plant and animal diseases) pursued their independent work, not only in separate buildings but in different parts of town.

It was by happenstance that Allan Pier, grocery shopping in Ames one Saturday, bumped into Dr. Hillman Nelson, who was head of the toxicology section in APHIS, and chanced to remark that he had some interesting samples from Michigan which he was testing for mycotoxins.

Dr. Nelson replied, "That's odd. We're working on some animal feed from Michigan too."

Had it not been for that random conversation the two pieces of research would have continued independently within the same government department—Pier wishing he had more samples of 402 to work with, Nelson with a large sample on hand, not knowing that the possibility of mycotoxicosis was already being explored. In any event, Rick Halbert, more than five hundred miles away and with no idea of the complexities of the U.S. Department of Agriculture's organization, developed the anxious habit of telephoning Dr. Pier two or three times a week to inquire about the research, expecting him to have all the answers.

In Dr. Nelson's department the investigation had been handed over to a veterinary toxicologist named Dr. Allan Furr. He was an enthusiastic, outgoing man; clever, outspoken, dedicated to his work and as single-minded about it as Ted Jackson. The problem of the 402 feed instantly intrigued him. He too thought first of mycotoxins and, not knowing about Pier's tests, did some of his own. He also drew a blank but did not accept that this ruled out the possibility.

"We did not have a test for all the known mycotoxins," Dr. Furr explained. "Also I am sure there were mycotoxins around which had not been identified. We certainly don't know all there is to know about them. But mycotoxins were then in vogue, and if I had been allowed to take the research further than I did I might have been as guilty as anyone in reaching a conclusion that some kind of mycotoxicosis was involved."

Furr used the second batch of feed which had come from Jackson on an experiment with six hogs and two steers. For the first few days they ate a moderate amount of the 402, being offered nothing else. Then they refused it altogether, so Furr blended it with standard feed, fifty-fifty, and the animals ate a little. He diluted it further to tempt them to eat again, but without much success. One of the pigs died unaccountably and quite horribly, bleeding from every orifice. The rest were put to death after three weeks, which, it was later realized, was too soon to show much effect of the toxin. Pathology tests showed some liver degeneration but nothing definitive.

Furr was so absorbed with his search for the elusive poison that he made several tests concurrently. At one stage he telephoned Rick Halbert and volunteered cautiously that maybe he had an answer. His analysis had shown too much urea in the feed, enough to cause toxicity. Rick had already ruled out urea poisoning by the analysis he had commissioned from the laboratory at Kalamazoo.

"That can't be," he told Furr. "Are you sure of your tests?"

Furr ran the test four more times, and on all four there was no significant level of urea. "That taught us both a lesson," Rick said later. "Never believe a single data point. Statistically you can have very little confidence in one measurement."

Or, as Furr explained, "When you take a lab sample you take a very small amount, and if that small amount has one little pellet of urea in it this could throw the whole result off."

His next thought was pesticides. Researchers have a standard procedure for discovering this kind of contaminant. Using a machine called a gas liquid chromatograph, Furr did a test known as a pesticide screen.

The main part of the machine is a large metal box with a variety of dials and switches on the outside and, inside, a sophisticated mechanism which takes a mixture of compounds apart and breaks it into its components, sorting out molecules in terms of size. The mixture to be tested—in this case a feed sample—is first thoroughly ground up to ensure uniformity and then mixed with a solvent.

This solution is then injected by a small syringe into the machine where it enters a thin coil of glass known as a column. There is a flow of gas through the column and a detector at the end of it. At a constant temperature, regulated by the scientist, the compound passes along the column where it is separated by the material inside this glass coil. Molecules of the lightest weight are separated off the compound first; those of the heaviest molecular weight, last. Every separation is registered by the movement of a tracing needle on a revolving cylinder of graph paper. At the end of the exercise—which in the case of pesticides like dieldrin, chlordane, and DDT takes twelve to fifteen minutes—the scientist has a graph showing a series of peaks and valleys. The length of time which it takes for every molecule to come off the column and to be recorded as a peak is crucial to the scientist in identifying the chemical. The graph of any specific mixture always has the same peaks and valleys in the same order, which provides a means of identification as individual as a fingerprint.

There is only one problem with this kind of research, the same problem that Rick encountered when he sent his first feed samples to the laboratory at Kalamazoo. The scientist must have some idea of what he is looking for. When Michigan was contaminated there were nearly two million organic chemicals—about thirty thousand of them used in commerce. For measuring the impurities in a sample, a gas liquid chromatograph is one of the most useful tools in a scientist's laboratory, but when he injects a compound into it he must already have made an informed guess about the family of chemicals into which it falls. Otherwise he would have to keep varying the material in the columns and the temperature through interminable permutations, and he might still fail to find an answer.

A scientist looking for commonplace pesticides would leave the machine on at less than 200 degrees centigrade for about twenty minutes, and if no printout appeared on the graph paper in that time he would quite properly determine that the sample did not contain a pesticide. He might run the same test with a different sample of the same compound a few more times, and if the paper still came out blank he would know that he had exhausted that possibility—which is exactly what happened with Allan Furr. After checking and counter-checking, he was positive the problem couldn't be caused by any substance like dieldrin, which had been giving a lot of trouble to farmers at the time. Polybrominated biphenyl was such a rare man-made chemical which only a handful of American scientists had ever handled that neither Allan Furr nor anyone else in his laboratory would have recognized it from the printout, even if it had appeared on the machine. But PBB did not start coming off the column within twenty minutes because it has an unusually heavy molecular weight. If a researcher wanted to get it off the column in that time span he would have to raise the temperature in the box to a higher level than is required for a standard pesticide screen. Or, he would have to leave the machine on much longer. When Furr failed to get a result in the expected time, he merely ruled out pesticides and started worrying about what other line of research he might take.

Then one of those strange things happened, perhaps accidental, perhaps intuitive, which would be the first break in Rick Halbert's long and frustrating search. A technician in Furr's laboratory left the gas chromatograph switched on while he went to lunch. There was no specific reason for doing this. It could have been carelessness. Or it might have been done, casually, with the thought that it was a good idea, before using the machine for a different test, to let it run for a while with nothing in it in order to clear the column of any residue of the previous sample. This is everyday procedure in laboratories, but with a difference. The roll of graph paper should first be removed, and this time it wasn't. The technician returned from lunch to find the machine still running and the graph paper piled up on the floor. He could have thrown it away without looking at it, but he didn't. And Furr, studying it with curiosity and concern, noted that after seventy-six minutes the needle had started drawing peaks. These were not the stalagmite shapes of a pesticide printout but long slow peaks which, in Furr's description, resembled the Rocky Mountain range. He telephoned Rick Halbert and told him.

"What does it mean?" Rick asked.

"I don't know," Allan Furr replied. "I never saw anything like it before. It might be significant. It might mean nothing. Why don't you send us a tissue sample and I'll run the test again with that."

A tissue sample, in this case, was a polite laboratory expression for a hunk of dead animal. This was a job for Ted Jackson, who took samples of liver, kidney, and fat from a diseased cow which had died calving and arranged for them to be sent to Ames. Meantime, Furr reproduced the same conditions with the chromatograph using tissue samples from the animals in his own experiment. Again came the long slow peaks which reminded him of the Rocky Mountain range. A few days later Jackson's samples arrived and produced the same result.

Still Furr didn't know what to make of it. He was not a chemist and there was no chemist in his section, nor in Dr. Pier's. From the other end of the telephone, in Michigan (and he continued to call Ames several times a week), Rick imagined the National Animal Disease Laboratory to be a huge shiny complex with every modern piece of research equipment imaginable, the ultimate in efficiency. He couldn't have been further from the truth. This sub-division of the U.S. Department of Agriculture, intended to serve the entire country, was an ill-equipped, under-staffed collection of scattered laboratories, poorly coordinated, with neither the expertise nor the funding to respond to the kind of crisis which faced Rick Halbert.

"Our instrumentation was about three generations senile at that time," Dr. Delmar Cassidy, who was both Nelson's and Furr's chief at Ames, recalled later. "Much of our equipment was obsolete. I didn't have a fully qualified chemist on the staff. I just had a biologist who had taken chemistry in college and who was doing his best. We didn't know what we had on that gas chromatograph—it could have been glue from a piece of wallpaper that came out of Mr. Halbert's house—and we didn't have the equipment or the expertise to find out."

Dr. Pier's section was just as badly off. His being a research, as opposed to diagnostic, laboratory, it lacked even a gas chromatograph. He, too, had to operate on a national basis within parochial limitations. "I would like to have had more people on my staff, more equipment," he said. "But all the money came out of one pot and we had to spread it around, and everybody was not always happy about the way it was spread."

The organization at Ames was typical of a government bureaucracy on a tight budget. Every section tended to be run like a small fiefdom, jealously guarding its own small privileges and possessions, a law unto itself, rarely coordinating with the work of others. Bureaucrats may flourish in such conditions, but ambitious scientists become bitterly unhappy. They need freedom (which requires money and leeway to exercise their imaginations) to develop individual research, not the restrictions of time and equipment which faced Allan Furr. But in the National Animal Diseases Laboratory at Ames the standard reward for ability was promotion within the system. If a scientist showed such promise that the organization wished to keep him, he might eventually be promoted to head a section, where his salary would be increased (based on the number of people he supervised rather than his qualities as a scientist) and where he would not only be frustrated from pursuing his own research but would be a daily witness to the frustrations of those who worked beneath him. It was the Peter Principle in excelsis: a place where the reward for a scientist with flair was promotion to the level of his own inefficiency, his skills not suited to the constraints of a job as bureaucrat. And there he might stultify, unable to pursue his proper work. Allan Furr was an extremely hard worker, a lone research type, who did his best to break out of the mold. And Delmar Cassidy, who was as much a victim of the system as anyone else at Ames, let him go as far as he dared, conscious that every extra day Furr devoted to the concerns of one farmer, Rick Halbert, other research work with wider application was piling up in the under-staffed laboratory.

"We all felt badly about Furr," he remarked many months later.

"He came to us with a Doctor of Veterinary Medicine degree and was working for his master's degree. He was very energetic, very bright, very competent, and rather blunt—a bit disturbing to his colleagues, but some colleagues need disturbing."

Allan Furr was certainly blunt in the telephone conversations he had with James McKean, the veterinarian for Farm Bureau Services. "We were not bashful about telling him that we felt the feed was the problem," Furr recalled. He was also determined to pursue the research, despite the restrictions of his laboratory. He knew that the next step in trying to identify the strange contaminant in Rick Halbert's feed would be to test it on a mass spectrometer—a complex and costly piece of equipment, well beyond the means of most research laboratories at the time—which examines the peaks made by the gas chromatograph, and determines the molecular weight of each one. Using this equipment, an experienced analyst can come much closer to a diagnosis.

There was no mass spectrometer at the National Animal Diseases Laboratory but there did happen to be one in Ames—in the chemistry department of Iowa State University. Furr knew about it because, along with Hillman Nelson, he attended a weekly seminar in veterinary toxicology at the university. These were small informal meetings presided over by Dr. William Buck, who was professor of toxicology at the university's College of Veterinary Medicine. As Furr became involved with the problem of the Halbert herd, he discussed it in this group. First he told of the dramatic drop in milk production and the cows' reluctance to eat. Later he reported how the feed had killed a pig in his own experiment. Another week he told of the late peaks on the gas chromatograph. It was like a serialized mystery story and Furr's veterinary colleagues were fascinated.

"We wanted to go all out to help him find out what it was," said Dr. Buck. When it became obvious, therefore, that a mass spectrometer was needed, the request was made to the university's chemistry department.

It was not the kind of favor that an outside researcher would normally ask, and Furr indicated that his employers were none too pleased about his request. As Rick Halbert remarked, "You don't just walk into a laboratory and say to a chemist, 'May I use your mass spec?' These are expensive machines which have to be cleaned up by experts every time they are used. It's a complicated job and you would probably have to ask weeks ahead."

The university essentially told Allan Furr that his sample could be tested, but not immediately. He then took a small amount of feed across to the chemistry department to wait its turn.

Meantime, Rick wanted to make quite sure that the late peaks emerging on the gas chromatograph were not due to the residue of some other compound in Furr's laboratory or, as he put it, "some gunk in the column." This time he went to a private, highly respected laboratory, the Wisconsin Alumni Research Foundation (WARF) in Madison, Wisconsin. Several feed tests were made there at Rick's request, including two whose results were of particular concern to him. The first was an analysis of the feed which showed that it contained very little, if any, of the magnesium oxide which Rick had felt to be so important to the 402 recipe that he had asked for its quantity to be increased. It was odd to discover, months after the addition was supposed to have been made, that the ingredient appeared to be missing.

The other test struck Rick as more significant. The WARF researchers produced the same unidentified late peaks on the gas chromatograph that Furr had found.

Rick reported this to Donald Armstrong when, at Armstrong's suggestion, there was a second meeting between the senior officers of Farm Bureau Services and the Halberts. It was held at the offices of the Battle Creek feed plant on February 22, which was too soon, Rick and his father were told, for even a preliminary report on the Agway experiment. This eight-week project had only just begun. But it emerged during the inconclusive talk (ostensibly to share research information) that Farm Bureau Services had also been contracting to WARF for additional feed tests, similar to the standard experiments that had been run before, checking for the presence of metal contaminants and pesticides, and finding nothing extraordinary.

Angrily, Rick demanded, "This feed kills calves and rats. Can't you admit there is *something* wrong with it?"

Reporting the exchange later, he said, "They would admit to nothing. We asked them repeatedly if other farmers were having similar problems, but they always said no."

When he questioned why the experimental mice died at Geagley Laboratory, Rick said he was told that this result was not conclusive because "this is cattle feed, not mice feed," regardless of the fact that mice will survive on almost anything. Ted Jackson had tried a similar experiment, feeding 402 pellets to rabbits. In the picturesque phrase of a young veterinary assistant of Jackson, "it killed them better'n a hammer," but Jackson received a similar response from Farm Bureau Services: the pellets were not made for rabbits.

A long time afterwards it would come to light that at three staff meetings at Farm Bureau Services between October 8, 1973, and January 7, 1974, Dr. McKean reported a number of unrelated farmers' complaints about his organization's cattle feed, several of these dairy herds showing symptoms similar to those of the Halberts' herd. The feed had been purchased at four scattered Michigan cooperatives—Allegan, Coopersville, McBain, and Stanwood—all of which had bought it from the Battle Creek plant. Some was 402, and all the other suspect mixtures had some ingredients in common with 402. In his report about complaints from McBain and Stanwood, on October 29, 1973, McKean had told his colleagues that "we don't know yet what caused the problem—bad ingredients, equipment problem, or human error," which certainly suggested a liability of some sort by Farm Bureau Services. Yet there is no indication that McKean or his superiors considered the possibility that there was a common explanation to all the different farmers' feed problems.

Every one of the complaining farmers averred he was told by staff members of Farm Bureau Services that the problem in his herd was unique (a strong suggestion that his husbandry was at fault), and for some time every one of them believed this. In the case of the Halbert herd this view was reinforced at the top management level of Farm Bureau Services by a memorandum which McKean sent to Donald Armstrong on February 22, 1974, the day of Armstrong's second meeting with the Halberts. The memorandum was headed "Halbert Dairy Problem" and it described the laboratory tests which had eliminated the possibility of feed contamination by heavy metals and pesticides. It reported Dr. Pier's comment that "his sample of feed apparently contained an unpalatable compound of unknown origin." And it concluded with a reflection on Rick's farm management, which must surely have influenced Donald Armstrong. McKean wrote:

"After perusal of all the tests run, history, and clinical signs, I am unable to conclusively indict the feed. I have a feeling that Rick's feeding and management programs have contributed to the problem. If palatability is the only problem with the 402 pellets, the serious effects seen in the herd since September cannot be totally laid at its door. I will continue to explore every available avenue to determine the problem source; but, to date, have found no feed-related problem which will explain the signs observed."

At home that evening Rick studied copies of the gas chromatograph results WARF had compiled for Farm Bureau Services. Two things puzzled him. In his analysis of the feed the laboratory technician had entered the quantities against the ingredients he isolated; there seemed to be about five times more lead than would be normal in cattle feed. Searching through his veterinary texts Rick noted that this much lead, five parts per million, could cause toxicity in young animals; he wondered again if this was the answer. Also, under the heading PCB (polychlorinated biphenyl), someone had typed the single word "interference."

"I called Jim McKean and asked him what it meant," Rick recalled. "He said he didn't know but would try to find out. He did not seem to think it was important."

Even with his scientific background, Rick was unable to make sense of the various clues: the virtual absence of magnesium oxide in the 402 mixture, the presence of too much lead (which laboratory tests must have been picking up from the Firemaster mixture), the late peaks on the gas chromatograph, the hint of some substance like PCB. Only someone who had handled PBB could have made anything of it, and at that time there were very few scientists outside Michigan Chemical Corporation who had even heard of the compound.

Of the even smaller number who had experimented with it, one was a toxicologist in the Netherlands, Dr. J.J.T.W.A. Strik. Another was Dr. Joel Bitman, who worked in the Animal Science Research Division of the U.S. Department of Agriculture's research center at Beltsville, Maryland. Both men read about the existence of PBB in May, 1971, when a full page advertisement by Michigan Chemical Corporation appeared in *Chemical and Engineering News,* announcing the new flame retardant which the company had developed. The advertisement included a coupon which serious researchers were invited to fill in, requesting a sample of PBB which they could use in their own experiments—an accepted procedure in the scientific profession when a new compound comes on the market.

Polybrominated biphenyl was particularly interesting to scientists who had already begun to realize the widespread toxic effects of the related chemical, PCB, which had been polluting lakes and rivers and poisoning wildlife. Strik had charted its consequence upon his country's birds; Bitman had been concerned about its effects upon eggs and poultry. Both men wanted to find out whether PBB held similar hazards.

Strik's request for a sample, through the company's European importer, was filled promptly. He used it in an experiment on Japanese quail and discovered that PBB was ten times more toxic than PCB, causing severe liver damage. His analysis of the compound and its effects formed an important part of his Ph.D. thesis at the State University of Utrecht, published in the late summer of 1973—the same time that the Halbert herd was poisoned. Early in September Strik mailed a copy of his thesis, *Experimentele leverporfyrie bij vogels,* * to the head office of Michigan Chemical Corporation in Chicago. The text was in Dutch with a short summary in English, bound together in a slim blue-covered volume with a bright red photograph of a diseased liver on the cover. Strik made the gesture out of courtesy, thinking the chemical company would be interested in his findings—findings which were to be very significant when it came to assessing PBB toxicity in people and cattle. If it ever arrived, however, his strange-looking little book failed

* Experimental Hepatic Porphyria in Birds

to intrigue. Months later no one at Michigan Chemical Corporation recalled receiving it.

Bitman fared worse. Michigan Chemical Corporation was reluctant to send him a sample of PBB. He argued back and forth for weeks and at one point was told that the reason he had not received a sample was because "this product is not for agricultural use"—words which would prove bitterly ironic. Bitman refused to accept this. He complained to his professional society and to the company which managed advertising for *Chemical and Engineering News,* charging that Michigan Chemical Corporation's advertisement contained a misrepresentation: it promised to supply a sample of PBB to any bona fide researcher but it hadn't honored his request. As a result of this prodding, the advertising company reported back to Bitman that, as he related the message to his superiors, "it was intentional that we did not receive samples because we were the U.S. Department of Agriculture and they might run into toxicity problems if they sent us a sample."

This only made Bitman more insistent. In a detailed letter to the magazine's editor he pointed out:

"A more important issue is at stake here. The public is entitled to know, and it is the responsibility of the U.S. Department of Agriculture and the scientific community to find out, whether polybrominated biphenyls . . . are *safe*. . . .

"It is important for all of us, including the Michigan Chemical Corporation, to learn more about any chemical which may pose a potential threat to man and his environment."

Eventually he received the sample. Whatever knowledge he gained from it, it was in the nature of government bureaucracy that Allan Furr, doing one kind of veterinary research for a USDA station in Iowa, should have heard nothing of Joel Bitman's work (also in the area of dairy animals) for a USDA station in Maryland, although both laboratories were supposed to serve the entire country. If Furr had had this information, if he had had access to the mass spectrometer in Bitman's laboratory, if Bitman had seen the drawing of the Rocky Mountain range which came out of the machine in Furr's department and compared it with his chart of PBB, the pieces of the jigsaw puzzle could have been put together by the end of February.

Instead, in late February or early March of 1974 there was a second massive contamination of cattle feed mixed at the Battle Creek plant. It was not 402 this time—that had been taken off the market—but a similar dairy mixture also requiring magnesium oxide and even more popular with Michigan farmers. When an analysis of a residual sample was eventually made, the PBB content was found to be so large that contaminated mixing equipment alone could not have been responsible. Firemaster, at full strength, had to have been dumped directly into this feed as surely as it was dumped into Rick Halbert's. Once again some worker at the plant must have taken sacks of Firemaster from the place where Nutrimaster was normally stacked and tipped the contents into the mixer.

And once again this heavily poisoned feed went out to dairy farms in different parts of the state.

5

Family Tragedy

FEBRUARY was a terrible month for the Halbert family. The herd was continuing to deteriorate and it seemed certain that these cows would never recover and the next generation was also doomed. Of the few calves that were born alive, some died within a few days, others looked sickly. The overgrowth of hooves on the older cattle had become even more grotesque and their coats were so thin that skin showed through the hair. The financial losses were enormous: it would have been less costly to have let the farm employees go, slaughter the herd, and close down the operation. The results of Rick's four months of research were still inconclusive. And far from helping a farmer in trouble, which was supposed to be its reason for existence, the feed company's parent organization, Michigan Farm Bureau, was silent.

All this was happening at the same time as the Watergate scandal was unfolding in Washington, and Rick could not help making the comparison:

"The things we heard from Farm Bureau Services were similar to what the press and people were hearing from the Nixon administration —mainly, We didn't do it, it is somebody else's problem.

"It seemed to be a one-way proposition. You share your information with them, they share nothing with you. You call them, they do not call you. You ask if others are having problems and they say no. At first it is a natural response to believe them. You want to believe. But soon it begins to come back to you from a number of different directions. All the signs point to them and their feed."

The professional tragedy was compounded by personal problems.

Two of the Halbert children were ill, one with pneumonia, the other with an ear infection. And Sandra was literally sick with worry. She developed a stomach ulcer as well as strange numbing pains in her chest and hands which one doctor described as "nothing organic, just a nerve thing." At one point she even related her health problems to a psychiatrist.

"It was no use," she said. "He had no way of grappling with the consequences of an amorphous business disaster. I couldn't face the thought of going any place, not even out of the house, without taking a tranquilizer."

At this bleak, bitter time of year with their livelihoods crumbling about them, the Halberts felt as though they were living out a Gothic tale. The catastrophe which encompassed the farm seemed also to be stretching out to their family and others close to them. One day one of the farm employees was working on a ladder when, for no apparent reason, it broke. It was a fairly new ladder and should have been good for years. The worker escaped with a twisted ankle, but Rick couldn't help wondering again about evil spirits and poltergeists. Then there was a mysterious, fatal accident in which Rick was closely involved. He could barely bring himself to speak of it, even long afterwards, and when he did his voice would drop to a whisper and his face turn white.

It happened to his cousin, Karl Halbert, who had worked on the family farm for seventeen years. He was the most dependable of men, careful and deliberate in his habits, a person who fussed conscientiously over details that others might get by with skipping. He was a bachelor in his late thirties with no worries or health problems of which anyone was aware, probably the least likely person on the farm to do anything foolish or forgetful.

On this particular February day Rick was working in the dairy barn and Karl was driving a heavy tractor with an empty manure spreader in tow, taking it from Ted Halbert's old farm to the barn, a distance of about one and a half miles. There was a lane between the fields which Karl would normally have followed. Suddenly one of the farm hands ran into the barn shouting to Rick, "Hey, there's a tractor out there going through fences!"

Rick knew Karl should have been driving it, and his first thought was that his cousin must have jumped out of the tractor for some reason and that the gear must have slipped from "park" to "drive." But it was unlike Karl to leave the brake off and the engine running. Rick began walking quickly towards his father's farm, an uneasy feeling in his stomach. When he had covered about two hundred yards he saw the tractor. It had been following the lane and then, abruptly, had run randomly across the fields, crashing a couple of fences. Following its crazy tracks in the snow Rick came upon the body of his cousin, lying on the hillside.

He had been thrown and run over by the massive wheels. A broken rib had punctured his liver and he had bled to death rapidly from internal injuries. An autopsy revealed nothing that could account for the accident.

"It was inexplicable, irrational," Rick said.

The family could only conjecture that while driving the tractor Karl suffered a blackout and lost control of the engine. But how and why? Not long before this happened he had worked alongside Ted Halbert, moving the mass of 402 pellets from the bin in which they had been stored. He had stood over them, shoveled them, breathed the dust. It is medically feasible that he could have ingested enough of the toxin, in his lungs and through his skin, for it to have caused a momentary dizziness, a feeling of disorientation—symptoms of which many contaminated Michigan farmers later complained. There will never be a way of knowing. The Halberts discounted the theory since Ted, a much stockier man than his rangily built nephew, suffered no ill effects from doing the same work with the poisoned feed. If the younger man had passed out as the result of it, why hadn't Ted? It did not do to dwell on it, then or afterwards, although Rick would never forget how he felt in those darkest days of personal disaster when he suddenly came upon his cousin's body, outstretched in the snow.

6

A Bureaucratic Decision

RICK had become a chronic telephone user. Regularly he called various officials at Farm Bureau Services, at Michigan Department of Agriculture, at Michigan State University, at WARF, at the USDA Station at Ames, and wherever else he felt he might glean information or enlist help. He developed the canniness of an investigative reporter, picking up a hint from one source, checking it against the word of another, making the kind of tantalizing comments and half-suggestions designed to draw out an unguarded remark which might provide the lead he sought. He gave very little away himself. It wasn't his nature. Besides, no proud farmer talks about his losses, especially when he doesn't know the reason for them and isn't sure whom to trust. His telephone bill soared to about two hundred dollars a month, which he saw not so much as a further drain on his dwindling resources but as a modest capital investment, which, if he persevered, might lead him to a cure for his wretchedly sick herd. Although he was under self-imposed pressure to get the research work done, he had to be selective about whom he approached. The animals were dying, and funds were running out.

His conversations with Allan Furr raised his spirits. Here at last was a pure researcher, fascinated with the problem and determined to dig to the root of it. Rick learned early that little but frustration was to be gained in making a formal approach to a government institution, even though this was usually the place with the expertise. One's request for help, no matter how urgent, got lost in the works. Or, more likely, it was dismissed because it came from an unknown outsider and not down the bureaucratic pipeline. Although Rick resented this, he well understood the reasons.

"The federal government has no mandate to help individuals," he rationalized. "It only has a mandate to help recognized groups, like war veterans or senior citizens. If you fall into an accepted category and your group has a problem, the whole system is mobilized for you and you are likely to get results. That's the way a bureaucracy has to work. We all have the mistaken impression that government can take initiatives and swing into action when something new and unusual happens. But it can't. It is quite helpless to assist individuals unless it acts in an institutional way. The government cannot run a private clinic for people's difficulties."

Even so, it bothered him that the U.S. Department of Agriculture showed such a lack of curiosity and coordination in the research it did for him. There *must* have been some communication between the veterinarians at the Michigan Department of Agriculture, Doctors Grover and Carter, and Dr. Ronald Scott, the USDA veterinarian at Lansing through whose office the Halberts' feed and tissue samples were directed to Ames.

In fact, all these experts, as well as Doctors Pier and Furr in Ames, had had detailed reports from Ted Jackson. Grover and Carter also knew by that time—although Halbert and Jackson didn't—that some other Michigan dairy farmers were complaining of feed problems. This piece of information never reached the USDA department heads at Ames, who might have mobilized the wider resources of their agency had they realized they were dealing not just with Frederic L. Halbert, a lone farmer with a peculiar problem which might be his fault, but with a widening group of Michigan dairymen who had all given their cows Farm Bureau Services feed. Instead, all the work fell to Allan Furr, a veterinary toxicologist without even a chemist to help him, doing the job out of personal interest and not because his superiors insisted or even wanted, doing it in the knowledge that they begrudged the time he spent on it and lacking an awareness of wider contamination which would have made all the difference in his research. In their lengthy telephone talks Furr and Halbert, sharing frustrations, felt they had formed a bond with one another. Allan Furr, earthy and plain-spoken, with a salty wisdom and a style which was bluntly direct—a man who met obstacles head on—and Rick Halbert, a generation younger, a loner, a complicated character who functioned on many levels and whose instinct in this case was to approach his goal cautiously, even deviously, testing every step of the way before he trod. They shared few qualities except scientific curiosity and persistence. It is hard to judge what they would have made of one another if they had ever met.

They almost did meet around the last week of February, 1974. Rick had been telephoning Furr every few days, "trying to keep the project on the front burner," he explained. Rick knew that while there was small chance of getting a government department to help an individual, it was occasionally possible, with luck and assiduous effort, to create an exception. If you could find a sympathetic person within the bureaucracy and stimulate his interest on a one-to-one basis then he might be inspired enough about your cause to fight for it within the system. Rick saw this happening with Allan Furr, although he was disturbed by the frequency with which Furr hinted of budgetary constraints. Clearly he was telling Rick (who was still visualizing Furr with free access to a grandiose laboratory): don't get too hopeful, there are narrow limits to what we can do.

However, in one conversation Furr made an offer which encouraged Rick enormously. He would make the long drive from Ames to the Halbert farm with a pick-up truck; he would inspect the farm and examine the sick cows; then he would load the truck with a large quantity of the suspect feed, which he would use in some longer-term animal experiments at Ames. The first batch of feed which Ted Jackson had sent was not enough for the detailed tests Furr had in mind.

They made a date, a Friday, and in Rick's mind it was firm. He waited all day at his farm but there was no sign or word of Allan Furr. He felt terribly let down. Early Monday morning he telephoned the Ames laboratory. Furr sounded bleak. He was sorry, he had been taken off the project and could do no more for Rick. No, no one had been assigned the job in his place and it hadn't been relegated to the back burner. It wasn't on any burner at all. In a separate telephone call he gave the same news to Ted Jackson.

All three men were angry and terribly disappointed. Furr told Rick he felt sure the order to drop the project had come from high up in the USDA bureaucracy, in Washington.

"Would it help if I called Mulhern to plead our case?" Rick asked.

"I wish you would," Furr replied.

Dr. Francis J. Mulhern was the administrator of the APHIS (Animal and Plant Health Inspection Service) section of the U.S. Department of Agriculture, the ultimate authority over the various APHIS research stations across the country. Rick telephoned his Washington office twice but this did no good. He was told, in effect, that the service had reached the limit of money and manpower which it felt able to expend on a problem apparently unique to the Halbert farm. Again, Rick caught the inference that whatever was wrong with his cattle might well be his fault. On one of his calls he managed to speak to Mulhern directly. "His attitude seemed to be that our difficulties could not be very serious because there had been no other complaints from Michigan," Rick reported. "Also it was indicated to me that the people at Ames had more important things to do."

Three years later, testifying before a U.S. Senate Committee, Dr. Francis Mulhern gave more detailed reasons for stopping Furr's research. By that time the decision had taken on considerable importance. If Furr had been allowed to continue, and the university's mass spectrometer test had been made, any competent chemist could have identified the compound as PBB—little known though it was—within days. If that had been done, if state and federal authorities had taken prompt action, the second massive contamination of feed at the Battle Creek plant could have been prevented, and the state's subsequent PBB problem might have been reduced by half. The flow of poisoned feed to the farms could have been stopped, and huge quantities of highly contaminated meat, milk, and eggs (the poison was also in poultry feed) could have been kept off the market and off Michigan dinner tables.

Almost all the meat and milk produced in Michigan is consumed within the state—although at the time of the PBB disaster some cattle went to Canada, and a large batch of contaminated chickens was reportedly sold to a manufacturer of canned soup with nationwide distribution. Most of Michigan's consumers were therefore exposed to heavily contaminated food from the fall of 1973 to the late spring of 1974, when Rick Halbert's persistence finally led to identification of PBB in the cattle feed. Some of the worst of the resulting human and animal misery might have been prevented if Allan Furr had been allowed to complete his research. "We were within a gnat's eyebrow of making a diagnosis," he commented, months later. Yet without the hindsight which later made their decision seem so preposterous, Furr's superiors—operating, as they were obliged and conditioned, within the constraints of a bureaucracy—made what seemed to them a rational decision.

As Dr. Mulhern told the Senate committee, the research work for Halbert's herd did not come under any heading for which he could allocate funds. There was no way of judging how long it would take or how much it would cost. It might never prove anything which would benefit the public. And there was the criticism that other research work, of the kind that the laboratory at Ames was routinely expected to do, was piling up and being neglected.

Afterwards Allan Furr's superior at Ames, Dr. Delmar Cassidy, insisted that it was he who recommended stopping Furr's research with the Halbert feed.

"It was a very painful decision," Cassidy said. "I was concerned about Halbert and his family. He was being wiped out financially and I did not have enough competent people or the equipment to help him. Even if Washington had given us ten million dollars for the project, it would have taken too long to hire a chemist and get the experimentation going, and Halbert was in desperate trouble. His whole livelihood was going down the drain. He needed help fast and we couldn't give it. It would have been wrong to let him think that we could. That was the basis of my decision, not money. It was a moral judgment. I had had to make such judgments before and would do so again. The only thing I regret about it in retrospect was that so many people misunderstood my reasons. But even knowing how it turned out, I still think I did the best thing. I sleep well at night."

Cassidy knew that skilled research staff and refined scientific equipment, including a mass spectrometer, were available at the USDA research station in Beltsville, Maryland—the place where, ironically, Joel Bitman had experimented on the sample of PBB which he had obtained from Michigan Chemical Corporation with such difficulty. Prompt referral of Rick Halbert to Beltsville would have produced immediate results, since Bitman and his close colleagues were among the few men in the country familiar with PBB. Later Cassidy said he assumed that one of the scientists under his authority and senior to Furr would have advised Halbert to try Beltsville—a curious assumption, since the staff at Ames and the staff at Beltsville handled different kinds of agricultural research and tended to regard one another as employees of rival institutions.

"Perhaps I was remiss in not calling Halbert personally," he said.

The decision to take Allan Furr off the research would not have been so devastating if only that call had been made. Two months of the very worst contamination could still have been prevented.

Some time afterwards, when the USDA hierarchy realized how this disaster could have been minimized, Cassidy's section was granted a graduate chemist and a $250,000 mass spectrometer. Even so, Cassidy remarked that "the money had to be wrenched from the bureaucracy to make us a more capable group."

As soon as he was pulled off the job Furr confided his frustrations to the next meeting of Dr. William Buck's group at the university. There was no doubt in the mind of the anyone at his seminar, Buck said, that the directive to discontinue the work came from USDA's hierarchy in Washington and was not based on Cassidy's recommendation alone. They had various theories which ranged from interdepartmental rivalries at high government level, specifically between the USDA and the Food and Drug Administration (which would inevitably become involved if a toxin was discovered) to the political influence of American Farm Bureau's Washington lobby. These theories were probably wrong, but in their anger Furr's colleagues were suspicious.

"We were all very aggravated about the situation with Furr," Buck said. "If he had been given the freedom to go ahead he would have known fairly quickly from the mass spec that the substance was a brominated biphenyl. Once he found that he would have started yelling at the FDA, and it wouldn't have taken the FDA's local investigators long to tell it was coming from Farm Bureau's feed. Even from the gas chromatogram we all knew it was something like a biphenyl because that's the way these substances act. We would have had to wait only a few more days to use that mass spec at the university. We all felt badly let down."

Buck, who had once worked for APHIS himself, was scornful of a government department which would crush the spirit of a researcher as keen as Allan Furr. "That organization, clear up to Washington, is a horrible bureaucracy in which all the boat-rockers and noncomformists have been weeded out," he remarked.

Although none of the officers of Farm Bureau Services admitted this to Rick Halbert, some of them were also suspicious about the magnesium oxide content of the 402. There had been complaints from other farmers including Peter Crum of Coopersville and Jerome Petroshus of Allegan, both of whom had cancelled orders for 402. The magnesium oxide, moreover, was an obvious suspect because it was the one ingredient in such high quality feeds which did not appear in the more widely sold mixtures. In December of 1973 Farm Bureau Services sent a sample of it to Kar Laboratory of Kalamazoo for analysis and were reassured that it was all right. The sample was taken from a randomly selected bag at Battle Creek plant and it really *was* magnesium oxide. Similar bags containing PBB must have been close at hand. If only one of them had been selected. . . . Instead, the staff at Battle Creek plant now had good reason to believe that whatever might be wrong with the 402 mixture, it wasn't the magnesium oxide. When they were unable to get trace minerals from their normal supplier during the early part of February, 1974, they used additional magnesium oxide in several of their other feed mixtures to make up the deficiency. At this time, however, as subsequent analysis showed, the bag from which they poured this ingredient actually contained PBB. Many of the farmers whose cattle, sickened on the 402—by then withdrawn from the market—turned instead to these other mixtures, numbered 405, 407, 410, and 412. Such was the nature of the second massive contamination, the one that might have been prevented if Allan Furr could have finished his investigation or if Delmar Cassidy had made that telephone call.

Ted Jackson was giving far more time than a country veterinarian can profitably devote to the problems of one client. After the research was halted at Ames he sent detailed reports to the WARF Laboratory, and a summary of the result of his own tests on the Halbert herd to Dr. James McKean of Farm Bureau Services. He worried about the problem continually; wrote letters and made telephone calls to anyone he thought might help. Colleagues chided him that he was neglecting his other veterinary work for the sake of a single herd of cows, and all this added to the frustration he already felt over his inability to diagnose Rick Halbert's problem. He was angry about the people at Ames and impatient when his letters went unanswered.

Even the letter which Furr had promised failed to arrive. Furr had told Rick he would send a list of laboratories capable of completing the feed analysis. But when the list did not materialize, Rick decided not to pursue it.

"We should have screamed bloody murder when the people at Ames gave up," Rick admitted. "But when you are an individual dealing with a government department it isn't always a good idea. You make yourself very vulnerable. Later on, farmers who were outspoken and made themselves visible came in for a lot of ridicule."

Rick tried a different approach. Through McKean, he attempted to persuade Farm Bureau to "pull some strings in Washington to keep this project going." Again this was unsuccessful. It was McKean's private thinking that Rick had probably aggravated Mulhern enough already: Rick had been calling APHIS headquarters even before Furr's research was halted. Weeks elapsed before Rick received a formal notification from Ames stating that "We regret not being able to reach any conclusion since we do not have adequate personnel, equipment and experimental animals to pursue the problem in depth"—without a hint as to who else was qualified to do the pursuing.

Rick also felt that officials at Farm Bureau Services were being deliberately uncommunicative. One day the company's risk manager, Ken Jones, called at the Halbert farm to discuss his company's insurance liability for the sick animals. No decision was reached but in the course of conversation Rick asked, "How are those calves doing, the ones in the Agway experiment?"

"Haven't you heard?" Jones responded, apparently surprised that Rick hadn't. "Two of them died."

"After he left," Rick recalled, "I called up Jim McKean at Farm Bureau and asked him the same question: 'How are the calves doing at Agway?' "

Rick understood him to say that they were doing all right. "I told him Jones was here, and he told me they died. There was this long silence. Then McKean said, 'I was not able to tell you that.' "

Rick added, "I received much glee from that conversation."

By "fishing around" Rick had made another discovery: Farm Bureau Services had had complaints about dairy feed from at least two other Michigan farmers, Peter Crum and Jerome Petroshus. McKean admitted this but hastened to add that he was sure their problems were quite different from those of the Halberts'. Rick, however, telephoned them both.

Crum, an immigrant from the Netherlands, spoke with a heavy accent, and Rick could not grasp much of what he was saying. Petroshus, unsure of who was telephoning, told Rick something of the difficulties he was having with his herd but did not discuss them in detail. However, Rick gathered enough to realize that here were two more pieces to the puzzle he was trying to solve. Had he been able to elicit their stories in detail he would have been even more fascinated.

Petroshus's farm in Allegan was thirty miles west of Rick's. Crum's place in Coopersville was sixty miles to the northwest. The two men bought 402 feed from different Farm Bureau cooperatives, and neither of them dealt directly with the Battle Creek plant as did Rick. Crum's problems began in the summer of 1973, when all but 3 of the cows in his herd of 120 developed mastitis. Some of his calves were born dead and, as in the Halbert herd, some cows died calving because they lacked the strength to deliver. Crum suspected the 402 pellets and complained to the Farm Bureau co-op at Coopersville. The co-op first sent a Farm Bureau nutritionist and then its veterinarian, Dr. McKean, from whom Crum understood, as Halbert had done, that he was the only farmer having problems. This was not evasion on McKean's part. He was not at liberty to discuss one farmer's problem with another and felt that Crum's complaint had a different cause from Halbert's. McKean called at the farm several times and Crum developed the uneasy feeling that "there was something wrong which he didn't want to tell me." McKean, on the other hand, was uneasy about Crum for different reasons. He knew, from local gossip, that this farmer was in financial difficulties and naturally suspected this might explain the poor state of his herd.

On December 4, 1973, one of Crum's best cows dropped dead. This upset him so much that he insisted that his own veterinarian—who happened to be Benjamin Heckhuis, the practitioner who first alerted Ted Jackson to the fact that he, too, had a client with a mysterious malady in his herd—call in one of the state veterinarians employed by the Michigan Department of Agriculture. Dr. Grover arrived and did an autopsy within a few hours of the cow's death. Watching, Crum was appalled by the sight of the animal's blood on the snow. It was thin and runny; it should have been thick and crimson. Grover took a number of tissue and feed samples back to his laboratory at Lansing; as he drove off he commented, "I think you have a feed problem." It was several weeks before the result of his analysis arrived in the mail, and all that Crum could make of it was that the percentage of protein in the pellets was lower than it should have been.

The day after the autopsy Crum heard from his veterinarian of an experiment at the Department of Agriculture's Lansing laboratory in which a number of mice died after being fed 402 pellets. Crum wondered if his feed had been used in the experiment: in fact it was Rick

Halbert's. Heckhuis might not have known about this test had it not been for his chance conversation with Jackson at the veterinarians' convention from which he had just returned. He reacted by rushing to Peter Crum's barn at six-thirty one December morning telling him that the experimental mice had died and he should stop feeding the pellets to his cows.

Crum was due for a new delivery of the 402 that day. He went directly to the Coopersville co-op and cancelled the order. This annoyed the co-op manager because Crum's feed order had already been loaded on one of his trucks. It was still waiting there when word came through from Farm Bureau's co-op in Allegan that one of its customers, Jerome Petroshus, could use a load of 402. So the truckload of contaminated feed, all five tons of it, was diverted to Allegan and delivered to Petroshus's farm.

Previously Jerome Petroshus had received some deliveries of 402 which were contaminated with PBB and some which were not. Hence he was alternately pleased and dissatisfied with the feed. After a September delivery he noticed that his cows were refusing to eat and he returned part of that delivery in October. A Farm Bureau nutritionist said he could not find anything wrong with it but conceded that it may not have been as palatable to the cows as usual. When that batch was mixed, he explained, the Battle Creek mill had run out of apple pomace; so this ingredient had been omitted. Now that it had been put into subsequent batches he was promised that the cows would again eagerly eat the feed. Petroshus was persuaded, and when the feed destined for Crum was dumped at his farm, he gave it to all his animals—cows, calves, and even ponies. They had about one good feed and then, again, all refused it. Petroshus returned the rest of this load to Farm Bureau Services on December 29. His herd, like Crum's and Halbert's, went downhill fast.

Crum, meantime, had suffered such serious financial reverses from the state of his herd that he decided to retire from farming. His cows had either died or dried up and he could no longer meet his loan payments. He had been a farmer all his life, first in Holland and since 1963 in Michigan; the decision hurt him terribly. He felt that his herd, which had dwindled from 172 cows at the beginning of 1973 to 98 at the year's end, was good for nothing but the slaughterhouse, so he refused to sell these cows with his 170 acres. He thought it dishonorable to burden any farmer investing in his land with these useless animals.

Towards the end of February Crum sent all ninety-eight cows to the Allendale Beef Company to be butchered. If he had had any doubts that they might not be fit to eat, these were dispelled by the fact that every animal passed the tests of federal and state inspectors, including Dr. Grover. It was noted that some of the cows' livers were twice the normal size, and the story spread among local farmers that Crum had had too much urea in his feed. However, all the meat market inspectors were satisfied, and it is reasonable to assume that all of Peter Crum's contaminated cows were sliced up and sold in Michigan supermarkets. His farm was bought by a farmer named Gerald Woltjer, and his is another part of this story.

Around the time his animals were slaughtered, Peter Crum heard what had happened to the load of 402 pellets which he had refused. He did not know Jerome Petroshus but, taking his teen-age daughter, Mia, who often acted as his interpreter because of his awkwardness with English, Crum drove to Allegan. He turned up on Jerome Petroshus's doorstep unannounced, described his experiences, and urged Petroshus not to feed the pellets to his herd. He was too late. Petroshus had used the feed and was about to send two of his cows to be butchered. At the slaughterhouse, remembering what Crum had told him, he took particular note of their livers. They were so enlarged and unhealthy looking that they had to be thrown away.

"But," Petroshus commented, "somebody ate those cows."

Before long the rest of his animals had to be destroyed—those which had not died in the meantime. For the next thirteen months Jerome Petroshus was in a desperate situation with no livestock on his farm. Eventually he got back in business; but Peter Crum was never able to return to farming, nor to speak of the experience without emotion.

He said, "It will bring me sorrow to the last days of my life."

7

More Pieces of the Puzzle

EVERY year towards the end of March, Michigan State University organizes an event known as Farmers' Week. Described in the trade press as "MSU's annual bash for farmers" it offers an up-to-date crash course on agricultural developments, from agronomy to economics. It is also a chance for farmers of this heavily agricultural state to discover how their operations stand in relation to others. Many of them would not miss it.

In March of 1974 Rick Halbert was unwilling to share news of his troubles, but he drove to the university's vast campus in East Lansing intending to hear some lectures. "Instead," he said, "I got into a nitty gritty discussion with Jim McKean. I told him that it just didn't seem possible to me that mag oxide wasn't the feed ingredient causing trouble, or that other farmers weren't complaining."

This discussion led to a session between the two men which took up an entire Wednesday morning in McKean's office in the opulent Lansing headquarters of Michigan Farm Bureau. McKean was reassuring about the magnesium oxide. He said it came from Michigan Chemical Corporation and that it was a food grade product, the same type that was used in antacids widely sold in drug stores for the relief of human gastric discomfort; so there couldn't be anything wrong with it. This, however, did not explain why so little magnesium oxide had shown up in the WARF analysis of the cattle feed, a finding which still bothered both men.

Rick returned to his question about whether other farmers could have received feed mixed at the same time as his. McKean produced his file of production schedules and Rick learned that, on the day his suspect feed was delivered, a batch had also gone from the Battle Creek plant to a Farm Bureau Services cooperative in Yale.

"Up to that point," Rick said, "it was understandable that if ours was the only farm to have that particular feed, there would have been no complaints from other farmers. But when I realized this wasn't so, I *knew* there had to be other people crying in the forest."

Yale is in a region of Michigan popularly known as the Thumb. The main body of the state, its lower peninsula, is shaped like a mitten for a giant's left hand—the wrist sinking into the states of Indiana and Ohio, the left side of the hand lapped by the waters of Lake Michigan, and the thumb and forefinger extending into Lake Huron. The Thumb has good farm country with rich black earth, but it is isolated from the mainstream of Michigan life and thinly populated. People in other parts of the state, particularly the town dwellers, regard it as almost another country. They talk about events and places "up in the Thumb" as if referring to Alaska.

When he had returned home from Lansing, Rick remembered that he knew an agricultural agent who had worked in the Thumb for the Michigan State University extension service. He phoned the man and asked if he knew anyone in the Yale area using Farm Bureau feed.

The agent gave him a name, Robert Demaray, and Rick made another of his probing telephone calls. At first he did not reveal his own problems, but subtly led up to questions about Demaray's experiences with Farm Bureau feed. These experiences were different from Rick's and less bizarre. Bob Demaray said his cows had some reproductive tract infections and were not reacting to treatment by antibiotics. "We were not seeing any of that," Rick remarked. "But he had given his cows only two pounds of the 402 a day with their regular feed and we had given twenty pounds." The condition shared by the Halbert and Demaray herds was an unusual number of abnormal births. Cows had gone beyond their delivery date and then, with difficulty, produced calves which were stillborn or deformed. The dissimilarity between other symptoms in the two herds added to Rick's confusion but turned out to be immensely significant. There were, in fact, two kinds of PBB contamination: the dramatic and virulent, experienced by cows like Rick's, which had very heavy doses of the toxin, and the low-level contamination in cows which received such minuscule amounts of PBB that it had to be calculated in tenths of parts per million of the total feed. Cattle faced with the heavily contaminated feed found it so unpalatable that they went hungry rather than eat more, but the others, on a less contaminated diet, did not seem to notice a disagreeable taste and continued to ingest it over a long period. Endless arguments have been waged in Michigan over the effect of low-level PBB contamination

—how serious it was and how the long-term effects of small quantities of PBB steadily building up in an animal's system compared with the more immediate poisoning of herds by massive doses. One thing soon became certain: in their various ways, tens of thousands of animals in both categories were doomed.

Bob Demaray told Rick Halbert that before he linked his herd problems to the 402 he had had such good results with this feed that he had recommended it to a neighbor, Arthur Laupichler. Laupichler's experience was even more tragic. Telephoning him, Rick discovered that Laupichler received his batch of the feed in January of 1974, at just about the time Farm Bureau Services determined that it was suspect and should be taken off the market. But the message did not reach the Farm Bureau Services cooperative at Yale in time to stop delivery of the three-ton order.

Laupichler used the feed as a top-dressing for his own homegrown high-moisture corn, adding it by hand, but after a few days he noticed his cows backing away from the new mixture. By the ninth or tenth day, his best milk producers, who were given the highest concentration of 402, refused to eat any of it.

Laupichler complained to the cooperative and asked for the remaining feed, more than two tons of it, to be removed. The feed was picked up without argument, which surprised him a little. He was not told that 402 was being recalled. About a month later he began to notice symptoms similar to those in the Demaray herd—uterine infections and abnormal births.

"I got together with my veterinarian," Laupichler recalled. "He was pulling his hair out worrying what it was. I thought it must be the 402, because it was the only new feed ingredient which we had not raised on the farm. So I got back to Farm Bureau in March and shortly afterwards they sent Dr. McKean along with a local salesman. They tried to tell me I was doing everything wrong—that I had not been dousing my rations, that I had been over-feeding or under-feeding mineral supplements, nitpicky things that didn't hold water." Laupichler made this second complaint to Farm Bureau after Rick Halbert telephoned him. "He started asking me questions about my animals," Laupichler said. "At first he did not tell me what was wrong with his. But I said to him, 'Hey, what's going on here? You must know something I don't know.' Then we got down in earnest to compare notes."

The similarities were a revelation to Art Laupichler. He too had been told by the staff at his local co-op that he was the only farmer complaining. He had been unable to check whether this was the case because "the farmers in my area are far apart, so we do not communicate very much." In fact, none of the Yale complaints reached McKean's desk until early April, following his Farmers' Week meeting with Rick. After his telephone calls to the Yale area, Rick remembered a remark that his father had made back in January: "When this is all known, it may turn out that there are hundreds of farmers with the same problems as ours."

Rick commented, "I thought it was impossible, but it turned out to be prophetic."

Crum, Petroshus, Demaray, and Laupichler were but a few of the PBB-afflicted farmers who had thought themselves alone. In the eastern part of the state, around Fremont, several dairymen were having inexplicable difficulties with their herds. One of them, Paul Greer, was a Fremont attorney who, having built up a prosperous small-town practice, had recently bought a five-hundred-acre farm to operate as a hobby. He had put in a manager who was improving the quality of the land and building up the herd when milk production suddenly diminished, and some of the best cows started aborting in early pregnancy. Greer began to regret his venture, especially when some of his legal colleagues started teasing him: "It just proves that a lawyer shouldn't go into farming. He doesn't know what he is doing."

Some of these gibes came from a young corporate lawyer in Grand Rapids, Gary Schenk, who had met Greer in the fall of 1973, when they had represented different interests in a case involving a trucking company. The dispute, settled out of court, brought them together several times over a period of six months, Greer becoming testier all the time because he was troubled about his farm. What he did not tell Schenk was that two dairy farmers in the Fremont area had come to him for legal advice about similar problems in their herds. Both men, Myron Kokx and Blaine Johnson, had bought feed from the Fremont co-op and felt this was to blame. Whereas Greer might be accused of agricultural inexperience, Kokx and Johnson were substantial, successful farmers who had maintained large dairy herds for years. Greer used the same feed as they—a mixture made up at the co-op using Farm Bureau ingredients. All three men thought the culprit to be moldy corn and had complained to the Battle Creek plant, never thinking to question the feed's magnesium oxide content—which, in their case, was not magnesium oxide, but PBB.

While Paul Greer maintained a professional silence about all this to Gary Schenk, the younger attorney was keeping a similar confidence of his own. For some time he had been legal adviser and vice president of a small family corporation, Westmac of Howard City, which was in the business of chicken farming, and this company was now in trouble. Egg production had dropped dramatically, and chickens were dying. Before death, the birds moved in an uncoordinated fashion as though their neurological systems had been affected. Westmac bought its feed directly from the Battle Creek plant; John Williams, the corporation's president, suspected this was the source of his problems. He had been promised an investigation by Farm Bureau Services—a promise which had also been made to Paul Greer and his clients. But Schenk was becoming dissatisfied with the slowness of results and had begun drafting a complaint against Farm Bureau Services, alleging that it was responsible for an unknown toxin in the chicken feed. Schenk knew nothing about dairy farming, and it never occurred to him that Westmac's unproductive chickens could have anything in common with Paul Greer's unproductive cows.

There were other pieces of the jigsaw puzzle. In the veterinary department of Michigan State University at least one senior student, Peter Van Vranken, was perplexed by the number of cases from the Fremont area which came into the referral clinic where he worked as an intern.

"You would go down a ward and all the animals would be from Fremont," he related. "They were individual cases from different farms and it was strange there should be so many from this one area because Fremont was two or three hours away. They were mostly bulls and valuable calves, all with a lack of appetite, and there were different suspicions about every one—things like infections and parasites. Nobody could figure it out, but nobody would admit he did not know."

Some weeks later Pete Van Vranken became Ted Jackson's assistant in Battle Creek. "I never thought to tell Ted about the Fremont cases. It didn't cross my mind that they could have the same thing wrong with them as Halbert's cows."

The experts were baffled. The fact that a number of farmers had complained of feed problems was not unusual. Farm Bureau Services served hundreds of Michigan farmers, and those who bought the more expensive, high protein feed such as 402 were those with the highest producing, most carefully bred cows. Such animals, like all other purebreeds, are more delicate and susceptible to infection than most. Their gastro-intestinal systems are easily upset, and frequent calving makes their reproductive organs vulnerable. A good dairy farmer is more watchful of his high-producing cows than of any other animals on his farm, since these are the main source of his income. Any digestive upset in a cow shows up instantly in reduced milk production. Any infection which needs treatment by antibiotics means that the milk must be thrown away—and this is throwing away cash. On a large farm, a bull or a heifer could be unwell for a day or two before the farmer would notice, but not a lactating cow.

Jim McKean knew from experience that as soon as a farmer senses trouble among his best milkers to which he does not have a ready answer, he is likely to blame the feed, just as a mother tends to blame the infant formula when she can't understand why her baby is cranky. The farmer may be right, or he may be doing something wrong himself. If, as in this case, the feed from Battle Creek *was* at fault it could be for a variety of indigenous reasons such as moldy corn, too much urea, an inappropriate mixture—all the things which had been suspected for the Halbert herd. It was part of McKean's job to discern the difficulty and put the farmer on the right track. Traveling around the state he encountered feed problems all the time, some more baffling than others. He did not immediately recognize a common factor in Rick Halbert's complaints and those from Coopersville, Allegan, and Fremont. The symptoms varied from one farm to another because the herds were different, the farmers had different feeding practices, and the cows had ingested different quantities of PBB at different stages of their breeding cycles. Even after they acknowledged that there was something wrong with the 402 mixture, McKean and his Farm Bureau colleagues were still thinking in terms of a toxin familiar in agriculture, such as a pesticide, and not one which might be alien and unknown. It added to their confusion that the Fremont farmers had never used 402, and that many of the farmers who did use it continued to be satisfied. In fact when Farm Bureau Services began recalling 402 in January, some customers resisted the company's attempts to take back their remaining supplies. They were, of course, the fortunate ones who had real magnesium oxide in their feed.

"They said that on 402 their cows were doing the best they had ever done, and they did not want to return the feed," McKean recalled. "That kind of colored my thinking."

He wondered whether local soil or climatic conditions had something to do with the cattle blight. Since he had not yet heard about the complaints from Yale it seemed to him that all the farmers having trouble were in the western part of the state. And there were still not enough of them to suggest an epidemic. The problems that were being relayed to Jim McKean were no greater in number than he regularly dealt with in his everyday work. They were, however, different, more serious, more stubborn, and cause for deep concern.

"I could not understand why animals in some herds were falling apart, and other animals getting the same feed were thriving," he said. "When there was a problem, it was never exactly the same from herd to herd. There were some similarities but for a long time I had no reason to think the complaints were connected. The fact that we had not yet found an answer to the Halberts' problem worried me every passing day. And yet I thought we were protected from something like this happening again because we had checked out the magnesium oxide and recalled all the 402.

"Faced with all the unknowns that we had I don't see how we could have proceeded in any other way. Now I find it easy to criticize myself because I know things today that if I had known then, or even suspected then, things like the magnesium oxide content, well of course I would have acted differently. In our meetings at Farm Bureau we kept coming back to that and we spent the better part of one night going over what we knew and did not know about it, and asking, 'Where do we go from here?' "

In February, still not satisfied with his earlier tests, McKean questioned the company's supplier of magnesium oxide and salt, Michigan Chemical Corporation. He was referred to the parent company, Northwest Industries in Chicago, which he telephoned.

"They put me on to a sales desk and I asked the man there where the magnesium oxide was made and how could I be sure it was pure. He gave me a whole thing about it being a food grade product, manufactured in a different warehouse from the other chemicals, and bagged separately. I had no reason to think I was not being told the truth."

That is why McKean reassured Rick about the quality of the magnesium oxide in their discussion during Farmers' Week. And why Rick, subsequently, thought McKean was not telling all he knew. McKean sensed this and was offended.

"Halbert and I were supposed to be working together," he said. "That was my instruction from Armstrong and I followed it to the letter." His research methods, however, were different from Halbert's. McKean proceeded steadily and, he felt, logically from one step to the next. Halbert pursued several courses of inquiry at the same time, hoping that at least one would prove fruitful. "He was like a bull in a china shop," McKean grumbled, "striking out in all directions." Both men were under intense pressure. Rick could see his family's business operation sinking into bankruptcy, and he thought McKean was not trying as hard as he could to help him. McKean, exhausted from long days on the job and with equally stubborn complaints from other farmers, felt that some of Rick's demands were unreasonable. It did not make matters easy for either of them that, by now, McKean had become the main contact at Farm Bureau with whom Rick had to deal.

So they parted after the Farmers' Week meeting, Rick still not satisfied (except for his new knowledge that there might be other farmers crying in the forest) and Jim McKean somewhat ruffled.

McKean was also left puzzling over a piece of information which he saw no reason to share with Rick. In February he had sent to the WARF laboratory samples of tissues from Peter Crum's cows after they had been butchered. He had taken them himself from the slaughterhouse. He also sent for analysis samples of the feed made up at the Fremont co-op about which Paul Greer and his clients complained. In both cases the printout on WARF's gas liquid chromatagraph showed the same mysterious late peaks as the printout for the Halberts' 402 had shown.

For the next few weeks McKean continued to assume that the foreign substance causing the peaks, whatever it might be, must be a contaminant of the corn, which was a common ingredient of both feeds. Not until early April did he discover that the Fremont feed also contained magnesium oxide—or, what was supposed to be magnesium oxide.

8

Needle in the Haystack

RICK'S strained relationship with Jim McKean resulted in some duplication of effort. In March McKean told Rick that he was having further tests run on the feed but did not say where. Rick made a similar decision but did not at first tell McKean. So without informing each other, both men were asking for follow-up studies from Donald Hughes, manager of the pesticide residue department at WARF, who assumed that the two were working closely together and sharing the results of their research. He did, however, note a difference in their approach. Rick frequently telephoned with suggestions and requests. Farm Bureau Services gave Hughes the impression "that they saw this as an isolated case so did not feel an enormous urgency."

Rick had sent WARF a second sample of 402 as well as one of 412, the Farm Bureau Services feed he used as a substitute after his cows sickened, and asked if the laboratory technicians could be more specific in categorizing the peaks which had showed up on the gas liquid chromatograph. The job really needed a mass spectrometer, but, like the USDA station at Ames, WARF lacked this sophisticated piece of equipment. Without it, Rick felt it might still be possible to get a general idea of the type of substance which was causing the drawing of the strange late peaks on the machine's graph paper. This was a matter of hooking up different kinds of detectors to the machine, noting whether the material being tested produced a response, and seeing if it could be matched with a known compound.

Suspecting that the feed contaminant might be a pesticide, the WARF technicians had used an electron capture detector. If they had been testing for biological material they would have used a sulphur or phosphorus detector and, in this case, got a negative reading. In the course of the test they observed that the peaks not only came off the machine late but remained constant, denoting the presence of halogen molecules. Don Hughes was also able to give Rick an estimate of the compound's unusually heavy molecular weight.

"Boy, that really perked me up," Rick said. "Finally we had something to work on. Up to this point I had thought that the peaks could have been incidental to the problem, caused by some harmless substance in the feed like Vitamin A. But the fact that they were halogenated and that the compound had a high molecular weight told me that in all probability we were dealing with something which was not only very toxic, but man-made. Halogenated organics of this weight rarely, if ever, occur in nature." There are four types of halogen: bromine, chlorine, flourine, and iodine. "I naturally thought of a chlorinated material," Rick added, "because chlorine is used much more often than bromine in industry. Don Hughes injected chlorinated standards into the machine and turned the temperature way up so the peaks would come out fast. But the peaks from the feed sample didn't match. So it couldn't be chlorine and we couldn't think what else. But at least we were able to demonstrate that there was something man-made in 402 which was biologically incompatible with our animals, and by late March Farm Bureau had accepted this. Up to then they were still telling us that our problem might not have anything to do with the feed."

Now there was no doubt about the next step—to find a mass spectrometer which would identify the substance causing the strange peaks on the graph. Jim McKean had privately reached the same conclusion but pursued it differently. On April 9 he commissioned WARF to farm out the work to the University of Wisconsin, with which it had close relations. He was told that this involved a contract arrangement which would take a little time. Rick took a more complex course. He had discovered that there was a mass spectrometer at the USDA Research Station in Beltsville, Maryland, and rather than make a direct approach and risk a rebuff, he started asking his various contacts in Michigan State University's Department of Agriculture if they could give him the name of anyone at Beltsville who worked in the field of pesticide residues. "By now this had become my standard method of operation," Rick said.

Typically, he was still tackling other lines of research concurrently. One of them led him to telephone Texas, where Rick tracked down a USDA toxicologist, Dr. Harry Smalley, one of the few who specialized in pregnancy problems of farm animals. "Farm Bureau had substantiated our claim that the feed kills calves. I felt we also had to prove that it interferes with the reproductive process." Most of Smalley's work had been with sheep, but he seemed interested and asked Rick to send him a quantity of feed. Rick went through the routine of requesting the USDA for a government Bill of Lading to make the shipment across state lines, but was denied this without explanation. It was an unexpected frustration, since there had been no difficulty about making a similar shipment to Ames. Undeterred, Rick packaged the feed in fibre drums and made his own arrangements for it to be trucked to Texas. "I called Dr. Smalley often, as was my custom," Rick said. "He gave the feed to pregnant ewes. Some had still births and the others did not give birth. These were necropsied and it was found they had absorbed the fetuses. The control group had normal lambs. But Dr. Smalley did not have these results until June and by that time we knew the explanation."

Back in March, Rick tried yet another tack. In a cautious and deliberately incomplete fashion he reported his difficulties to the Detroit office of the Food and Drug Administration. Rick had pondered this move ever since the USDA research at Ames was arbitrarily called to a halt. He knew that this FDA office with its extensive research facilities could probably complete the work. Or it could bring pressure on USDA. He also knew that FDA was primarily in the business of protecting the public food supply and that if it suspected a cattle disease it could force his milk off the market. He might gain by making the contact or he might be immeasurably worse off. Most of his friends advised against it. As he himself described it, "If you really want to find out what's going on, you go to someone who will help you, not to someone who is likely to throw you in prison." Believing that the move could also present a threat to Farm Bureau Services, Rick said that he "threatened Jim McKean with it."

"I mulled it over for a while and then decided to bite the bullet and suffer the consequences. I called the FDA in mid-March and talked to a Mr. John Dempster in Detroit. I approached him in a circuitous way. I wanted to apprise him of the problem rather than confront him, so I said we were having difficulty with a feed company and that some of our animals had been affected. I said there was perhaps too much lead in the feed. I did not tell him about the mice experiment, our calf experiment, or the Agway experiment. I generalized because all we had at that point was a number of leads, and I wanted to see what the FDA would come up with. One thing you quickly learn is that if you want quality research you do not point the way. You don't make suggestions like, 'Leave your chromatograph on all day. . . .' "

The next day a FDA inspector called at the Halbert farm and took samples of the green pellets still stacked in the barn, the feed that was currently being used, hay, corn, milk from some of the sickly cows, and salt which had come from Michigan Chemical Corporation. Rick was impressed by the speed of this response. Here at last is a government agency which works fast, he thought. But after the FDA inspector had driven off to Detroit with what Rick called "his truckload of goodies," Rick heard little else from FDA except a report that the agency had followed up the only hint he had given, had tested the Halbert milk for the presence of lead, and had alerted Michigan's Department of Agriculture to the possibility; they had, however, found the milk to be within normal limits.

"Basically I found that FDA is an organization that tells you nothing," Rick concluded. "I called the Detroit office repeatedly and asked them how they were coming along, and I finally concluded they were not particularly interested in our problems.

"Still I went on calling, and as they weren't getting anywhere I started giving them hints. I told them about the late peaks on the chromatograph and how we had learned that the substance did not contain sulphur or phosphorus but was halogenated. The only response I got was to be told that FDA was not a research organization but a regulatory agency. So my efforts with FDA got off to a fast start but blew out very quickly. It was my perception that they were not interested unless I could tell them what it was."

Meanwhile, he was still trying to find a way of using the mass spectrometer at the Beltsville Research Station, unaware of the arrangement which Jim McKean had made for the University of Wisconsin to do the work. After some digging around, Rick discovered the name of a contact at Beltsville, recommended by a veterinarian at Michigan State University. "I telephoned this man and asked if he knew anyone working in the area of toxic residues," he recalled. "He told me about two researchers in the pesticide degradation laboratory, Dr. Joel Bitman and Dr. George Fries. This was early April. I called their lab and discovered they were both attending a professional meeting in Los Angeles and I would have to wait at least a week to talk to them." It was left that Fries would return Rick's call when he got back to Maryland.

Although this was another frustration, Rick, without knowing it, had hit the jackpot. Bitman and Fries were the only researchers anywhere who, having obtained samples of PBB from Michigan Chemical Corporation, had tested the effects upon farm animals. Their narrow field of research was concerned with the effects upon animals of toxic chemicals, such as pesticides, which inadvertantly entered the environment, and it was their aim to discover whether residues of these chemicals lingered in animal food products and could be harmful to humans. That is why, having worked with PCB, which was known to pose these hazards, Bitman had insisted on obtaining a sample of the new but related chemical, PBB, when Michigan Chemical Corporation introduced it in 1971.

Of all the chemists in the country who had access to mass spectrometers and who might have become known to Rick, the two men working in the USDA research station at Beltsville, Maryland, with the special knowledge they had gained and the equipment they had on hand were literally the only ones in a position to make an immediate identification of the toxin in his feed, and not only that, but also to have a good idea of its effects upon a dairy herd and upon the people who ate contaminated dairy products. They were the needle in the haystack, but neither Rick nor they realized it. A combination of persistence and blind chance had led him to them, but he might have hammered on the doors of scientific establishments for years and never found an answer.

"It just wasn't a coincidence, Halbert finding Fries," said Ted Jackson's assistant, Pete Van Vranken. "It was an incredible fluke—like going all the way to California and bumping into a guy you knew in high school." Yet when Fries eventually returned Rick's telephone call, his reaction to the story of the troubled herd was so cool as to seem almost disinterested.

Rick had been referred to George Fries because Fries had done a lot of research with dairy cattle. But, he explained to Rick, it was not USDA policy to become involved in the comparatively small, local problems of a single farmer. Also, from Rick's description of his herd, this did not seem to be a problem which fell in his field. In a sense, it wasn't. Rick was talking about a massive contamination; George Fries was used to thinking in terms of residues. The effects of each are quite different. Recognizing the inevitability of chemicals in the environment, Fries's special area of concern was to discover what quantities of certain chemicals an animal like a dairy cow was able to tolerate without becoming sick and without any chemical residue showing up in its meat or milk. Rick's animals were sick already. Furthermore, Fries had recently been diverted from his work to investigate another farmer's complaints and had found that "they didn't add up to much." That man, Fries determined, was another of the chronic complainers with whom USDA was burdened from time to time. He couldn't help wondering whether the same was true of Rick Halbert.

On the telephone Rick tried desperately to light a spark of enthusiasm in Fries. From his experience as a chemist he understood the difficulties only too well: "All researchers have a very narrow field of interest and if you don't hit that you are out of luck." He urged Fries to check his story with others who had investigated it: Dr. Allan Furr at the USDA research station in Ames, Iowa; Dr. Donald Grover, veterinarian for the Michigan Department of Agriculture; Dr. Donald Hillman, animal nutritionist at Michigan State University; Donald Hughes, who was in charge of the research at WARF.

George Fries was impressed by the list. He himself had the reputation of being a skilled and impeccable researcher but he was far too busy to waste time on nonessentials. Of the names Rick had submitted the most familiar to him was that of Donald Hillman, whom Fries had known at MSU when he was studying there for his doctorate. He telephoned Hillman, who passed on the opinion he had formed several months back: that the condition of the Halbert cows was probably due to the effects of moldy corn. Since this scarcely seemed like a residue problem, Fries again told Rick that he doubted if he could help. In his standard fashion, Rick then tried to tempt the other man with a little more information—about the halogenation and the indications of a man-made chemical—and urged him to talk to Allan Furr and Don Hughes. Still on his guard, reluctant to become involved in litigation which might develop between a farmer and his feed supplier, Fries telephoned Furr. He also checked with two of his former colleagues at MSU who assured him that Rick, his father, and his brother were widely regarded as excellent farmers.

On Wednesday, April 17, George Fries again telephoned Rick. He said that maybe he had been too hasty in judging the problem to be outside his field of research, and asked Rick to mail him a small sample of the feed. Immediately Rick put about two ounces of it in a package and hurried down the road to the Bedford post office, a small rural station a mile south of his barn, where he mailed it, special delivery. On Friday the nineteenth he made one of his routine telephone calls to Jim McKean at Lansing to ask if there was any more news of Farm Bureau's research. Rick will always remember McKean's response.

"He told me that WARF had just relayed to him some interesting information from the University of Wisconsin. The university researchers had just tested the feed on a low resolution mass spectrometer, and although this is not the kind which shows the fine detail, they had found the halide and identified bromine. It was a tremendous revelation. I nearly fell off my chair."

A halide is a combination of two elements, one of which is a halogen. In PBB the halogen, bromine, is combined with biphenyl at enormous heat and these two chemicals form a highly toxic compound. Rick had continued to suspect that the halogen with which he was dealing was a chlorine and, over and over, had gone through the list of chlorines used in industry which could be contaminants—possibilities like PCB and the chlorinated naphthalene which he had considered months before. The discovery that it was bromine instantly narrowed his search.

"I hung up and called George Fries right away. I was very excited because I had the feeling we were almost there. I asked him if he had received my package and he said no. So I told him about the bromine and described to him the chemical fingerprint on the gas chromatograph as it had been told to me—the shape of the peaks on the graph, their relative sizes, mass numbers, and the order in which the major and minor peaks fell.

"There was a long pause at the other end of the phone. He must have been digging through his files for a matching graph. Then he said the material sounded familiar to him, that he thought he knew what it was."

George Fries was remembering the experiment he had done with PBB two years earlier. He had at first fed it to chickens and then had fed the chicken manure to cows (who will uncomplainingly eat the manure for its nitrogen content) to see if the cattle suffered because of the residual PBB. No ill effects were noted, but Fries had not expected to find any because the quantities given were so small. The Halbert cows, he noted later, had received a thousand times as much PBB as the cattle in his experiment.

"I worked with a bromine which had late peaks like you are describing," he told Rick on the telephone. "It's a commercial fire retardant, polybrominated biphenyl."

"Who makes it?" Rick asked.

"Michigan Chemical Corporation," The name was not easy to forget, considering the trouble Fries and Bitman had had in securing a sample.

Rick was elated. "Then we have the answer. Michigan Chemical Corporation supplies Farm Bureau with magnesium oxide, and we know that there wasn't the usual eight pounds per ton of it in our batch of cattle feed."

"That's probably it," Fried responded. "I bet some idiot made a mistake in the warehouse and you got PBB instead."

Cautiously he added, "Of course this is only a guess and I still have to wait for your sample."

Rick, however, had no doubt. There could be no other explanation, he thought. Fries was almost as certain but, knowing that such a heavy dose of chemical contaminant in the human food chain would involve federal agencies and, inevitably, legal action, he knew he must have irrefutable evidence. And that meant making his own test of the feed on a gas chromatograph and comparing the printout with the one in his files. While he waited to establish this final link, Rick rushed into action.

First he telephoned Farm Bureau Services and described what had happened. Next he telephoned Michigan Chemical's corporate headquarters in Chicago. Rick's impression of the reactions to his startling news was that "Michigan Chemical was incredulous. Farm Bureau Services' attitude, I think, was one of glee. They felt they had found somebody to take the crown of thorns from their head."

The package of feed was delivered to George Fries's desk about noon the following Monday. He shook a little of it into a solvent and injected it into a gas chromatograph. The charts matched, but to be absolutely sure Fries confirmed the test by mass spectography. There was no longer any doubt.

After making several calls to Washington, Fries again telephoned Rick. "I hope it won't cause difficulties for you," he said, "but I have had to notify our people in Washington that there may be a residue problem in your state."

"It was the understatement of my life," Rick commented later. "We did not just have a problem. We had a disaster."

PART TWO

IN SEARCH OF PROOF

They thought it was like Watergate and they called it Cattlegate. But there was no big cover-up. It was not what was done that was wrong, but what was not done by a number of people in authority who did not realize the magnitude of the problem.

—HARRY IWASKO, state official

We were mired in a swamp of ignorance.

—MAURICE REIZEN, state official

9

Chemicals to Save Lives

IF it were possible to slice the mitten-shaped lower peninsula of Michigan across the middle, laterally, the rock strata beneath the soil would be seen in the distinct shape of a dish. Clinging to the sides of the dish is a layer of salt deposits, relic of a time about 250 million years ago when this land was saturated by sea. Salt is close to the surface on the east and west coasts, but in the center of the state, the bottom of the dish, it lies deep beneath a layer of brine which is rich in minerals. Many elements of the ocean are there, but in greater concentration: Michigan's subterranean brine yields sodium chloride, calcium chloride, magnesium chloride, and bromine, and forms the basis of a constantly expanding chemical industry. The sodium chloride is processed into common salt. Magnesium goes into pharmaceuticals, is an ingredient of gunpowder, and forms part of the insulation for electrical wires. Calcium chloride is used for dust control on roads. Bromine and its compounds have a variety of industrial applications from insecticides and fire retardants to an "anti-knock" additive for leaded gasoline.

The Dow Chemical Company (Rick Halbert's former employer) has been at Midland, Michigan, since 1897, when Herbert H. Dow pioneered a process of extracting bromine from the subterranean brine; the company he founded has become one of the largest bromine producers in the world. Less than thirty miles to the southeast, Michigan Chemical Corporation was founded, years later, on the banks of the same Pine River where it would sink wells to similar brine reservoirs. It soon became the largest employer in St. Louis, Michigan, which, with a population of only four thousand, could boast only one other claim to public attention: its geographical situation. "St. Louis," reads the green and white sign at the city limit, "Middle of the Mitten." There above the center of the dish, tucked out of sight behind the white frame houses of the main highway, beyond the war memorial and the Dandy Drive-In Restaurant, developed an untidy conglomeration of concrete sheds, brick warehouses, drum-shaped storage tanks, rusty iron towers, and tall gray chimneys which formed the premises of the Michigan Chemical Corporation.

At one time it was a small subsidiary of a subsidiary of the giant Northwest Industries. But, stimulated by the increasingly stringent fire safety legislation of the late 1960s, it expanded by developing a wide range of fire retardant chemicals, based on bromine—the most effective flame-retarding element. These new products, the company advertised, would "fill the growing need for fire safety in today's world." All were marketed under the trade name of Firemaster, and distinguished from one another by stock letters and numbers which were shorthand for the chemical formulae involved. Firemaster LV-T23P was the company's name for tris (2,3 dibromopropyl) phosphate, a flame retardant developed for incorporation in man-made fibers prior to spinning. Popularly known as Tris, it was widely used for five years in children's night clothes, until someone discovered that it could cause cancer and genetic damage. Firemaster PHT4 was developed to make substances like Naugahyde and tires inflammable. Firemaster BP-6, designed for incorporation with hard plastics, was Michigan Chemical Corporation's version of polybrominated biphenyl—which soon became known throughout the state as PBB.

The cancer-causing potential of Tris was an unexpected setback for the fire retardant industry. Otherwise, it seemed, these new chemicals were a benefit to mankind, greatly reducing the risk of injury and death from fires. The federal legislation which required all children's sleepwear to be flame retardant was the direct result of lobbying by medical authorities who were gravely concerned about the high incidence of dreadful burns suffered by children in home accidents. The child undressing by an electric heater, the infant climbing up to reach a forbidden cookie jar on a shelf above the gas stove could be incinerated in an instant in flammable cotton pyjamas. After the U.S. Department of Commerce set mandatory flameproof standards for children's nightclothes in 1972 the incidence of death and injury to children from fires in the home dropped by almost 50 percent. At that point, before anyone realized that there was a different kind of hazard from wearing fabrics impregnated with the new chemicals, the manufacturers of flame retardants had good reason to portray themselves as guardians of the public safety.

Meantime, in the late 1960s and early 1970s, demand for higher standards of flameproofing was spreading through other American industries. This was spurred by the disquieting discovery that the United States had a fire death rate substantially higher than that of any other industrialized country: in 1972, 12,000 Americans died and 300,000 were seriously burned in fires.

"No one really knows what we were doing differently from other countries but there was no doubt: we had the worst fire problem in the civilized world," said Dr. Frederic Clark, director of the Center for Fire Research at the National Bureau of Standards. "We became aware of this in the late sixties. Before then our fire statistics were not being reported to a central authority or being compared with others. It has been speculated that one reason for our excess of fires may be the fact that this country uses more energy than any other country, so there is an increase in the sources of ignition—more potential for faulty switches, loose wires.

"A lot of fires are caused by careless smokers. So the thinking has been to make more and more fabrics and furnishings and plastics flame retardant. But when a standard for flameproofing is imposed—a standard which says, for example, that no material used inside an automobile may burn at more than so many inches a minute—it is up to the manufacturer to determine how he will meet it. He can use a material which is inherently flameproof. Or he can use a different material, possibly a cheaper one, and add a chemical fire retardant."

The chemical companies filled this need. It had been known for some years that some of the halogens, especially bromines, were good fire retardants. The challenge to researchers was to find a way to use these chemicals without changing the composition of the material with which they were to be bonded. Bromine and chlorine by themselves would probably rot fibers, discolor and soften plastics. In time they might also lose strength. But by the addition of another chemical they could be stabilized. The challenge to the industrial chemist was to find an appropriate formula, arriving at the most effective brominated or chlorinated compound which would be compatible with the material to be made flameproof. There were endless permutations. A formula which would be right for clothing textiles might not work for upholstery fabric. The fire retardant that would bond with the polyurethane in automobile upholstery would be wrong for plastic seat covers, and the chemical used in seat covers could not be incorporated with the hard plastic of a dashboard. These were considerations which governed the development of different fire retardants. It was known that many were highly toxic and there was some concern about potential health hazards to workers handling the new compounds, but in the controlled conditions under which most chemicals are made—or are supposed to be made—most manufacturers felt these could safely be overcome. But, as with the manufacture of earlier halogenated compounds—pesticides such as dieldrin, mirex, kepone, and DDT—hazards developed which no one anticipated. The classic case was Tris. It had been tested to discover whether it was a skin irritant, and it wasn't; whether when incorporated in a fabric, it could be inhaled or ingested, and it couldn't. Not until it had claimed a large section of the market for children's sleepwear did anyone discover that the chemical could be absorbed through the pores of a perspiring, sleeping child.

All this was seven years in the future when, in 1970, Michigan Chemical Corporation introduced PBB. At that time fire retardant chemicals were generally regarded as a boon, and the company was proud of its widening range of Firemaster compounds. This was one of its special projects because no other chemical company in the United States had produced polybrominated biphenyl. Before doing so, Michigan Chemical followed the standard practice of having an independent laboratory test the compound not only to determine whether it would be commercially effective but also to ensure that it would not be dangerous in use. Although this test was required by the Federal Hazardous Substances Act and had some effect of the chemical industry policing itself, the motives were not entirely altruistic. Manufacturers need to be as sure as possible, before they market a product, that there is the least possible risk of being sued by those who use it. For a judgment on the acute toxicity of PBB, Michigan Chemical Corporation went to Hilltop Research Inc. of Miamiville, Ohio.

The Hilltop study was completed on May 22, 1970. From the results of tests on rats and rabbits, it advised Michigan Chemical Corporation that its version of polybrominated biphenyl, known as Firemaster BP-6, "is classified as non-toxic by ingestion or dermal application, is not a primary skin irritant or corrosive material, is not an eye irritant, and is not highly toxic by inhalation exposure." Delighted, the company went ahead with plans for large-scale production of this new fire retardant, which was admirably suited for incorporation with the hard plastics used for telephones, electric typewriters, calculators, hair dryers, television sets, automobile fixtures—a whole range of everyday objects which may be at risk from fire as the result of overheating. It quoted the Hilltop study to potential customers and also added this warning, dated December 28, 1971:

> The acute [single dose] toxicity of BP-6 is relatively low. In common with other brominated and chlorinated hydrocarbons, however, BP-6 is oil and fat soluble. If ingested or inhaled in small quantities over a period of time, we would expect BP-6 and similar materials to accumulate in fatty tissue and in the liver, which certainly is undesirable and possibly could be dangerous.
>
> Accordingly BP-6 should not be incorporated in any product that might

be expected to come in contact with food or feed, and reasonable care should be taken to protect plant workers from dust and fumes of BP-6.

This warning from the corporate headquarters of Michigan Chemical Corporation in Chicago failed to filter down to the workers at the plant who actually made Firemaster BP-6. Charles L. Touzeau, the plant manager and the company's chief officer in St. Louis, testified later that he did not see the warning notice, nor did he know of the dangers of this type of PBB until after the cattle feed contamination was discovered. By this time his men had been working with the chemical for three years. Some of them said that at a meeting of their bargaining committee, Touzeau remarked, half in jest, that BP-6 was "so safe you could even eat the stuff."* Consequently, they were not worried about its toxicity. They were more worried about the personal discomforts of working with this dusty, sticky substance which clung to their clothes. Through the grievance committee of their union branch, Local 7-224 of the Oil, Chemical and Atomic Workers' Union, the men in the BP-6 section made this written request on February 21, 1974:

"We the BP-6 dept. feel we should be furnished coveralls and shower time due to the products we work in, such as HBR acide, BP-6, bromain, toluene, and biphenyl dust. These are both irritating to the skin and very distructive to clothing."

"We were concerned because the area was quite dusty," said Carl Husted, the local's chief steward. "We were not concerned about the product, because we thought the product was safe. Most of us ate our lunches right in the area. It was quicker than using the canteen service, and if you think the chemical you are working with is harmless, why not? The toxicity of BP-6 was known at the head office in Chicago, but not by the people at the plant. When we worked with Tris we did not know how toxic that was either."

The men's written request for shower time and protective clothing was answered the same day, in a somewhat impatient tone.

"To my knowledge there has been no violation by Management regarding this matter," responded Richard Jeffries, assistant to the plant manager. "I feel this grievance is not in order."

The men working with BP-6 wore the same hard hats and safety glasses that were worn throughout the plant and presumably this was thought to be adequate protection. But three months later, a few days after Rick Halbert's efforts led to identification of PBB in the cattle feed,

* Touzeau probably gained this impression from the Hilltop study, the one document on BP-6 which he had seen. Referring to acute—and not chronic—effects it stated that BP-6 "is classified as non-toxic by ingestion or dermal application, is not a primary skin irritant or eye irritant, and is not highly toxic by inhalation exposure."

the men were given not only shower time and coveralls to be laundered by the company, but also respirators, which they hadn't requested.

Many of them soon gave up wearing the respirators because the masks became sticky and sweaty on the inside. Now that they were more comfortable in their coveralls, the men in the BP-6 section saw no reason to make themselves uncomfortable with dust masks. Even after they learned that cattle and farm families were ill from ingesting PBB, Michigan Chemical's workers still did not fear a health risk to themselves. Despite the extravagant joke across the bargaining table, they knew PBB wasn't meant to be eaten, not even by cattle. But they saw no objection, except an aesthetic one, to breathing and handling it. "When you are busy working to provide for a family, that's where your thinking begins and stops," said one man in the BP-6 section.

When Michigan Chemical Corporation began commercial production of Firemaster BP-6, two of its rivals explored the idea of putting their versions of polybrominated biphenyl on the market. After studies with experimental animals, both decided against it on grounds that there were likely to be too many health hazards for people exposed to the chemical. Both studies were less optimistic than the Hilltop report. Chemists for DuPont, the largest American chemical company, found PBB caused liver enlargement in rats, that it concentrated and remained in body fat, and that it would probably build up there to dangerous levels. They recommended against DuPont manufacturing the compound because of its toxicity and the likelihood that it would accumulate in the environment. They also felt that it would be extremely difficult to protect their own workers from the health hazards of any amount of brominated biphenyl.

Researchers for the Dow Chemical Company came up with similar findings. Concerned about the possibility of PBB leaching into the environment, Dr. Perry Gehring, a Dow toxicologist, advised his company that it had "a high potential for producing chronic toxic effects." His formal recommendations concluded, "In light of the experiences with polychlorinated biphenyls, the continued development and use of this compound as a fire retardant appears very unwise." Both the Dow and DuPont studies were published and available to other companies in 1972. DuPont's received particularly wide circulation, being presented at a national meeting of the Society of Toxicology in March of that year. This was a thorough and detailed report, pointing out that while the acute oral toxicity of PBB was low (one of the main points in the Hilltop report) the long-term effects of small quantities in the environment could include liver damage and abnormal births.

Michigan Chemical Corporation continued to increase its output of PBB. After manufacturing a few thousand pounds experimentally in 1969 and 1970, it began full commercial production the following year and continued until November 1974—seven months after the chemical was identified in the cattle feed. Throughout this time, it was the only American producer of PBB, turning out almost twelve million pounds.

Polybrominated biphenyl, or PBB, is not a precise name for the chemical which contaminated Michigan. The formula for PBB is relatively simple but can be varied in a number of ways, and Firemaster BP-6 was one of these variations. It called for the addition of six bromine atoms to every biphenyl molecule, but other versions of PBB could be arrived at by adding more or less bromine. The version with which Dow experimented had eight bromine atoms.

But to begin with the biphenyl—purchased by Michigan Chemical from two other manufacturers, Pilot Industries and Monsanto. This is a white crystalline substance, created from benzene. Benzene, obtained from coal tar, is a compound of hydrogen and carbon, its molecule consisting of six atoms of each element. The carbon atoms are arranged in a hexagon called the benzene ring or benzene nucleus. The six hydrogen atoms are attached to the outside of this ring, one to each of the carbon atoms. A biphenyl molecule is composed of two benzene rings attached to one another by a bond between a carbon atom on each ring.

Biphenyl was first developed in the 1920s, as a heat transfer fluid in the refining of lubricating oils. Polybrominated biphenyls are formed by substituting bromine atoms for hydrogen on the two fused rings that have become a biphenyl molecule. The prefix "poly," meaning many, implies a certain latitude as to how many of the ten hydrogen atoms in the biphenyl molecule are replaced. Hexabrominated biphenyl, which is BP-6 in its purest form, has six bromine atoms and four hydrogen atoms in every molecule. In commercial production this exchange of atoms is achieved by adding a catalyst, usually iron, to bromine and biphenyl, which are "cooked" together in a reactor at very high heat —the higher the heat, the more the bromination and the more chemically stable the end product. The final result is a whole new molecule with properties quite different from those of its components.

Polychlorinated biphenyls (PCBs) are arrived at by the same process, but by using chlorine instead of bromine. Both chemicals fall into a class of relatively new toxic synthetic compounds known as halogenated hydrocarbons (hydrogen and carbon compounds bonded with a halogen), which have been invented to solve some of the problems of our overpopulated, industrialized society. Most of the earlier chemicals of this category were developed as insecticides, fungicides, and weed killers; many of the later ones were fire retardants. They have also taken the form of soft plastics (vinyl chloride), transformer fluids, refrigerants, and aerosol propellants. All of them have a propensity for creating problems which were unforeseen when they were developed. One of the first, DDT, has killed wildlife along with the unwelcome insects, upsetting the delicate balance of the environment.

Ironically, it was in Michigan that some of the earliest disasters with halogenated hydrocarbons happened in the United States. Rachel Carson, the biologist whose book *Silent Spring* created the first awareness of the devastating damage these chemicals could cause, reported that Michigan was one of the first states to receive large-scale air spraying by a chemical of this category. The spraying was done over Detroit, aimed at destroying Japanese beetles. It destroyed birds and domestic animals too. Virtually all the robins disappeared on the huge forested campus of Michigan State University after it was sprayed with DDT in 1954 to halt the spread of Dutch elm disease. Puzzled, researchers at the university thought some strange infection had stricken the wildlife. Dissecting bodies and eggs of robins they were shocked to find traces of DDT. These birds had ingested it from eating worms which had fed from the mulch of fallen leaves. Earlier in the year the leaves had been growing on trees sprayed with the insecticide—a chemical so persistent that numerous rainstorms had not washed it away. The earthworms had concentrated the DDT in their bodies: the same cumulative process that DuPont and Dow researchers predicted might happen with PBB.

Time proved them right. Three years after Michigan Chemical Corporation began dumping its liquid wastes from the PBB process into the Pine River, the state's Department of Natural Resources found low levels of the chemical in the water downstream—and much higher levels in the fish. This is a phenomenon known to scientists as biomagnification, whereby residues which accumulate in organisms low in the food chain multiply in intensity as they are transferred to higher organisms. Hence the earthworms survived while the robins did not. And, like the robins, an animal or human does not need to ingest these chemicals directly to develop several of a variety of ugly symptoms, many of them irreversible.

DDT, mirex, dieldrin, kepone, endrin, heptachlor, chlordane, aldrin, PCB, vinyl chloride, hexachlorobenzene, Tris, PBB—the list of halogenated hydrocarbons created in the laboratory lengthened rapidly during the industrial development which followed the Second World War. If only they could be strictly contained, all of these chemicals could be beneficial. But since this is almost impossible, many have leached into the environment with disastrous results. Some have been identified as carcinogens.

Some are known to damage the liver, the immune system, the central nervous system, or the genes. All can produce a combination of some of these ills. Stored in the body fat, halogenated hydrocarbons assault the liver and other organs when the fat is metabolized. In pregnancy they cross the placenta and enter the fetus. Some have a potential for causing birth defects and most of them can appear in mother's milk. The effects of these tenacious toxins can be carried through the food chain, as it was to the Michigan robins when they devoured earthworms—and as it was to Michigan people, twenty years later, when they drank the milk and ate the meat from contaminated cows.

Oblivious of all this, the men at Michigan Chemical Corporation's St. Louis plant manufactured PBB, their minds as devoid of personal concern as if they were workers in a pickle factory, dumping ingredients into a vat of ketchup. One of the older employees in the BP-6 department, Thomas Walsh, described the process in much those terms:

"You have a big reactor, glass lined, like a seven-hundred-and-fifty-gallon kettle. You put in the biphenyl from fifty-pound bags. Biphenyl looks like Epsom salts and sometimes it solidifies, so we would have to beat the bags with baseball bats to break it up. You open the manhole on the reactor and manually dump it in. You then close the lid and pump bromine in from storage tanks. This is a red liquid and when it hits the air it produces an orange gas. It is very caustic.

"You turn on the steam in the reactor and this melts the biphenyl. There is an agitator, like the agitator in a washing machine, which mixes the bromine and biphenyl together. The reactor is run at very high temperatures for several hours. The bromine goes in in stages until the formula is correct and everything is really balanced out. If you have too much bromine, a catalyst has to be added to burn off the gases. At the end you are left with an amber colored liquid which solidifies to an opaque crystal as soon as it hits the air. It is pumped from the reactor on to a cooling belt and it looks like peanut brittle."

The recipe did not always turn out according to plan. "The stuff would solidify while we were pumping it from the reactor on to the belt," Walsh recalled. "If there was a gap in the steam tracer line, it might build up and cause a block. But if we would open the reactor valve wider to release it there would be a splatter. I must have scraped tons of it off the floor. You could not move in that building without getting it on you. It stuck to your clothing like hot resin. We were all carrying it home with us."

One of Walsh's colleagues complained, "After it came out of the reactor it would just get over everything. It would be exceptionally sticky and the pipes would clog up. When you would take a pipe apart and try to get it unclogged it would gush and be all over the floor." Another man in the BP-6 department told of the difficulty of keeping the product confined to the cooling belt: often it would overflow. He remembered one occasion when the belt snapped and PBB "poured all over the place."

This was in the same area where the men ate their lunches. At the end of the workday they tracked home PBB dust on their shoes and put their contaminated clothes in the family wash.

"We were all pretty dumb," said Ron Orwig, a union official, who worked in another section of the plant. "It's not that we weren't concerned. We were concerned—about making a living and putting bread on the table. And I would think to myself: this chemical can't be dangerous, or the company would give us protective equipment." Orwig did not blame the plant's manager, Charles Touzeau, for failing to warn them of the hazards of PBB. Touzeau was respected throughout the plant as a decent man, and his workers refused to believe that he would mislead them.

"He didn't lie to us," said Orwig. "He just didn't know. But the bosses in Chicago knew all right."

Local 7-224 of the Oil, Chemical and Atomic Workers' Union, which represented the 240 manual workers at the plant, did not have the expertise in its ranks to ask the management probing questions about the nature of chemicals its members were handling. There was no union official at this level with that kind of educational background, and it was a long time before the leadership at the union's national headquarters became directly involved in the welfare of its members at Michigan Chemical. The local had a Health and Safety Committee whose chairman, Charlie Gross, was a shipper in the bromine warehouse. Gross noted that dust was "all over the place" in the BP-6 department, but admitted that he had had no training in public health and did not know what precautions should be taken by chemical workers.

"We had always been very concerned about working with the known toxic substances," Carl Husted explained. "For that we insisted on coveralls and rubber gloves. But this new product, BP-6, was not known to be toxic. Not by us."

The stuff which so innocently resembled peanut brittle could be put to yet another profitable use, Michigan Chemical Corporation decided. Its original form was developed to bond with the ingredients of hard plastics. But, ground to granules, it could combine with softer substances like polyurethane, the material of which upholstery and mattresses are often made. In search of this expanded market, the company formulated another version of PBB and called it Firemaster FF-1. Essentially this was 98 percent BP-6. The other 2 percent was a calcium polysilicate agent commercially known as Floguard, the purpose of which was to convert the crystals to granular form. After this treatment, which did not materially change the chemical properties of BP-6, the product looked quite different. The amber colored crystals were transformed to a crumbly grayish-white powder, which closely resembled the magnesium oxide processed in another section of the plant.

The conversion of BP-6 to FF-1 was not done on the premises. After treating an experimental six hundred pounds of the crystals and being satisfied with the result, Michigan Chemical, lacking the equipment and manpower, contracted with the Cincinnati Chemical Processing Company of Batavia, Ohio, to do the work on a regular basis. Starting in November 1971, at much the same time production of BP-6 began, drums of it were shipped to Batavia for processing, along with a supply of fifty-pound heavy brown paper bags on which was a stencilled identification "Firemaster FF-1" plus the lot number. There was no color-coding on these bags. Michigan Chemical had designed bags for BP-6 with a distinctive red marking and a warning about the need to wear protective gloves when handling the contents, but decided against going to this expense for FF-1 until it was sure of a good market for the product.

There wasn't. While demands for Firemaster BP-6 increased, the FF-1 version of polybrominated biphenyl was not a commercial success. Michigan Chemical sent its last shipment to Batavia for processing in July 1972, then discontinued the item. Over the eight-month period, the Batavia company had sent some FF-1 directly to Michigan Chemical's customers and the bulk of it back to the St. Louis plant, the first delivery in drums and all the rest in fifty-pound bags. There it was stored along with bags of BP-6 awaiting delivery to customers. Some of it would never be sold. In the opinion of Michigan Chemical Corporation's former president, Dr. Paul Hoffman, the procedures for keeping food-grade products, such as magnesium oxide, and the fire retardant apart "should have prevented any misshipments." The practice of producing items for human or animal consumption in the same plant as toxic substances was, at that time, widespread in the U.S. chemical industry.

"There's nothing unusual about a chemical manufacturer producing at a common plant site . . . materials that are destined for food and feed use as well as those for industrial chemical applications," said Dr. A. Fred Kerst, the scientist in charge of research and development at Michigan Chemical's corporate headquarters in Chicago. He cited the rival Dow Chemical's Midland plant, which turned out more than five-hundred products ranging from aspirin to weed killer.

The comparison may not have been fair. Some aspects of Michigan Chemical's operation were so sloppy that one can only hope they were unique. After the cattle contamination was discovered, the company's officers refused to talk to the press except through their attorneys; almost three years later, however, an unusual insight into management conditions was given in testimony from several senior officers of Michigan Chemical Corporation who were subpoenaed to appear in one of the state's circuit courts. One of these men told how workers had so little knowledge of the chemicals they were handling, that "they would have treated all the products the same." Theoretically, the toxic chemicals and those destined for the food chain were kept well apart. In practice, this did not always happen. At one point, PBB was stored in the same warehouse as salt, and there were several occasions when suffocating bromine fumes from the BP-6 section were so pervasive that workers were driven away from the salt-loading dock.

The two reactors in which BP-6 was made were also used for the other toxic fire retardants and were merely steam-cleaned by the operators, never professionally cleaned, so that the starting water for one operation was contaminated with residues from another. This was of no great concern when none of the chemicals was destined for human or animal consumption. However, the brine used in most of the plant's operations was constantly recirculated through its bromine towers so that water contaminated with Tris residues, for example, became a potential contaminant of brine used in the manufacture of magnesium oxide. A memorandum from a senior chemist at the company's corporate headquarters in Chicago to his superior expressed concern about this brine being used to wash feed ingredients, but there was no serious attempt to clean up the operation until Rick Halbert telephoned with the news that Michigan Chemical's PBB had poisoned Farm Bureau feed. Then, said Touzeau's second-in-command, William Thorne, "all hell broke loose."

Thorne and his assistant, Richard Jeffries, admitted that the standard of housekeeping at Michigan Chemical had been a constant concern. There was PBB dust spilled from broken bags in the warehouses; dust tracked from one department to another on the shoes and clothes of workers. Jeffries complained that some men failed to clean up their work areas at the end of shifts, and after his inspection rounds he wrote them notices like these:

> Bagging area is starting to be a safety hazard due to excessive amount of dust present.
>
> FF-1 warehouse pile very poor. Broken bags. Empty drum.

The FF-1 was kept in the same warehouse as BP-6, but as BP-6 production increased the space was inadequate. At that time the excess bags of both fire retardants were stored together in another warehouse along with rock salt. The FF-1 hung around for months. Bags became torn, their powdery contents spilling out. Sometimes men maneuvering fork loaders accidentally ran into the bags with their machinery and neglected to follow the instructions to seal the tears with masking tape.

In June of 1973, almost a year after Michigan Chemical had decided to discontinue FF-1, whole bags and broken bags of the compound were still taking up warehouse space. Recognizing that there was small chance of ever selling them, Thorne ordered Jeffries to get rid of the stuff. He wanted the salvageable FF-1 dumped back into the reactors and reprocessed into BP-6, and all the residual dust cleaned up. Jeffries counted the bags, checked the number against the inventory, and found that about ten of the fifty-pound bags were missing. He reported the discrepancy to Thorne.

After a further fruitless check through all the warehouses, it was obvious to Thorne that there was no way to determine precisely how many bags of FF-1 had disappeared. Some of the compound may have already been reprocessed into BP-6. Some had probably been lost with what Thorne called "normal bag attrition"—swept up or trodden to other parts of the plant after it had leaked on the floor. Thorne was not much concerned about the loss. There was no market for FF-1; he wanted to get rid of it anyway, so he listed the missing bags as being "in process," meaning that their contents were assumed to have gone back into the reactor.

"The physical condition of the bags in the warehouse was so bad," Thorne recalled. "There were ripped bags; they were disarrayed. They had been setting around for so long, when he [Jeffries] told me he was short my assumption at that point was that it was a normal bag attrition of the material in the warehouse that accounted for the shortage. There was no concern in my mind at that time at all that there had been any misshipment." Hence, Thorne did not report the discrepancy to Touzeau.

The misshipment, later investigation showed, probably happened on May 2, 1973, when between ten and twenty of the fifty-pound bags of FF-1 were loaded on a truck along with a delivery of salt and dispatched to the Farm Bureau Services mixing plant. Jeffries missed them a month later. If the inventories at Michigan Chemical had been more precisely kept and if the warehouse had been neatly maintained, perhaps neither he nor Thorne would have dismissed the loss of FF-1 so lightly. An alarm would—or should—have gone out, records searched, deliveries checked; moreover, if the word had reached the Battle Creek feed plant promptly there would probably have been time to prevent the feed and the cows and the people of Michigan from being contaminated.

"It was a human error," Touzeau reflected five years later. "I find it difficult to believe that in any operation you can completely eliminate the possibility of human error. Our procedures were such that I would not have thought it could happen. But it did happen and I'm terribly sorry. I am sorry for the people at the plant because it was an emotional thing for them, and I'm sorry for the farmers and all the other people who suffered."

Several factors, most of them unrelated, caused the human error to be so easy to make. There was the sloppy condition of the warehouse, the failure of Michigan Chemical's corporate headquarters to communicate to its staff at the St. Louis plant the dangers of the materials they were dealing with, the corporate decision to delay the expense of ordering distinctive bags for a toxic substance until its promoters were sure that the product was a commercial success. Added to these disastrous misjudgments was the uncanny coincidence that at the same time as Firemaster FF-1 was being phased out, Nutrimaster, which resembled it, was being introduced.

Farm products were a minor part of Michigan Chemical's operations; nevertheless, the discovery that additional magnesium oxide in cattle feed could increase milk production provided the potential for the expansion of one of its markets. For some time it had supplied magnesium oxide to feed companies in bags labelled Magmaster, or with the chemical's generic name. It was a very small selling item. But in the spring of 1973, with the new knowledge of magnesium oxide's potential for dairy farmers, Michigan Chemical decided to stimulate sales by renaming the feed-grade version of its product Nutrimaster. This time it made an early decision to use color-coded bags and ordered a quantity with a blue diagonal stripe. Unfortunately, there was a paper shortage; the order was slow in being filled. The pre-printed bags did not arrive until late summer. Meantime, Michigan Chemical packaged Nutrimaster in the same large brown bags with the same kind of black stencilled letters in which they packaged Firemaster FF-1.

The compound of errors having been made, Charles Touzeau puzzled over the irony, the incredible coincidence, of workers at Farm Bureau Services making an equally disastrous blunder—and of the two blunders dove-tailing. He wondered why no one, but no one, at the Battle Creek plant noticed that the bags were stencilled Firemaster instead of Nutrimaster. But it was explicable. Not only could the stencilling have been indistinct or the lettering crinkled into the folds of the paper, but also, and more important, most of the men at the feed plant identified magnesium oxide with the label it had had in the past—simply magnesium oxide. Only recently had a few of them learned that the bags with the fancy new name of Nutrimaster contained the same substance. This whimsy of Michigan Chemical for replacing generic descriptions with sales-appealing titles was unfamiliar to them. Magmaster, Nutrimaster, Firemaster, it all sounded much the same. They would never get around to using the new name, no matter what the wording on the bag. "Pass the mag oxide," they would call to one another when they needed that ingredient for the mixer.

Rick Halbert's telephone call to Michigan Chemical's corporate headquarters in Chicago set off a flurry of activity. Just as Rick knew in his bones that Fries was right in his tentative conclusion about PBB being the feed contaminant, so the company's executives took Fries's expert opinion seriously. Like Rick, they acted without waiting for Fries to confirm his judgment by scientific analysis of the feed.

It was the afternoon of Friday, April 19, 1974. Dr. A. Fred Kerst, vice president of the corporation and its senior chemist, had been looking forward to a quiet weekend. Instead he was despatched to St. Louis to take charge of an immediate investigation into the cause of the mix-up. Lynn Hahn, who headed the analytical group of Kerst's research and development department, went too. Charles Touzeau was told to be on hand and to be prepared to close the plant for a few days. Touzeau passed the message down to subordinates. A courier was sent to the Halbert farm to pick up feed and milk samples.

The company's research and management men met at the St. Louis plant on Saturday morning. That, in Thorne's description, was when all hell broke loose. They tried to arrive at a precise count of how many bags of FF-1 were missing. Touzeau, who had only just heard of the discrepancy, estimated thirteen or less. A check with the inventories kept in Chicago revealed an alarming figure: 112,000 pounds of PBB (in either the BP-6 or FF-1 versions) could not be accounted for. Another search through the records showed that some sales had not been counted, bringing the 112,147 pounds down to 9,198 pounds. Touzeau thought even this was an exaggerated figure. He pointed out that some of the chemical was in the remaining inventory and some of the unsaleable FF-1 had already been reprocessed in BP-6, so was possibly counted twice. There was never any way of knowing for sure how much FF-1 was delivered to Farm Bureau Services. The most widely reported estimate was Touzeau's: thirteen fifty-pound bags. Rick Halbert's veterinarian, Ted Jackson, was one of many who disagreed with this. He pointed out that if all the sixty-five tons of 402 feed which went to the Halbert farm was contaminated (and there was no evidence to suggest it wasn't) then, since the formula called for eight pounds of magnesium oxide per ton, more than ten bags of PBB would have gone to the Halbert farm alone. And obviously a great deal more went to other farms.

Michigan Chemical's officers were not sure what form the missing PBB took. Was it BP-6 or FF-1? They were also perplexed about how the misshipment happened. Had Farm Bureau Services received magnesium oxide which had been heavily contaminated with PBB at the plant? Had PBB accidentally been put into Nutrimaster bags? Or had Firemaster in its own bags been shipped out instead? In view of the cross-contamination at the plant, it is not surprising that the first possibility was considered. The second was judged unlikely so far as FF-1 was concerned, because this was bagged in Batavia. For the third alternative—a direct misshipment—no evidence could be found. The company's weekend investigation therefore concentrated on the possibility of cross-contamination at the plant. Lynn Hahn, making tests in the St. Louis laboratory, found plenty.

That Saturday afternoon, April 20, 1974, he tested the first feed samples from the Halbert farm and found a PBB level of 2,700 parts per million, an enormous degree of toxicity. A feed sample from Farm Bureau Services was 4,500 parts per million.*

Hahn also tested magnesium oxide and salt from Michigan Chemical's warehouses and found traces of PBB. Puzzling whether this was the result of contamination in the plant or contamination of his testing equipment from airborne particles, he sent out to a local store for a sealed container of Morton table salt and a bottle of distilled water. He tested samples of both on a gas chromatograph and again got readings denoting the presence of PBB. The reading for Morton salt showed 23 parts per million. Hahn knew that the salt could not possibly be contaminated. The tests were really showing something quite different: that Lynn Hahn was working in a laboratory where the background level of PBB was so high that it was impossible to keep his testing equipment clean. This is a difficulty often confronting chemists but at St. Louis it was so bad as to make testing futile. Hahn telephoned George Fries and asked his opinion.

"He thought we had a serious background problem," Hahn testified. "He said he had run into the same problem. Even if you wash your glassware you can still end up with a detectable level . . . in materials that should not have any." What Hahn was finding was far more than a mere detectable level. It was irrefutable evidence of serious cross-contamination throughout the plant.

The plant was closed from April 21 to 23 while it was thoroughly cleaned. On returning to work, the men were not told about the feed contamination but were advised of the improvements in working conditions—protective measures against the toxins they handled. They were not given a reason, and assumed the management was responding to their union complaint. If by this time any of them had heard about the cattle contamination, which was unlikely, they did not make the

* Immediately after the contamination became known FDA set one part per million as the highest permissible level in milk and meat. This was reduced to 0.3 six months later.

connection with Michigan Chemical. A college student who had worked at the plant the previous summer returned to the BP-6 department in June of 1974 and noticed a changed atmosphere about the place, an unexplained carefulness and concern. "Obviously," he said, "it hadn't reached the papers, but there was a feeling that something had happened and they [the management] were kind of worried."

The story had in fact been given some publicity, but not in the small-town newspapers which were read in St. Louis. So far as is known, the first news story on the contamination appeared in the *Wall Street Journal* of May 8, 1974, and was the result of a call by Rick Halbert (for whom the telephone had become an extension of himself) to the paper's editorial department in New York. It was an unobtrusive piece of copy on an inside page which was not likely to get many people alarmed, or even very interested. Headed "Contamination of Some Dairy Cattle Feed is Indirectly Linked to '73 Paper Shortage," it made the entire incident seem like a passing agricultural curiosity which had hurt Rick Halbert and a few other farmers, and which "may have been responsible for a sharp drop in milk production of some dairy herds, and possibly the death of some calves."

Rick was credited with the detective work which led to identification of the fire retardant. His veterinarian, who had put in almost as many hours, not only in research but in traveling back and forth to the farm to treat hopelessly sick cows, was not mentioned. ("Ted was real pained about that," commented Jackson's assistant, Pete Van Vranken.) The *Wall Street Journal* report stated that officials of Farm Bureau Services and Michigan Chemical were "a bit reluctant to discuss what happened," but quoted Theodore Girard, then president of Michigan Chemical, as saying that the herd contamination was a single incident in which "one or two or three" of the fifty-pound bags of PBB were inadvertently mixed with a shipment of magnesium oxide. Girard blamed the mix-up on a paper bag shortage which had prevented Michigan Chemical from getting its usual color-coded bags for magnesium oxide. "The substitute bags," he said, "somewhat resembled the bags used for the flame retardant."

From this developed the story, perpetuated from one newspaper to another, about the origin of the mix-up. It was not quite accurate but it made good reading. It related how, because of the paper shortage, Michigan Chemical was unable to use either the blue-marked bags for Nutrimaster or the red-marked bags for Firemaster, but temporarily put both products in plain brown bags with black stenciling. (Michigan Chemical never took the precaution of using color-coded bags for FF-1, only for BP-6.) The story went on to tell how an illiterate truck driver, loaded the wrong bags at the chemical company and unloaded them at the feed plant. The truth was less dramatic but just as devastating. The trucking was done by St. Louis Freight Lines, an independent company, and there was no evidence of its employing an illiterate. In any event, as Charles Touzeau sadly admitted, the loading was done by a fork lift operated by Michigan Chemical men: their union would not have permitted an outsider to do this job.

One of the Michigan Chemical employees who worked on the loading dock, Leo Aldridge, testified, "Everybody back there has made mistakes at times. Like you put too many of one thing on the truck, or something like this here [referring to the despatch of Firemaster to the feed plant]. Usually nobody will say nothing. The truck drivers say nothing. You put too much or put the wrong stuff or not enough."

Although the date of this misshipment had been calculated as May 2, 1973; although Rick Halbert's cows began to sicken that September, the Oil, Chemical and Atomic Workers' Union insisted that the Firemaster could only have been despatched to Farm Bureau Services between October 5 and 17, 1973, when the union members were on strike. During those two weeks supervisors and foremen ran the plant. "And," said Ron Orwig on behalf of the local, "they don't know nothing about it, so a mix-up could have been easy."

10

In Quarantine

THE last piece of the puzzle fell into place on April 30, when an inspector from the federal Food and Drug Administration called at a small feed mill at Mendon, twenty miles southwest of Battle Creek. With Dr. George Fries's confirmation of PBB's presence in the cattle feed, the Detroit office of FDA (which Rick Halbert had felt to be so uninterested in his problem) had sprung into action and was checking stocks in feed outlets around the state.

For months farmers in the Mendon area had complained of cattle problems. The troubles of one Mendon farmer, William Coomer, went back to July 1973 (two months before Rick Halbert's), but the Farm Bureau Services salesman who heard Coomer's repeated complaints did not pass them on to his head office until the fall. He assumed that Coomer's cows had been eating moldy corn and that the problem was a local one. Nine months later, the FDA inspector discovered what had really happened.

The Mendon mill, like many others around the state, mixed its own feed, buying some ingredients from Farm Bureau Services. One of these ingredients was magnesium oxide—which meant that Mendon and any other mill which had inadvertently received PBB could have gone on turning out contaminated feed for almost four months after the Battle Creek plant recalled the 402 mixture. Checking through the stock at Mendon, the FDA inspector found an opened fifty-pound bag, supposedly of magnesium oxide. Some of the contents had been used and the heavy brown paper was folded over at the top, which made the stencilled name difficult to read. Dampness had caused the substance in the bag to become hard and lumpy. Someone would have to break up these lumps manually before the materials would be fit to go into the mixer, so the bag had been pushed into a corner and forgotten. When the FDA inspector examined it, he saw that the bag was stencilled Firemaster FF-1, plus a lot number. In all other respects it was identical to a Nutrimaster bag.

His discovery confirmed the suspicions of Michigan Chemical's own researchers. It was the FF-1 variety of PBB which had been delivered instead of Nutrimaster. BP-6 looked so different from magnesium oxide that one could not have been confused for the other. The "Mendon bag," as it became known, provided other vital evidence for subsequent researchers, for it was this material (FF-1), not "pure PBB" (a phrase widely and erroneously used in Michigan, not even literally applicable to BP-6) which poisoned animals and food throughout the state. The contents of the Mendon bag were pure PBB only in the terminology of industry, where it was meant to be a fire retardant. In terms of a toxin whose chemistry must be analyzed precisely in order for scientists to evaluate what it might do to its victims and how, eventually, it might be flushed from their bodies, the substance wasn't pure at all. Like the laboratory equipment at the St. Louis plant, Firemaster FF-1 had been cross-contaminated with particles of other chemicals.

Bill Coomer had been a regular customer at the Mendon mill, where his cattle feed was made to order, and magnesium oxide (or what he thought was magnesium oxide) was an ingredient of Coomer's formula; but no one told him about the discovery of PBB in the "Mendon bag." For nine months he had had terrible problems with his herd. A load of feed which he received in July 1973 must have been heavily laced with PBB. After one feeding from it, most of his cows refused a second. Some became bloated, and one—a formerly healthy animal, due to calve in two weeks—suddenly dropped dead. Coomer complained to the feed salesman, and together they concluded that if the problem wasn't moldy corn, either the magnesium oxide or the mathionine in the mixture was causing trouble. Mathionine is an amino acid, often incorporated in the feed of lactating cows to prevent their forming too much fat from lack of exercise. The salesman, Lance Coplan, tasted the mathionine, thought it had an odd, bitter tang, and suggested omitting this ingredient from the next load of feed. Believing he had found the solution to Coomer's problem, Coplan did not check the magnesium oxide.

Subsequent deliveries of feed to Coomer, without the mathionine, must also have contained PBB. Eventually, when his entire herd had to be destroyed, some of his cows were found to contain even higher levels of toxin than the Halberts'. One tissue sample from Coomer's herd tested at 8,000 parts per million of PBB. Yet more than two weeks passed after the cause of the feed contamination had been established (and more than one week after discovery of the Mendon bag) before Bill Coomer learned about it, and then by accident.

A feed salesman from a rival company, hearing of his dissatisfaction with Farm Bureau Services, had called at the farm in the hope of recruiting a new customer.

"Did you see the report in the *Wall Street Journal*?" he asked.

"No," Bill Coomer replied.

"Wait a minute. I'll get it from the truck."

He produced a well-read copy, turned to page 14 and there Coomer read the news story inspired by Rick Halbert.

As soon as the salesman left, Coomer called Farm Bureau Services, where a staff member confirmed that some feed from the Battle Creek plant had been contaminated. The discovery of the Mendon bag was not mentioned.

Many other small cooperatives around the state received direct deliveries of magnesium oxide from the Battle Creek feed plant, and in the second half of 1973 several of them must have received PBB instead. The Fremont co-op, which served some substantial dairy farms in the area, was delivered at least eight fifty-pound bags of it, according to Paul Greer, the farmer-attorney whose cows were among the first to be contaminated.

"It had hardened so much they had to put it in the grinder before mixing it with the feed," Greer discovered. "The co-op sent four or five of the bags back to the Battle Creek plant, but later no one there had any record of receiving it. It could have gone out again to some other customer."

Dr. James McKean was sent to Fremont to investigate Paul Greer's cattle problem. It was one of several feed complaints which he was following up, along with Rick Halbert's, but as he had observed, every one varied enough to make him think that they were unrelated. Halbert had been feeding his cows the 402 formula made up at Battle Creek, while the Fremont farmers were using a locally produced dairy mixture. McKean may have lacked perception and detective skills, but he was not wanting in honesty when he told Greer that he knew of no other complaint quite like that of the Fremont farmers. It was by then February 1974, McKean knew that the 402 had been withdrawn from sale, so there was little danger of any of that having been diverted to Fremont. However, he had just discovered, or was on the point of discovering, that a gas chromatogram of the Fremont feed produced the same late peaks as Halbert's 402, but he had been unable to determine why. He did not suspect anything wrong with magnesium oxide because he had recently been assured of its purity by Michigan Chemical Corporation and he was theorizing that perhaps some local condition of soil or climate was responsible. Farm Bureau Services did, however, feel there was now enough justification for Halbert and Greer to submit claims for damages, and advised them to do so, in the knowledge that the company carried liability insurance.

"Our milk production had dropped to less than half," Greer said. "The cows were aborting, trying to deliver dead calves. The calves were dropping like flies. We were shooting them every day. We lost ninety percent of our calves, and buried them on the hill." Greer's farm was one of the most highly contaminated in the state, but, with no inkling of the cause, he said he "blamed everything and everyone but PBB." He blamed the young man who was managing the farm for him, he blamed his son, who was helping, he blamed himself for getting into this expensive hobby of agriculture and not confining himself to the law. Sometimes, to forget his troubles, he drank an extra martini; then he was even more ready to shift the blame. It was a tragi-comedy which was played, with variations, in dozens of Michigan farmhouses that winter and throughout the succeeding year: the farmer convinced that his management was not at fault, yet mystified by the worsening state of his herd, growing suspicious that his wife or his sons or his hired hands—whoever did the feeding and milking—had to be neglectful, or worse. And those being blamed, knowing they were doing their best, turning their own anger and distress back on the farmer until he began to doubt his abilities as well as theirs.

In many cases this situation was worsened by the fact that some farm families who had eaten their own contaminated produce had begun to suffer physically and psychologically from the contamination. They were not "themselves," and the strengths of mind and body so essential to farm work were faltering. This did not happen to Greer or Halbert or on many of the farms which suffered such high-level contamination that its results were immediately devastating. In that sense, they were fortunate. Most of them saw enough wrong with their animals to stop eating their own meat and drinking their own milk early in the course of the disaster—although the little they had was probably too much for safety.

Throughout the winter of 1973–4, Paul Greer and his neighbors in the Fremont area had no idea that there were any other Michigan farmers suffering. But on May 8 Greer was leafing through the *Wall Street Journal* when he came upon the news story which had caused Bill Coomer, almost two hundred miles south, to drop his farm chores and pick up the telephone.

"So it was Farm Bureau's feed all the time," Greer muttered. "The bastards."

He did not know it, nor did Coomer or Halbert, but the Michigan

Department of Agriculture was already testing the bulk milk tanks in dairies around the state for evidence of PBB. It was a crude analysis because every tank contained milk from a number of farms, and if only a few of those farms were highly contaminated and the rest all right, the dilution factor could prevent the contamination from showing up. The inspectors were looking for a PBB level of one part per million or more, which was the lowest level their equipment was capable of testing, and far higher levels than this must have passed inspection because of the unrefined nature of the test. Conversely, when PBB did show up in a bulk tank test, it probably indicated a highly contaminated herd. Such farms were tracked down by the Michigan Department of Agriculture which also searched the records of feed mills to trace farmers who could have bought contaminated feed.

On Friday May 10, 1974, eight Michigan dairy farmers were served with quarantine notices and told they could no longer send their milk to market. They were Paul Greer, Blaine Johnson, and Myron Kokx, in the Fremont area; Rick Halbert near Battle Creek; Bob Demaray and Art Laupichler in the Yale area; Jerome Petroshus in Allegan; and a beef farmer, Carroll Robinson, of Grant. Anyone pinpointing these farmsteads on a map could see how widespread was the contamination—across southern, eastern, and western Michigan—and that was barely the beginning. Five more farms were quarantined the following Monday, three on Tuesday, including Bill Coomer's. By the end of May the total reached thirty-nine. One was a few hundred miles north of the rest, in the state's Upper Peninsula, owned by a beef cattle breeder who had bought a load of feed on a recent trip south. The more contaminated farms MDA discovered, the more its inspectors realized they were barely scratching the surface of the problem.

Robert Rottier, a Fremont farmer who was quarantined in May, made this prophetic comment, "I believe that we affected farmers are in a disaster situation, through no fault of our own, and the best way to come out of it is to destroy all the sick animals and start all over again." He was right, but no one in authority was listening.

Rick Halbert was devastated by the quarantine. Throughout his months of research he had been a most careful planner, analyzing every move before he made it. Yet this was a consequence he had not remotely envisaged. He had assumed that once he was able to prove that the blame lay with Farm Bureau Services he could collect some compensation money, use it to restore his herd, and get back to full milk production. Instead, he was arbitrarily cut off from his milk market, the only source of income to the farm during months of overwhelming losses.

"When George Fries identified PBB in the feed my one thought was, This is it, this is finally it. I thought our troubles were almost over, that the insurance company would see the light and we could soon get back into full production," Rick said. "I had no idea that PBB was in the milk. But when Michigan Department of Agriculture quarantined our farm, that turned the whole thing around and we were in even more trouble than before. It is one thing to have PBB. It is another thing to have no income. It was costing us eight hundred dollars a day in feed and labor to keep our animals, and once we were quarantined there was no way of earning a penny to offset it." Up to then the Halberts had been sending all of their milk to market.

All the quarantined farmers experienced the same sense of shock. Once the cause of their herds' sickness was identified they had expected quick financial relief. Instead, they were being frozen out of business. There was no precedent for a disaster like this, no ready formula for the kind of government help which is available when the cause can be blamed on the Almighty. Nobody wanted to shoulder responsibility for the series of human errors which had begun in the chemical company, been compounded in the feed plant, and had spread through scores of dairy farms in rural Michigan. Nor did state authorities want to broadcast the fact that contaminated milk and meat must have been sold for months in urban supermarkets across the state.

"As time went on, the situation became more and more unworkable," said Rick. "We had to find a way of feeding animals of no value —keeping them alive as evidence—while Michigan Chemical Corporation and Farm Bureau Services squabbled and sat on their thumbs. We were more or less at the mercy of the two of them. The state did not come up with a solution, and either we had to go on feeding our useless animals or let them starve to death."

With this much to worry about, the quarantined farmers gave little thought to the potential harm done to their families by months of consuming their own contaminated meat. Milk was of less concern to them, since few modern dairy farmers drink their own milk. Pasteurization and bottling is an industry in itself, so most farm families sell all their milk in bulk and buy milk for home use at the local store. Meat is another matter. Most dairy farmers do their own butchering, store the proceeds in their home freezers, and rarely buy meat at retail prices. This habit was so deeply ingrained in the Halbert family that as late as May 1974, the month after PBB was identified as the contaminant, they butchered two animals.

"We had eaten a very little of the meat when we had thought to have it tested," Rick said. "We found a PBB content as high as ten parts per million."

Every day the Halberts flung truckloads of contaminated milk into the woods on their farm. It had to be disposed of somehow, and other quarantined farmers found similar solutions. "At the bottom of my farm there was a large gully," Paul Greer related. "I let the neighbors dump their milk there along with mine."

Not knowing the potency of PBB and the strange way in which it adhered tenaciously to soil, none of them realized that they were contaminating the land.

At the Battle Creek feed plant Paul Mullineaux, production manager, was doing his best to find out how his company's part in the disaster happened. He had no inkling that anything serious had been amiss until the end of April, when Rick Halbert broke the news that George Fries had found a highly toxic chemical in the feed. In the preceding months Mullineaux had gathered only small gleanings of the various farmers' complaints, and to him they seemed unexceptional. He was used to farmers grumbling. Inevitably, among the feed plant's long list of customers there were always some dissatisfied, just as there had been some who had told him that 402 was the most successful feed they had ever used.

In late 1973, Mullineaux knew there had been suspicions about the apple pomace in 402, and that it had been checked for the presence of pesticides. He remembered Jerome Petroshus returning a load of 402 that December, and John Williams complaining about his chicken feed a month later. Since he was not included in many of Farm Bureau Services' regular staff meetings, Mullineaux knew no details of the other feed problems, including Halbert's. But a time came when the company veterinarian, Dr. James McKean, told Mullineaux that production of 402 must cease. He was clearly passing on an order from the company's headquarters in Lansing, and Mullineaux did not ask for a reason. He was not that kind of man.

Paul Mullineaux was properly proud of the Battle Creek feed plant. He had been in charge of it since January 1969, before the building was even completed, and some of his ideas were incorporated in the design. It was an efficient design, so well arranged that when the first PBB suit against the company eventually came to a Michigan court, Judge William R. Peterson made a point of praising it in his decision. "A wonder of computerized automation," he wrote. And so it was, with systems that weighed and moved precise quantities of ingredients by conveyor belt so that huge batches of feed could be mixed by the operation of levers and buttons. Only the ingredients which were needed in very small amounts, like magnesium oxide, had to be added by hand.

Mullineaux was a good manager, a man with an orderly mind who ran a tight ship. He had grown up on a farm and was superintendent of a feed company's plant in Indiana for nine years before joining Farm Bureau Services. Yet in middle age, with all this agricultural experience behind him, it was the five years he had spent as a staff sergeant in the U.S. Marines which had left the strongest impression on his character and bearing. Here he was, a quarter of a century later, shoulders squared, gray hair brushed briskly back, still carrying himself as though he were on the parade ground, never questioning an instruction, unwavering in loyalty to the powers which paid him.

Now his employers were in trouble because of something that went wrong at the feed plant, and it was his duty to find out when and why. Immediately after PBB was identified in Halbert's feed, while the Battle Creek plant was closed for a massive cleanup, Mullineaux asked his workers to try and reconstruct everything that had happened at the plant in the latter part of 1973, in the hope that someone might dredge from his memory some detail which might explain how the mix-up had come about. He urged them to be frank and to give him all the information they could. He cautioned them not to discuss the matter with any outsider—not to customers, certainly not to the press, not even to their own families. Mullineaux's warning was interpreted by the workers as a threat that if unfavorable gossip about the contaminated feed were spread, the plant might have to close down and that would be the end of their jobs.

This secrecy might have endured had it not been for an action taken by Mullineaux's employers which redounded in the worst way on themselves. Besieged by damage claims from farmers, Farm Bureau Services sued Michigan Chemical for $270 million. After many months of negotiation the case was settled out of court, reportedly for a mere $18 million; thereafter the two companies were unlikely allies in paying some farmers' claims and in defending themselves against others. Although their attorneys never seemed quite comfortable in each other's presence, they maintained a united front and, as time went on and the farmers' claims increased, would like to have erased from the record the evidence they had collected about one another when they were adversaries.

This evidence was in the form of pretrial depositions taken by both sets of company attorneys when each was trying to uncover the weaker aspects of the other's case. Some of it—which could perhaps answer questions which continue to haunt researchers—may never be made public. But some of the depositions by workers at the Battle Creek feed plant were filed, according to legal procedure, in the Missaukee County Clerk's office at Lake City, available to anyone interested enough to read through hundreds of thousands of words of typescript. The fact that these particular depositions went into the public domain was probably an accident. It would have been prudent for the documents to have been filed under seal, a nicety which the attorneys could have requested. But the request was not made; and some testimony by Mullineaux and his workers, containing the kind of detail he strove so hard to keep within the confines of the company, became, in the fall of 1975, just as available as if it had been placed on the shelves of a public library.

The depositions, taken earlier that year, told a revealing story of Mullineaux's attempt to piece together the events that led to PBB being dumped into the cattle feed. There was a mixer operator at the plant named Charles Szeluga, generally known as Chuck, who casually stepped into Mullineaux's office one day in July 1974 as he returned from taking out a truck delivery. What Chuck Szeluga then said—or, rather, what he afterwards said that he said—was devastating.

"Y'know, Paul, I got to thinking. It seems to me that at one time I saw in the plant out there two skids of mag oxide that there was something different about, and I called you and asked you should I put them together with the rest and you said yes."

This alleged conversation on the plant's internal telephone system happened a year previously, in July 1973, according to Szeluga. Mullineaux replied that he had no recollection of it. Szeluga racked his brains a little harder and recalled what had struck him as different about the magnesium oxide on the two skids. The bags on one skid were labelled Firemaster and the bags on the other were labelled Nutrimaster: it was these which he insisted Mullineaux had told him to put together with the rest of the magnesium oxide.

Mullineaux still denied any memory of the event and, long after the depositions were taken, continued to deny it under oath in court. Even so, his sense of rectitude was such that after Szeluga left his office Mullineaux telephoned his superior, Donald Shepard, manager of Farm Bureau Services feed division in Lansing, and repeated Szeluga's statement, damning as it was against himself. Although Mullineaux found it difficult to believe, he felt it was his duty to report the conversation.

Szeluga's story remained unsubstantiated for nine months. Then on a Monday morning of the following April, two days before his deposition was to be taken, David Hulet, a mixer operator, announced to Mullineaux, "Paul, I've been thinking all weekend. I believe I remember seeing Firemaster out there in the plant."

Controlling his impatience, Mullineaux responded, "Dave, you have been asked specific questions about this many times and you never recalled seeing Firemaster before. Now suddenly you remember it?"

"Yes, I'm sure of it."

"Dave, you be positive of what you are saying. Remember you are going to deposition and you must tell the truth. If you really believe you saw Firemaster in the plant you must surely say so. But be sure you tell the truth."

Among their fellow workers who testified there was a division of opinion as to whether Szeluga and Hulet had permitted their imaginations to color their memories. No one else at the plant could remember seeing Firemaster. The two men's stories were nevertheless widely accepted, and for ever after the "authentic" account of the feed mix-up, told and retold in press reports, was one in which a worker at the Battle Creek plant drew his overseer's attention to some bags labelled Firemaster instead of Nutrimaster, and in which the overseer allegedly responded something like, "It's all the same thing. Put it in the mixer." This made a neat addendum to the tale of the illiterate truck driver, and both stories were plausible enough to be regarded as accurate.

Six and a half years later, in a separate court case, a psychiatrist testified that Szeluga still suffered an intense depression which "disabled him from the enjoyment of life," believing himself to be personally responsible for the contamination of Michigan.

How ever it happened, the mistake was easy enough to make, given the dangerous practice of chemical manufacturers to call their products sometimes by generic names, sometimes by trade names of their own invention—magnesium oxide, Magnacite, Magmaster, Nutrimaster—and given the educational level of most of the men who did menial jobs at the Battle Creek plant.

"They're not hiring Rhodes scholars out there to mix the feed," commented Ted Jackson's veterinary assistant, Pete Van Vranken, who had clients in the neighborhood. "The bags of ingredients come off the delivery truck and are stored in the warehouse, a different pile for each ingredient. When it comes to dumping ingredients in the mixer you take the bags off the piles as you need them. If you ever ran a fork lift you'd know you don't stop to read the lettering. You just look at the color coding on the bags. I'm not sure I'd have caught the mistake myself."

One former worker at the plant, a man most likely to have handled the Firemaster, admitted that he had difficulty reading anything. His name was Ronald Jex, and it had been his job to unload the incoming trucks. He remembered one driver, "a white man called Shorty," who drove the truck that brought in Michigan Chemical's weekly deliveries of loose salt, often with bags of some other feed ingredient—Jex did not know what—spread on top. When the truck arrived, Jex would go to the control room and open the chute for the salt bin. He would take the bagged ingredients to the warehouse and ask whoever was in charge where he should stack them. When Jex's deposition was taken, this exchange took place between him and L. Roland Roegge, attorney for Michigan Chemical:

Roegge: How well do you read?
Jex: Not very good.

R: If you saw a word that said Nutrimaster, would you be able to read it?

J: No.

R: If you saw a bag that said Firemaster, would you be able to read it?

J: I believe so.

R: Why would you be able to read Firemaster and not Nutrimaster?

J: Because I have seen the word fire more than I did—

R: The Nutri—?

J: Yes.

R: But you say you do have some difficulty reading, is that right?

J: Quite a bit.

Later in the day's questioning, Paul Greer, representing some of the affected farmers, pursued the same line of inquiry. Jex had volunteered that in 1973 he unloaded "just about every bag" that came in on Shorty's truck. Greer asked if he could remember the markings on any of the bags.

"I can't say for sure," Jex replied.

"Why can't you say for sure, Ron?" Greer inquired. The question gently asked, disarmed Ronald Jex of his last shred of pretence. He answered,

"Because I don't know how to read."

Greer wanted to be sure he had this right. "I'm not trying to embarrass you, but isn't it true that you can't read properly, or enough to be able to see what was on those bags?"

"That's correct," Jex replied.

11

Recontamination Begins

EVEN with the benefit of a higher education, there was at least one college graduate in the warehouse who could not decipher the lettering on the bags. It was sometimes smudged and the trade names gave no hint of the product. Jack Galbreath had often complained about them. Picking up a bag with the blurred lettering Nutrimaster, he would demand imperiously, "What the hell is this?"

"Mag oxide," someone would tell him.

"Then why the hell doesn't it say so?" And, grumbling about the inanity of chemical manufacturers, he would—without further question—dump the contents into the mixer or load the bag on a truck for delivery to some farmers' co-op. Later he said if he had been told that Firemaster was yet another name for magnesium oxide, he would not have hesitated to load it as such.

Galbreath was an oddity among workers at the plant. A thirty-nine-year-old bachelor at the time of the feed mix-up, he had taken a menial job at Farm Bureau Services although he held a master's degree in education. After teaching chemistry for seven years, he was studying for a doctorate when he suddenly decided to leave the world of academics.

"I wanted out," he explained. "I didn't want any more secondary school teaching. I was disillusioned, fed up with the doctoral program, and if you are going to stay in education as such you need a doctorate, so I just plain quit."

He had a business interest in a bookstore in the nearby town of

Kalamazoo, wanted to stay in the area, and found it difficult to get a job outside his own field. Employers regarded him suspiciously as either under-qualified or over-qualified. Farm Bureau Services had not been so fussy. So there he was, earning, as he said, "a great big four dollars and twenty-three cents an hour," after promotion from the job of catching bags from the bagging machine to that of mixer operator. Perhaps it says something for the tedious nature of these jobs and the mechanical way in which men come to do them that even a worker with Galbreath's intellect and knowledge of chemistry could, for different reasons, have been just as responsible for putting the wrong ingredient in a batch of cattle feed as the almost illiterate Ronald Jex. When the men's depositions were taken, none of the attorneys' questions was able to elicit who was actually responsible, and it might have been unfair to pin the blame on any one man. Clearly there was some carelessness at the feed plant, but at least its employees believed that the ingredients they handled were edible and that a little too much of one could not do devastating harm. It was quite different from the carelessness of a chemical company which put euphemistic names on toxic products, failed to put "poison" signs on the bags, and was so casual about storage conditions that toxins came to be confused with items destined for the food chain.

Nevertheless, conditions at the feed plant were conducive to error. Galbreath said that sometimes the computer (which governed the flow of different ingredients into the mixer) did not work properly, and he remembered an occasion when "there was a mix-up of urea in the bins." Most of these odd mistakes seemed directly related to the very limited skills of some of the workers, and lack of knowledge about the products they handled. Mullineaux, with his orderly mind, must have been sorely tried.

In his deposition he told how he had instructed Dave Hulet and a twenty-year-old employee named Wayne Edwards about the precautions that had to be taken when running the mixer.

"The formula was given to them, how much to put on the scale. They would select the bin [by operating a switch on the computer] and run it until they got the weight by watching the dial." Mullineaux said he went over this very carefully with them, explaining how federal regulations required them to maintain a log of quantities used and how they should check the names of all ingredients on the bags. He instructed them to come to him with any queries. Many of these queries, he said, were about the contents of the bags, although the men soon became familiar with trade names. Mullineaux recalled that, aside from being called Magnacite, Magmaster, and Nutrimaster, magnesium oxide had "two or three other names I don't now remember," and mathionine had the trade names of Hydan and MHA. Jack Galbreath was justified in wondering aloud why the hell manufacturers were so obtuse.

Night shifts were often worked at the plant, and Paul Mullineaux could not be there round the clock. In his absence, workers would sometimes go to warehouseman Jack Weber. Ronald Jex related how when he unloaded the bags from Shorty's truck—the bags whose names he could not properly read—he would ask Weber where to stack them. But around the spring of 1973 Jack Weber took several weeks off to undergo an operation, and a laborer at the feed mill, Robert Wonegeshik, temporarily took his place. Wonegeshik, who left Farm Bureau Services after five months "because I felt like it," said he had had no training for the warehouseman's job and admitted there were two products he kept getting mixed up "because they was both about the same thing."

Asked if he remembered an ingredient named magnesium oxide, Wonegeshik replied, "I'm not sure on that. I think there was but I don't know. There was different kinds of magnesium stuff but I don't remember what the name was that they used in there. Seems like there was some that they used in there but I don't remember the name of it." Questioned further, he said he had never heard of either Nutrimaster or Firemaster. He was better able to remember the bags that had red markings and the bags that had green, not the words imprinted on them.

Another employee who worked as a bagger, William Nichols, testified that if he handled a bag with Nutrimaster or some similar inexplicable word on it he would ask the foreman or plant manager what it meant. But, he added, if he were working the night shift "there would not be anybody to ask."

Wayne Edwards, the young man who worked with Chuck Szeluga and Dave Hulet on the feed mixer, was impressed by the poor labeling of the magnesium oxide bags. "You had to probably pick up two or three bags and dig down and find out whether they said magnesium oxide. It was stencilled and blurred a lot of times." He indicated that if there were a bag whose printing was illegible and it was in the magnesium oxide section, he would assume it was magnesium oxide.

"Wouldn't you ask somebody about it before you put it in the mixer?" he was asked. At first Edwards replied that he supposed he would, but when the question was pressed he admitted, "If it was on the skid that said mag oxide I would assume it was mag oxide, being on the same skid, being that it came in that way." He added that had he seen the word Firemaster on one of the bags, "it would have meant nothing, only the skid. If I found a bag on the skid that said mag oxide I would assume that's what it was."

An attorney reminded him, "Didn't you say earlier that if you didn't know what was in a bag that was labeled with a name on it you had not heard of, that you would check what it was first?"

Edwards replied, "It would be doubtful that I would notice anything printed on the bag if it was on the skid with the mag oxide." He then added, "Anyway, you couldn't check through it bag by bag because they weren't labeled that well. There was no information whatsoever, hardly." Asked if he had ever complained to Mullineaux about the poor labeling, Edwards commented, "It wouldn't have done no good to complain to him. He couldn't have done nothing about it."

Once PBB was identified in the cattle feed, Mullineaux and his superiors did plenty. Early in May 1974 the plant was shut down for a few days and cleaned out. All the magnesium oxide was moved out of the warehouse and into a storage garage, along with the 402 pellets previously returned by farmers and now known to be contaminated. A few months later all this was buried at landfill sites designated by the state's Department of Natural Resources. Officials of Farm Bureau Services thought they were handling the problem efficiently, but they had no real idea of the nature, toxicity, and persistence of PBB. Removing every last trace from the mixing plant was like trying to rid a room full of furniture of every last speck of dust. They washed and flushed and scrubbed but still traces of PBB remained. It could not be entirely cleansed from the mixer, at least not until a great deal more feed had gone through, every batch incorporating within itself a residual—yet still measurable—amount of toxic residue. It was not known until many months later that PBB, being soluble in fat, clung to the soybean oil which was often used to flush the mixer; not only that, it also had unexpected electrostatic properties, causing it to adhere tenaciously to metal and concrete. Thus particles of it defied the thorough vacuuming and hosing which were supposed to have cleansed the plant, and remained in the mixing and bagging machinery.

Mullineaux's men were ignorant of these facts. (From the testimony, some of them seemed only dimly aware of the enormous consequences of the disaster in which they had participated. Jex remembered Mullineaux calling the men together and telling them there had been a mix-up in the feed, but he had no idea what kind of mix-up. He got the impression that the main reason for Mullineaux's annoyance was because he believed some of the men had been "goofing off." Jex felt this criticism was unjustified and wasn't at all sure what the fuss was about.)

Around this time, early in May, Rick Halbert was anxious to get rid of the mound of poisoned pellets which remained in his father's barn. He no longer needed them for proof of his claim; so, keeping a few hundred pounds back against the possibility of future research, he asked one of his farm hands to take the rest to Battle Creek. The man returned with a story which horrified Rick. He said that the load of pellets had been accepted and were taken inside the plant. Rick immediately picked up the telephone.

"You guys have got to be crazy," he exploded at one of the plant's employees. "You'll contaminate the place all over again."

"You've got the story wrong," Rick was told. "Your feed is not being kept in the plant. It was put into the garage for storage."

Rick checked back with the employee who had driven the truck. This man explained that although he did not actually dump the feed in the plant, he was instructed, on arriving there, to put it on a conveyor belt which took it inside the building. It had to go inside to be bagged before being stacked in the garage. The bagging operation involved running the feed through the entire mechanical system, from the bulk receiving area into the mixer and thence to the bagger. Obviously the plant workers still thought that any carry-over contamination from the PBB could be removed in its entirety by running flush material—oil and grain which would be discarded—through the machinery before its next use. Rick was appalled at this lack of understanding. He telephoned Farm Bureau Services a second time but felt he could not get anyone there to comprehend the implications of the error. The men at the plant probably thought they were using supreme caution in bagging the contaminated feed, rather than letting it lie around loose. They had no idea that they had recontaminated the mixing and bagging machines. "Whoever ordered feed from Farm Bureau that day certainly got a shot of PBB," Rick remarked.

The ironic twist to this tale, unknown to Rick, was that Farm Bureau itself was the next recipient of feed from the plant. Among its diverse agricultural operations, Michigan Farm Bureau had an egg marketing division which owned about 250,000 chickens at various depots around the state. In the past, feed for these birds had been purchased locally, but then it was decided to buy all the feed in bulk from Farm Bureau's own affiliate, the plant at Battle Creek.

Having bagged the contaminated Halbert feed, Mullineaux's men cleaned the machine by running through soybean oil and feed material which were afterwards thrown away. This was adequate when flushing the machinery of feed containing antibiotics, but hopelessly inadequate to remove PBB. The next item to go through the machine was a load of urea for Farm Bureau's chickens. As the direct result of this action, all these birds soon became contaminated with PBB and had to be destroyed.

Farm Bureau Services has often been accused of a contemptuous attitude towards Michigan farmers by knowingly sending out feed which contained trace levels of PBB, while assuring these customers that it was safe. The incident of the poisoned chickens demonstrated the innocence—and ignorance—of the officers of Farm Bureau Services. They had never had to deal with a crisis of this kind before; they had no staff toxicologist to advise them; and employees at all levels failed to understand that PBB could not be excised from the plant machinery in the same familiar fashion as antibiotics and moldy corn.

Farm Bureau Services has also been accused by farmers and their attorneys of trying to recover some of its investment by deliberately reprocessing PBB-contaminated feed with uncontaminated feed, relying on the dilution factor to bring the PBB content down to a "safe" level. This is standard practice in the industry if a feed mixture is found to have too much of one ingredient or another, and usually it does no harm, none of the normal feed ingredients being poisonous. Farm Bureau Services strenuously denied doing this with the PBB-laced feed. However, because of the very high residual levels of PBB in feed which continued to come out of the plant weeks after the place had been cleaned out, not just once but several times, many farmers refused to believe this denial.

The fact was, Farm Bureau Services simply didn't know how to get rid of the PBB. None of the standard solutions worked and, modern as it was, the mixing and bagging machinery was wide open to cross-contamination. Although the place was cleaned out nine times in 1974, it was revealed, long afterwards, that levels of PBB persisted in feed manufactured there until 1977—and the feed plant's management could not understand why. After the first two cleanups, at the end of April and in early May 1974, the PBB level was reduced to less than 0.5 ppm in a sample taken that August. If, as Farm Bureau Services insisted, contaminated feed was not being recycled, there had to be something dreadfully amiss about the way in which the plant was operated.

Minutes of a Farm Bureau Services staff meeting on August 5, 1974, made the bleak observation: "PBB throughout the plant at Battle Creek. Could contaminate animals." Meantime the farmers were being told that there was nothing to worry about. And, curiously, most of the federal and state agencies seem to have accepted the feed company's word.

Mullineaux said he had never heard of PBB before the FDA called him during the first week of May 1974 and asked him to send a man to pick up a bag of it at the Mendon mill. Doubtless none of his superiors had heard of it either. Aside from the limited research by George Fries and Joel Bitman, nothing was known of its effect upon animals. In that sense, the ignorance at Farm Bureau Services was justified. But by 1974 a great deal was known about halogenated hydrocarbons. Much had been written about PCB, DDT, dieldrin, mirex, chlordane, and others.

It was well known that these chemicals had killed wildlife and polluted soil and waters in the western world, that their toxicity was intense and long-lasting. There was no reason to suppose that PBB, a member of the same chemical family, would be less pervasive. It was this basic chemical knowledge which caused Rick Halbert to wonder whether the workers at Battle Creek were crazy when they ran his contaminated feed through the bagging machine. But he was unable to impress upon them the mistake they were making.

Innocently, the same error must have occurred several times before. When Jerome Petroshus returned some 402 to the Battle Creek plant, the previous October, it had undoubtedly been run through the bagging machine for easy storage in the garage. The same thing happened on a much larger scale the following January after the statewide recall of all 402 from farms and co-ops. Every time this contaminated feed was run through the warehouse machinery, the feed which followed it inevitably picked up PBB. The higher the fat content of this subsequent feed, the more PBB it was likely to incorporate. Consequently PBB contamination was much more widespread than was first suspected. It was in the lower-grade feed which went to beef herds. It was in feed sold for sheep and chickens. It was in medicated feed like Aureomycin Crumbles, which the Battle Creek plant made up two or three times a month for American Cynamid and sold under the latter's label. It could have been in any item that went out from the plant between the spring of 1973 and at least into the spring or summer of 1974—a period during which Farm Bureau Services must have manufactured at least 50,000 tons of feed. (The plant produced 52,000 tons of feed in the calendar year 1973 alone.) PBB must also have been in feeds produced by less sophisticated mixing machines at local farm cooperatives throughout Michigan, and it was already in some of the salt produced for dairy farm use by Michigan Chemical. The small cooperatives had even less technical ability to eradicate PBB from their warehouses than the feed plant at Battle Creek.

Almost a year later, in March 1975, Michigan Department of Agriculture inspectors found unacceptable levels of PBB in sweepings at three grain storage elevators. One was the Fremont co-op, patronized by Paul Greer and his neighbors. Another was at Caledonia, near Grand Rapids, and the third was at Stanwood, near Big Rapids. The department ordered the grain elevators closed for a few days and thoroughly cleaned, but even this belated precaution was not fully effective. In March of 1977 farmers in the Stanwood area were still suspicious that the feed they were buying was not completely "clean" and surreptitiously organized a test of some of the dust at the Stanwood mill. It was collected by a mill employee, Michael Creighton, whose brother Ron was one of the affected farmers. The farmers sent half a cup of the dust for laboratory analysis and discovered it contained trace levels of PBB. Michigan Department of Agriculture continued to insist that the mill was free of contamination, and Creighton was fired from his job. The reason given for his dismissal by the mill manager, Steve Carr, was that Creighton had engaged in a covert activity, that of helping himself to company property—dust!—without asking permission.

Creighton's discovery happened more than three years after the original contamination. Around the same time it emerged that the order to clean up yet another cooperative, McBain, had never been carried out properly. An MDA inspector, Harold Wester, testified in court that the manager of the McBain mill had refused to close the place when ordered, arguing that he was sealing the contaminated equipment with paint instead. Wester testified in March 1977 that his department had made regular tests of the McBain mill dust over the previous three years, and had frequently found small traces of PBB. Farmers trading at the mill were not told of these findings.

Over and over, the poison was spread among Michigan livestock. Scores of farmers like Paul Greer contaminated their own land by burying animals which were full of PBB. As time went on and the contamination spread, other farmers realized the dangers and sold their dead animals to rendering plants. While most of these plants refused carcasses from quarantined farms, many farms with highly contaminated cattle had not been quarantined. The rendering plants ground up the meat and bones of the dead animals, and the resulting mixture became a major ingredient of cat, dog, hog, and chicken feed.

Rick Halbert used this disposal method for several of his animals before he understood why they had died. Later he estimated that some of his dead cows probably had a PBB fat content of 5,000 ppm. If one of their carcasses had been recycled into feed it could have so contaminated 15,000 other animals that none of them would have been considered fit for human consumption, according to the FDA guideline.

Months passed before any of the state's urban residents realized that the poisoning of Michigan dairy herds threatened them too. Very little Michigan meat, milk, and eggs was customarily sold outside the state, since dairy and beef farmers could barely fill local needs, but this meant an enormous concentration of PBB-contaminated products within Michigan. Townspeople were as likely to be affected as rural communities. Typically, most of the milk produced on the Halbert farm was not sold around the village of Banfield or even in the nearest town of Battle Creek, but to the citizens of Detroit—people who bought it at supermarkets and who, even if they had heard about some farmers' problems with poisoned feed, would have been unable to relate them to the contents of those cardboard cartons in their refrigerators.

Even farmers were slow to understand all the disaster's implications, especially if they had not bought 402 mixture. They equated it with the familiar cattle diseases described in their veterinary manuals —alarming but controllable. A few were nervous about "catching" PBB.

"There was talk about it being infectious or contagious," said Rick. "Some farmers traded in their tractors because they were afraid the tires would spread it around. One right-wing family I knew thought it was a communist plot."

At the other extreme, Rick met a farmer at an agricultural seminar shortly after the PBB story broke. "He came up to me and wanted to know where he could get some Firemaster. I had a funny feeling about that request—that maybe things weren't going too well with him and he wanted to collect some insurance."

To the few farmers who had qualms about "catching" PBB, Farm Bureau Services and the Michigan Department of Agriculture were reassuring. It gave them to understand that the contamination was relatively small, localized, and under control. No official statement mentioned or even hinted that PBB had doubtless been entering the food chain for months before it was identified. On May 13, 1974, MDA put out its first news release on the subject, one which was only modestly reported in some of the state's newspapers.

It stated that fifteen dairy farmers had been quarantined because it was believed that a fire retardant chemical had "found its way into the milk supplies through contamination of cattle feed." The release added (and this was the paragraph most noted by the press):

> Only a very few of the state's eight thousand Grade A dairy farms are involved, and the feed and chemical in question have been recalled. Also, milk from contaminated herds has been removed from the market. Thus, there is little need for concern about public milk supplies.

That same week, Farm Bureau Services sent out a detailed memorandum to all its feed dealers in Michigan, urging them to contact every one of their dairy customers personally. An eightfold message was to be impressed upon these farmers, as follows:

> A. The only [contaminated] feed were two dairy feeds manufactured at Battle Creek previous to March 1, 1974, and that every pound of that feed has been withdrawn from the trade and that there is no danger of them being exposed to those feeds now.
>
> B. That all magnesium oxide has been recalled and a "stop sale" has been imposed and shall remain in effect until laboratory analysis indicates that the product is "clean" and safe to sell.
>
> C. That a new supplier of magnesium oxide has been obtained and only that product is being used by Farm Bureau at this time.

D. That the feed plant at Battle Creek is producing "100 percent clean" feed now and has been for some time.

E. That all feeds purchased from Farm Bureau Services warehouses are clean and no danger exists of exposure to PBBs from these sources at this time.

F. That eleven herds which were sampled have been found to be completely uncontaminated.

G. That there is no health hazard to consumers of processed milk.

H. That every effort is being made by Farm Bureau Services to act in the best interest of our dairy customers and the dairy industry in this matter.

The statement was duplicated by cooperatives and sent to farmer customers with further reassurances that the local mill from which they were buying feed was uncontaminated. Unknown to these feed suppliers, this was in many cases incorrect—as were paragraphs A, D, E, and G of the memorandum. A number of feeds other than 402 had been contaminated; the Battle Creek plant was still not producing entirely clean feed; there was a continued danger of contamination from any feed which went through its machines; and there had been a widespread health hazard to consumers of Michigan milk for at least nine months. There was some suggestion in this and other official statements at the time that PBB in milk was like the tuberculin bacillus: pasteurization destroyed it.

Meantime, Michigan Department of Agriculture's director, B. Dale Ball, had written to the state's governor, William G. Milliken, apprising him of the first farm quarantines. Again, the tone of the communication was reassuring. After reiterating that the contaminated feed had been removed from the market and that only a few herds were affected, Mr. Ball added:

> We can safely say that the public milk supply is cleaner, safer and more wholesome than at any time in history due to our rigid laws of inspection and enforcement, combined with our highly sophisticated laboratory equipment and analytical methods. These enable us to pinpoint and deal with contaminants such as polybrominated biphenyl, which only a few years ago would have gone undetected due to lack of the necessary sophistication in tracing them down.

It would almost seem from this communication that MDA (whose veterinarians had given up early on in Rick Halbert's long search) knew exactly what PBB was and how to deal with it. Lacking independent knowledge of the disaster, neither the press nor the governor questioned any of these comforting statements. The Michigan newspapers had not yet learned to question the word of state agencies, and the governor depended upon these agencies for specialized information.

While the MDA, with its very limited knowledge of PBB, was not deliberately lying to either of them, it was certainly concealing its doubts and believing what it wanted to believe in an area where little was known and nothing proven. As a department of the state government with two distinct duties to perform, it was torn. It existed to promote Michigan agriculture and to protect the public's food supply. Normally these interests were mutually supportive. Suddenly they had become dichotomous. Mr. Ball, who owned a substantial farm of his own, must have realized that any suggestion of PBB-poisoned food on the urban market could be disastrous to Michigan's dairy and beef industries. His office, however, had been advised by the federal Food and Drug Administration that no milk with the slightest measurable trace of PBB should be sold.

On May 9, four days before he approved the press release and wrote reassuringly to Governor Milliken, he had been warned by his deputy director, Dr. George L. Whitehead, that the number of herds suspected to be contaminated had risen from seven to forty-four in three days. Probably, Dale Ball estimated, this figure could swell to one hundred, and he expressed this opinion in a conversation with the governor. It did not need much arithmetic or knowledge of the dairy industry to deduce that contaminated milk had been going to market for months and would continue to do so from PBB-affected herds which had not yet been identified. Dale Ball's department was in a dreadful dilemma. Should it alarm the public, perhaps unnecessarily, and risk permanent ruin to the dairy farmers of Michigan? Or should it protect the dairy industry and pray that the public risk was minimal?

In his May 9 memorandum to his superior, Dr. Whitehead advised a public stance of extreme caution. "We are accumulating a large amount of information in regard to this cattle feed contamination problem," he wrote, "and it appears that the economic loss to the dairy industry could be disastrous. Because of the legal implications of this case, we have to be very guarded in the release of any information to the outside until we have your decision on what action we will take."

12

What the Experts Knew

MR. BALL decided to gather all available information about PBB. On Friday, May 17, he convened a meeting at which experts from federal and state agencies, Farm Bureau Services, and Michigan Chemical Corporation would share their knowledge of the chemical and plan strategy for dealing with the contamination. This conference was held at Michigan Department of Agriculture's Geagley Laboratory on the campus of Michigan State University at East Lansing. It lasted all day and was closed to the public and press.

Mr. Ball presided and there was a healthy representation from the Michigan Department of Agriculture and Michigan Farm Bureau. At various stages of the PBB contamination crisis, Michigan Farm Bureau insisted that Farm Bureau Services (its affiliate which sold feed and farm goods) was an independent organization with a separate financial structure. If it were bankrupted by the claims of people owning contaminated farms, the considerable resources of Michigan Farm Bureau —would remain intact: such was the corporate structure. Yet when the feed operation was threatened and the matter was to be discussed at a closed conference, Farm Bureau's president, Elton Smith, and its general counsel, William Wilkinson, were among the select few whom Dale Ball invited. Apart from Dr. James McKean, the Bureau's veterinarian, none of the officers of Farm Bureau Services was listed among the participants. Rick Halbert was not invited, nor was his veterinarian, Ted Jackson. No one at the meeting represented the farmers, except for some administrative officials of Michigan Dairy Foods Council and Michigan Milk Producers' Association. The guest list also included Dr.

Fred Kerst of Michigan Chemical Corporation, his sales manager, Howard Washer, and experts from the Michigan Department of Health, the state attorney general's office, and Michigan State University. Two federal government departments were represented: the Food and Drug Administration and the U.S. Department of Agriculture. One of the two USDA delegates was Dr. George Fries, the scientist who had identified PBB in the Halbert feed. Then there was Dr. Lynn Willett of the Ohio Agriculture Research Center, who had done a number of animal experiments with the closely related chemical PCB.

Despite the omissions, it was an impressive assembly of experts. Between them they should have been able to come up with a fairly informed assessment of the toxicity of PBB, how it could be expected to affect farm animals and people who ate contaminated dairy products, how widespread the contamination might be, and what resources might be needed to eliminate it.

Dale Ball, a rotund, affable man, opened the morning session with a comfortable speech of the kind that he was accustomed to making at farmers' conventions. It began with a rambling, anecdotal story, told against himself and intended to put everyone at ease. ("I'm sure that some of you people have heard of the embarrassing situation that occurred to me a few years ago, but since this is a meeting that has to do with milk, I thought I'd explain to some of you that haven't . . .") In thus setting the tone of the meeting, Ball gave no hint of the extremely serious topic to be discussed. It took him a while to get to PBB, a chemical so unknown to his department that it had been incorrectly inscribed on the agenda as PCC. He explained conversationally, "It was brought to our attention by Mr. Halbert, who lives down in southwestern Michigan and is a very alert gentleman. I talked to him yesterday and I think I'll go down and take a course from him, because he's really up on this problem."

Although Ball was reassuring that MDA believed that the dilution in milk tanks had protected the public from contamination, his deputy, Dr. George Whitehead, warned that the PBB might not only be in milk and meat, but probably also in "condensed milk, eggs, and dried eggs and I don't know where to stop." Felix Schneider, chief chemist at the FDA's office in Detroit, explained that manufactured dairy products such as butter, cheese, condensed milk, and ice cream were especially vulnerable because of their high fat content, and that quantities of these items were being checked—and if contaminated, seized—by the department. Encouragingly, he added, "Hopefully, we'll get the problem wrapped up and it will not spread into other areas. We have not been able to identify the fact that it has gone beyond the state of Michigan at this time."

Schneider did not raise the possibility of contaminated meat being on the market, because meat inspection in the state was not the FDA's responsibility. It belonged to the Michigan Department of Agriculture which, having convened the conference, did not see fit to raise the issue in this forum. According to a transcript of the tape-recorded session, the only discussion of the subject was this brief exchange between two unidentified participants:

> "What should the disposition of the cows be?" [a reference to the contaminated herds]
>
> "You mean where should they be put?"
>
> "I'd tell them, don't eat the beef for damn sure."
>
> "What if it has been sold?"
>
> "I think that the cases in which the meat was gone before anyone knew it are hopeless. I don't know how you could trace that if it had been gone more than a week. I think it's an impossible situation. Correct me if I'm wrong."
>
> "You mean into the commercial market?"
>
> "Yes."
>
> "I think that's gone forever and I'd hate to have too much conversation develop on that because it can't help anybody."

Swiftly the subject was switched, narrowed down, to the condition of contaminated cows and the health of families living on quarantined farms who consumed their own produce. Amazingly, experts at the conference did not seriously consider whether anyone else could have been harmed. On the contrary, several of them—including Felix Schneider of the FDA—emphasized that there seemed to be no danger to the general populace. It was assumed that those who might have ingested PBB would have done so in such diluted form that it did no damage.

Dr. Harold Humphrey, environmental epidemiologist at the Michigan State Department of Public Health, promised that his department would undertake a study of the few people in Michigan who were thought to be in danger—those families on quarantined farms. He speculated about the symptoms which might appear if they became ill. He pointed out that there was no toxicological data and no known antidote for PBB, and that the effects of being poisoned by it were likely to be long-lasting. The only way of attempting to predict them was by extrapolation from recent knowledge of the *yusho* incident in Japan. Little had been written of this outside of medical literature, but it was a devastating story.

In 1968 thousands of people throughout southwestern Japan were afflicted with hideous boils, dark brown pigmentation of the skin, sticky discharge from the eyes, deteriorating vision, dramatic weight loss, and bodily weakness. Lesser symptoms included numbness of the limbs, headache, vomiting, and diarrhea. This plague ravaged villages and appeared sporadically in urban neighborhoods, defying the ministrations of modern medicine. Eventually its cause was traced to a shipment of commercial rice oil, a staple item in Japanese kitchens. During its preparation, one batch of oil was heavily contaminated by polychlorinated biphenyl (PCB), which was used in pipes surrounding the huge vats where processed oil was heated to remove malodorous material. One day when the heat was even more intense than usual, tiny holes developed in the pipes and PCB leaked heavily. Since Japanese deep-fry much food in rice oil, many victims made themselves progressively worse by using the same contaminated oil time and again. This disaster has been described in medical literature as the *yusho* (rice oil) incident.

There were clear parallels between the *yusho* incident and the Michigan disaster. The chemicals were similar. The process for making each of them began with biphenyl. In one case it was chlorinated, in the other brominated. Both chlorine and bromine are halogens, and halogenated hydrocarbons were known to be both toxic and persistent. The chief difference was that in Michigan the chemical had first been diluted, or perhaps changed in character, by passing through the digestive tract of an animal before reaching the people. Nevertheless, Dr. Humphrey speculated that if any farm families became ill they might be expected to develop dermatological symptoms or peripheral nervous disorders, like the Japanese. He said that his department had no guidelines for treating PBB toxicosis but would follow procedures for PCB poisoning. Since there was no known cure, these procedures were limited to treating symptoms.

"How long will it take to decide if there's anything wrong?" he was asked.

"I don't know," Dr. Humphrey replied. But, speaking hypothetically, he went on to give the most accurate prediction of the day's conference.

Suppose, he continued, a person had drunk PBB-contaminated milk daily for several months. From what was known of this family of chemicals, the PBB would not be excreted. It would accumulate in the human body, probably in the adipose tissue. "As far as I can tell in talking with experts, there it might sit for a very long time. What might mobilize it in the future, I don't know. It might just sit there for ever, much like DDT sits in the adipose tissue.

"It has not been proven that DDT would cause any particular problems over the long haul. However, we do know that with *yusho* disease—and again this is not the brominated form, this is the chlorinated form—some symptoms were long lasting. In other words, the skin disorders did not just clear up and go, but rather the person had those problems for some time."

Dr. George Fries, who had flown in from Maryland at a day's notice, described his PBB experiments with cows. They had taken place a year earlier and were not very helpful in this case, he admitted, because the sample of PBB which his colleague had eventually obtained from Michigan Chemical Corporation had been fed to four cows in very tiny amounts compared with the doses which the Halbert herd had ingested. He estimated these to have been three hundred times as great as the quantities given in his experiment. This experiment was never aimed at studying a massive contamination, Fries explained. He had wanted to find out whether, if PBB entered the environment, the small amount of residue which a farm animal might pick up could do serious harm. He saw no evidence that it did, although he made an interesting observation about PBB's effect on chickens. When a few of them were fed up to 20 ppm they continued to produce eggs which hatched; whereas, if PCB were fed at the same level "hatchability was wiped out and production lowered."

There was no comment on this finding about the chickens, which turned out to be significant for the people of Michigan. If only PBB had had the same effect, hundreds of thousands of contaminated eggs would never have been laid, and human beings would have been spared this additional exposure.

From the public health aspect, Fries made an even more important statement about the effect of PBB on dairy cows. He explained that because PBB was soluble in fat, a cow would excrete it in her milk but that this would happen very, very slowly and at a steadily decreasing rate. He indicated that although enormously high levels of PBB had been found in the Halbert herd, milk from these cows must have been more heavily contaminated several months earlier, when the cows were still eating the toxic feed. When they were eventually tested for PBB, it was merely residual levels of the chemical which were being measured.

No one pursued this point either, and it was left unsaid that from the fall of 1973 to the late spring of 1974 all this milk—not just from the Halberts' cows but from an unknown number of other heavily contaminated herds—had been sold to consumers. Still, delegates at the conference expressed no concern that anyone other than a few farm families might develop some version of *yusho* disease. Instead the talk kept drifting back to Dale Ball's major worry: how soon the affected farmers could get their herds back into full production.

Dr. Lynn Willett of the Ohio Agricultural Research Center was then called upon since he had studied the effect upon cows of PCB and various chlorinated hydrocarbon pesticides. Again, he could only surmise what PBB might do, but he confirmed that because of its solubility the only way a cow could rid herself of the compound was through her milk. When she lost weight, as most contaminated animals did, the PBB which had settled in her fatty tissue would remain in her body, presumably shifting to the bones, the blood stream, or other tissues. And, Willett added, when her milk production diminished (an experience reported by all of the afflicted farmers), there was no longer any way that the cow could excrete the chemical. It would still be present in her body, even when she had the physical appearance of starvation.

Again, there was a brief discussion (and again not pursued) about how these veterinary facts might apply to humans. Delegates acknowledged that the PBB in a nursing mother would be excreted through her breast milk—and that all the rest of the contaminated human population, lacking this means to rid themselves of the chemical through fat excretion, would retain it in their systems for a long time. Concern was also expressed that the chemical would cross the placenta of a pregnant female and damage the offspring.

They talked about how they could get rid of contaminated milk and animal carcasses. They agreed that it would probably be safer for quarantined farmers to throw their milk into their woodlands than upon soil which might serve for pasture. They wondered how they could safely bury contaminated animals without PBB leaching into the water supply, and whether, if the fat were stripped from the carcasses, hides could be salvaged and the lean meat used for dog food. But how could they dispose of the contaminated fat? It was acknowledged that it would have to be incinerated at enormous heat, something like 2000 degrees Fahrenheit, to destroy the toxicity of PBB, and that there was no equipment to handle this on a mass scale in Michigan. It was also agreed that for farmers to send carcasses of contaminated animals to rendering plants would risk recycling the PBB into other animal feed. But no one thought to warn the farmers of this danger; hence many rendering companies continued to take carcasses from contaminated farms, unchecked. It was just assumed that if a "safe" burial site were offered to quarantined farmers, all would take advantage of it.

The real problem remained with the farmers. It was made clear that they must decide whether to destroy contaminated animals or try and nurse them back to health—and production. Dale Ball stated that under Michigan law his department did not have authority to condemn animals, only to keep contaminated food off the market. Some of his MDA colleages expressed concern that a variety of farm animals might be affected, since they were liable to eat cattle feed that might be left lying around. Three premises had already been quarantined because chickens, swine, and sheep had been contaminated in this way. Dr. Whitehead also warned of the danger of cross-contamination in a dairy barn where presumably healthy cows were milked along with sick ones whose milk had to be thrown away. The experts discussed what criteria should determine the feasibility of trying to nurse a valuable dairy cow back to health and how long it might take. A few months, thought some. Too long to be profitable, thought others.

The two delegates from Michigan Chemical Corporation were questioned about the composition of Firemaster. They averred that to the best of their knowledge their firm was its only manufacturer. Only a limited amount of the chemical was made, they said, and production ceased in July 1972. But Dr. Fred Kerst and Howard Washer were referring only to Firemaster FF-1, the granular variety of PBB which got into the cattle feed because it looked like magnesium oxide, and *not* to Firemaster BP-6 (the same compound before granulation). The latter, produced in far greater quantity, was still being turned out at the St. Louis plant. When its production ceased, six months after the Geagley Laboratory conference, Michigan Chemical Corporation had made almost twelve thousand tons of PBB over four years.

Dr. Kerst made some statements which could have been misunderstood by laymen. At Geagley Laboratory and at subsequent meetings with officials of state and federal regulatory agencies, he spoke of "pure PBB." Technically, there was no such thing. The cross-contamination at Michigan Chemical's St. Louis plant was widespread; furthermore, before the biphenyl was brought into the plant some of it contained small amounts of naphthalene, the stuff of which mothballs are made. As an experienced chemist Kerst must have been aware that when biphenyl is brominated, the naphthalene is brominated too; this could turn a relatively harmless substance into a potentially toxic contaminant of the PBB. The contamination did not matter in a fire retardant, but it may have made a great difference in the effect of the chemical on animals and humans.

Three years later, in court, Dr. Kerst admitted that at the time of the Geagley Laboratory meeting he knew of the brominated naphthalene in PBB. While he had no direct evidence of its toxic effect, he knew that when naphthalene was chlorinated it could cause hyperkeratosis in cattle—the mysterious X-disease, which Rick Halbert had suspected early on. Kerst's veterinary knowledge sprang from the fact that he was raised on a ranch farm and from childhood had had an understanding of cows. But at the Geagley Laboratory meeting he did not reveal that there was naphthalene in the PBB because, he said, he wasn't asked.

These scientific subtleties were lost on the agricultural experts who dominated the Geagley Laboratory meeting. None of them understood that the contaminant in the quarantined herds, which by then numbered nineteen, was in itself contaminated and may have been significantly different from the laboratory-produced sample of PBB which George Fries had used in his experiment on cows. This may or may not have mattered, but it should have been considered. Rick Halbert would undoubtedly have seized upon this point if the facts had been presented to him. He knew that in the process of making one compound a chemist inadvertently creates minute quantities of other compounds, and that total purity is ever elusive: the best to be hoped for is a chemical purity of about 99 percent, and there was no cause even to strive for this in a chemical not intended to go near food. As for the ingredients of PBB, a chemist would expect to find contaminants not only in the biphenyl but also in the bromine; moreover, there was the possibility of further contaminants as the two interacted. And as Kerst knew, the composition of the biphenyl varied since it came from two different suppliers. In the careful production of a laboratory sample all these possibilities are reduced. They increase dramatically when the compound is mixed in the same huge vats that have been used for other chemicals—and when it is exposed to yet more chemicals at the plant.

Years later Rick still wondered about this. "Maybe there was another chemical which helped to do the damage," he speculated. "In all that time nobody ever took our feed, extracted the PBB, and tested to see whether mice who ate it would still die."

This thesis was not considered at the Geagley Laboratory meeting and Rick had other worries on his mind. Based on George Fries's limited research, the thin hope was held out that if farmers went on milking their quarantined cattle, instead of letting them dry up and rest, the cows' body burden of PBB might be reduced by half in about eight weeks. It was traumatic for the farmers to continue feeding for maximum milk production and then to throw the milk away. With no compensation for the cost, Rick and his brother Mark were dumping eight thousand quarts into their woodlot every day. Soon the scrub growth there began to turn yellow and die. Neither man dared to dwell on the thought of what milk like this might have done to all the people who had been drinking it.

Aside from their staff veterinarian, Jim McKean, Farm Bureau delegates had little to say at the Geagley Laboratory meeting. McKean gave a brief history of the various farmers' complaints, along with the impression that his organization had reacted promptly by stopping production of 402 feed in December, recalling the outstanding supplies in January, and recalling all magnesium oxide in April, when it was realized that this was the basic source of the contamination.

"We feel that the lid is on this problem," he added. "We have some feed which is questionable but we are analyzing it right now. We had a man in the country meeting with our dealerships yesterday and today, and he should complete five of them by this evening. A survey of customers that might have received questionable feed is being made and additional laboratory work is being done. The names are being supplied to the dairy people. I don't know whether this will turn out to be significant or not, but when that one is done I'll feel fairly confident that hopefully we'll have exhausted our search and will have the problem delineated. So it appears that we are down the road as to getting everyone identified that could have been involved. At least I hope we are. We've run out of just about everything else. We've checked all of the feed that we have samples for. We retain samples for ninety days. We've checked all the feed samples that had magnesium oxide in it, or added to it, at the Battle Creek plant."

At the time McKean was talking, FDA inspectors were finishing a three-day inspection of the Battle Creek feed plant. They reported these deficiencies:

Farm Bureau Services did not take adequate measures to avoid cross-contamination of nonmedicated feeds with medicated feeds; its equipment design allowed a build-up of residual materials in various pieces of equipment (the reason why so much subsequent feed was contaminated with PBB); there were open bags of antibiotic drugs in the raw material storage area; there were unlabeled bags of ingredients; and the inventory records for one drug contained discrepancies.

On the basis of FDA's findings the feed company lost its permit to continue producing medicated feed. This was not restored until July 1975, more than a year later.

After the May 17 conference a one-page press release was handed out. Essentially it was drafted at the Michigan Department of Agriculture before the meeting took place, and amended slightly by delegates. Quoting Dale Ball, it emphasized that "Michigan Department of Agriculture has acted to remove milk from nineteen affected herds from the market, and tests of milk at retail outlets indicate no danger in the public milk supply."

The original draft also stated that the health of farm families who had consumed contaminated milk was now a major concern, being investigated by the Michigan Department of Public Health. Someone at the meeting felt that the phrase "major concern" was too alarming and might cause "every TV news media in the state to run out to those farms." The health department, after all, had no funds to do a clinical survey of farm families. Whatever it did for them would have to be limited and the money somehow squeezed out of the department's regular budget.

So the two words were stricken from that part of the draft but left in the following sentence: "Mr. Ball said a major concern is the disposal of contaminated feed and animals." The other revision involved changing a figure. Originally it was stated that an estimated twenty-five hundred animals were "involved" in the contamination. Someone, again unidentified, suggested reducing this to "one thousand lactating animals." Another delegate thought it would be best not to mention a figure. These comments ensued:

> "If you don't use a number, you're going to have to answer the phone."
> "If you don't know the number, then that's the answer you give them."
> "Say approximately fifteen hundred. You couldn't go wrong with that."

And fifteen hundred it was.

One speaker observed, "The thing that concerns me is that you can use adjectives to alarm the public very easily. Say a toxic chemical was involved and that doesn't alarm anybody. Say 'highly toxic' and it will alarm everybody who reads it. I think that in an objective reporting session you ought to avoid that sort of thing."

The final statement issued as the session ended was only five paragraphs long. There was little in it to excite the curiosity of a copy editor who would have received it on his desk as the slender Saturday morning newspapers were going to press. Most of the Michigan newspapers did not bother to print it.

13

The Road to Kalkaska

No one understood the magnitude of the problem, not even the farmers involved. Each believed himself to be one of an afflicted few and was thinking only of personal survival. Farm Bureau and the MDA regarded this as a purely agricultural crisis and everything they did was aimed at putting the stricken farmers back in business. Their chief concern was to reduce the farm contamination below quarantine levels so that dairymen could market their milk again.

Since PBB is accumulative and persistent, the contamination was thus allowed to spread. MDA officials registered annoyance that Farm Bureau did not tell them of the increasing number of dairy farmers' feed problems at the time they were happening (for months they knew only about Halbert), maintaining that their resources might have helped to trace the cause sooner. Even so, the two agricultural organizations continued to work in close cooperation. Although he no longer belonged to Farm Bureau, Dale Ball (whose office, across the street in Lansing from the state Capitol, was decorated with family portraits, citations confirming his own success as a farmer, and a plaque inscribed "Cows may come and cows may go, but the bull in this office goes on for ever") was an archetypal Farm Bureau member of a passing generation: bluff, hearty, and unsophisticated.

Nearby, at the state's Department of Public Health, PBB's threat to the populace was underestimated from the beginning—and eventually its director, Dr. Maurice S. Reizen, admitted this publicly. Taking his advice from these two state agencies, Governor Milliken acted for many months as though this were no more than a bad case of foot rot among herds in the hinterlands. Michigan's votes are predominantly urban, so it was comforting for the governor to be told that only a few herds were damaged and that there was no risk to the public health. For many months he had little to say about PBB.

Having satisfied themselves that the contamination was concentrated in Michigan, with no discernible threat to interstate commerce, the federal agencies determined that their responsibility was limited and that it was basically Michigan's concern. There must also have been some embarrassment in Washington's high places. Although Rick Halbert had pleaded with senior officials of the USDA and FDA to help him solve his herd problem, USDA had reacted by calling Allan Furr off the research when close to identifying the contaminant, and FDA had apparently done nothing.

Among all of these bureaucratic departments there was no provision for any agency to search out every Michigan farm to which suspected feed had gone, test every animal for the presence of PBB, destroy all the contaminated ones, and arrange compensation for the farmers. Ideally, every meat and dairy item in all of Michigan's stores and supermarkets should have been seized and tested. It would have cost millions of dollars and no one had the overall authority or funds. If this had been a contagious disease with a likelihood of spreading across the country, federal agencies would have been mobilized. Instead, it was like an industrial cancer—unenvisaged, insidious, embarrassing to those who caused it, and shameful to those who suffered from it. Even after PBB was found in the cattle feed, it was still hard for a farmer to admit to neighbors that there was something wrong with his scrawny, stumbling cows. The symptoms so often were mistaken as evidence of his own poor management. Many of the afflicted farmers tried to nurse their cattle back to health—and since few had the financial resources of a large dairy operation like the Halberts', some of those who had avoided quarantine continued sending contaminated milk to market, diluting it in their bulk milk tanks with the milk of healthier cows until it reached an acceptable level. They had to live somehow.

Right after the Geagley Laboratory meeting, Dale Ball tried to get a bill through the state legislature which would have eased the burden on quarantined farmers. It was the end of May, and the nineteen quarantined farms had increased to thirty-nine cattle herds, seven swine herds, two sheep flocks, and four poultry flocks. There was no end to the increase in sight, although officially MDA was stating its belief that most of the contaminated herds had now been quarantined. Meantime its research staff was working around the clock, seven days a week, checking milk and feed samples.

The bill drafted by Ball's staff aimed at releasing state funds to pay the costs of the immediate crisis. It proposed giving MDA the authority to condemn contaminated animals and to reimburse the farmers for their losses. At the same time it was proposed that MDA should be authorized to sue those responsible for feed contamination, in an attempt to recover the money which the state would have to pay to the farmers. Thus the farmers would have been kept financially stable while the courts determined liability. It was a measure which would have done much to restore a clean food supply and prevent the contamination from spreading, but it did not stand a chance in the legislature. Until the state could recover the money which it would be paying out to farmers, the costs would have to be passed on to Michigan's taxpayers. Further, Ball's proposal would have created a precedent which the governor opposed. He was against putting the state government into the business of indemnifying farmers for damages which the state had no part in causing.

"There was no hope of getting that bill out of committee," Rick Halbert complained bitterly. "This is a very urban state, and it pays little but lip service to farmers."

The measure was not only watered down in the legislature; its intent was completely changed. The act which the governor signed on July 2 merely authorized the director of MDA to approve a disposal site for the severely contaminated animals, and to maintain legal action against Farm Bureau Services and Michigan Chemical for recovery of state funds required to transport and slaughter these animals. This new law left the farmers in limbo, to claim what they could from the feed and chemical companies. For its pains in drafting the emasculated bill, MDA was accused by some legislators of attempting to bail out Farm Bureau and Michigan Chemical. Meantime, these two powerful businesses argued back and forth as to which should be held culpable, and the farmers, who were near bankruptcy, were ignored.

Michigan Department of Agriculture tried making direct appeals to the federal government. It argued to the U.S. Department of Agriculture that because the contamination was being contained in Michigan, Americans in other states were being protected. If it had crossed state lines, federal authorities would have had to bear the expense, therefore USDA should share with MDA, fifty-fifty, the cost of containment. This theory got the same reception as one which Dale Ball expounded to Governor Milliken: if an agency of the state government went to some expense to protect the public from an unexpected hazard, the public should share the cost. Neither the USDA nor the governor was responsive.

Ball's deputy, Dr. George Whitehead (a veterinarian), was given the major responsibility in MDA for handling the PBB crisis, and when he testified before a U.S. Senate Committee almost three years later, this is how he described the problems which faced his agency in 1974:

> It was an extremely critical year for the Department of Agriculture because everyone was working long hours to resolve a problem that we considered most serious, with little or no research data and insufficient funding. All requests for federal assistance, with the exception of FDA, were shrugged off. The USDA took the approach that it was not a disease, therefore it was up to the contaminators and the state to handle the problem. Letters to the U.S. Secretary of Agriculture and contacts with other USDA officials were quietly ignored. The EPA [Environmental Protection Agency] was not interested because it was not a pesticide, and university research was not begun because they said they had no funding for such research, and some said the problem was already resolved. In short, the research, veterinary and laboratory assistance we needed so desperately during this period was not available, and we continued to do the best we could with department facilities and personnel. . . .
>
> The situation was further complicated during this period by unusually prolonged claims settlements, and the affected farmers naturally became very frustrated. This has generated hundreds of lawsuits against the feed manufacturer and the chemical company. These farmers were, and some still are, in dire financial straits and had no place to turn for assistance. They called the feed manufacturer, the U. S. Congress, the governor of Michigan, the state legislature, and nothing happened because there was no provision anywhere to help the livestock owner who suffered from an accidental contamination of his animals. Then they turned to the Michigan Department of Agriculture, and some said that inasmuch as we had quarantined their animals and caused their problem, that we were responsible for what had happened to them.

Despite the efforts MDA was making, it still did not understand that the disaster could—or had already—spread beyond the farms. If it had understood, it would surely have made more noise at government level. It would have warned the urban public and those who represented it in the legislature. It would have alerted the health department. It would have gone on demanding funds for research and testing instead of complaining and making do.

About a month after PBB was identified in the feed, a group of state officials met in Dale Ball's office to try and organize some financial aid for the quarantined farmers. They did not succeed. One participant was Harry Iwasko, state assistant attorney general, whose special responsibility was commerce and agriculture.

"It was clearly very serious," he recalled, "but they thought it involved only a handful of farms. It was a relatively new idea to use magnesium oxide in cattle feed, so they assumed that only a few sophisticated and highly successful farmers would have bought this particular Farm Bureau mixture. At that time MDA was estimating that the extent of the problem would involve paying out about three and a half million dollars to the quarantined farmers. They seemed confident that that would take care of it."

Ball tried hard to get relief for these afflicted farmers. He badgered Elton Smith, president of Michigan Farm Bureau, for help in disposing of the contaminated animals and reimbursing the farmers. While waiting for action on his doomed legislative measure, he also asked the state attorney general's office for a legal opinion: did he, as director of MDA, have authority to condemn contaminated animals, and did the state have the responsibility to pay indemnity? The answer was no to both questions. Ball was told that his department could condemn animals only if it was necessary to prevent spreading disease. And since it would be illegal for MDA to condemn PBB-contaminated herds, the state could not be held responsible for the cost of MDA's illegal act. Various Michigan laws were cited by Attorney General Frank J. Kelley in support of this position. He was not being obtuse. There was nothing on the statute books to cope with a crisis like this and no way of putting it there without going through the legislature.

Elton Smith responded with a long letter, full of circumlocution, saying little. He couldn't advise the farmers whether or not to destroy their sick animals because he didn't know if or when an antidote to PBB could be found, hence he couldn't judge whether it would be economically worth keeping them in hopes of a future cure. If it was decided to destroy them, Elton Smith agreed that Farm Bureau would help to find a safe burial site. That was as far as he was prepared to go, except for a nebulous promise at the end of his letter to Dale Ball:

"We are a farmer-oriented organization," he wrote. "We recognize the problems being suffered by the affected farmers and are certainly sympathetic to their plight. Our insurance carriers have been instructed to follow through upon an expeditious settlement apparatus conceived by us, and our attorney is in daily contact with them to monitor the progress of this process."

Dale Ball cut short that argument by advising Elton Smith, as soon as the emasculated bill passed the legislature in June, that Farm Bureau would be charged an initial $100,000 followed by $60,000 a week for the cost of preparing and maintaining a burial site. There was still nothing for the farmers.

Even these plans for destroying highly contaminated cattle were held up by arguments. Incineration or rendering of carcasses was ruled out as impractical and unsafe, and, after a statewide search, geologists at Michigan's Department of Natural Resources approved a fifteen-acre section of state-owned land in Kalkaska County for a mass grave. The area, in the north of Michigan's lower peninsula, was thought to be ideal. It was remote. It formed a natural clearing in dense pine woods.

No one lived within two miles, and there was at least half a mile of state-owned land on all sides of the proposed burial site. There seemed to be no possibility of contaminating water supplies because the nearest river was more than three miles away and the water table was about a hundred feet below ground, sealed by a layer of impermeable clay at a depth of about forty feet. It was estimated that about five thousand cattle and smaller numbers of other livestock would be buried in lined trenches, twelve feet deep, over the space of a few weeks, and then the graves permanently sealed.

All the state officials involved believed this to be a safe plan, one which could bring only benefits to the citizens of Michigan. The risk of further contamination on the farms would be removed along with the diseased animals. Quarantines could be lifted and farmers could return to making a living. The public would be protected by the destruction of what was believed to be all the contaminated food in Michigan—since the dairy products which MDA had seized were to be buried too. With Farm Bureau paying burial costs, it was agreed that the afflicted animals would be trucked to Kalkaska, where MDA veterinarians would supervise their slaughter.

It all seemed so safe, so foolproof, that even Harry Iwasko—the assistant attorney general who had been skeptical of MDA's understanding of the disaster—scoffed at the suggestion by the nervous residents of Kalkaska County that the water supply could be polluted. He explained that when the animals' bodies eventually decomposed underground, "the only thing we are going to do is make the land more fertile."

Many months later, when the mass grave had to be expanded to hold more than six times the number of bodies that had been envisaged, it developed that Iwasko and all the other officials had been overly optimistic. The trenches became so filled with carcasses that this woodland cemetery became known as an animal Auschwitz, and traces of PBB were eventually found in the local water supply.

In the summer of 1974 the only people who worried publicly about this possibility were members of the Kalkaska County Board of Commissioners, and none of the state officials had thought to consult them. On July 2 and 3, 116 contaminated cattle were put to death and buried at Kalkaska. On July 8 the county commissioners took legal action to prevent the burial of any more, and on July 9, in Kalkaska Circuit Court, a judge upheld this restraining order.

MDA was at a loss to tell farmers what to do. Its experts were advising that the most ailing animals should probably be destroyed; at the same time they knew the risks of soil and water pollution if farmers buried them on their own land. There were perhaps more serious risks in keeping them alive, especially since twice every day they would be milked with the same machinery that might be used for other animals whose milk was going to market. Reluctantly, Ball advised these farmers that if they wanted to dispose of their sick cattle, they had better slaughter them and bury them on their own land.

"I did not feel this was a proper solution to the problem," he said, "but apparently it was the only solution available to them at the time." Ball offered the help of MDA veterinarians to ensure that the animals were killed humanely. One who took advantage of this was Bill Coomer, who had 178 hogs killed and buried on his farm at Mendon.

The worst aspect of this for the farmers was that it put the onus on them to determine whether or not contaminated animals should live or die. It was a decision that they would be obliged to face, over and over, in the months to come—even after a state burial site was approved. At no point did any federal or state agency take the responsibility of inspecting herds, animal by animal, and making an impartial judgment as to which could be salvaged and which should be sacrificed. No one was ever given the authority. Garry Zuiderveen, a farmer who shot a number of his cows, was unable to speak of it for ever after without emotion in his voice and tears in his eyes.

"We should never have had to make that decision," he said. "It was the darkest day in my life when I shot those cows. A farmer is an immensely proud person. Anything wrong with his herd reflects on his husbandry and his herdmanship, and the sickness of those animals is a reproach to him. It hurts him to see them suffer, and even when they are not getting better he still hopes he can help them. I can still remember the pain in my gut when I would go in that barn in the morning, wondering if they had made it through the night. . . . And yet there was this economic necessity. It was impossible to maintain those cows and throw their milk away. I could not have made any kind of living. I had been losing money already, and when you start going backwards on a farm, you slip back so fast it's unbelievable. So the decision to shoot those cows was forced upon us, and we should never have had to make it. It should have been left to people like MDA or FDA who were not so emotionally involved."

On August 21, after hearing reassuring expert testimony, Circuit Judge Charles Wickens overruled the restraining order which the Kalkaska County Commissioners had obtained against MDA. He said he was satisfied that the mass grave was environmentally safe and that the state had a responsibility to ensure speedy disposal of the contaminated cattle. Six days later the extermination was resumed.

It was a pitiful sight which worsened as months passed. By August more than a hundred farms had been quarantined, and from then on thousands of animals were taken to Kalkaska. Farm Bureau had taken some of them off the farmers' hands and kept them in holding areas, feeding them enough to sustain life in the six weeks that the judge's decision was pending. Other cattle were despatched directly from the farms. All of them were mangey, emaciated, and bewildered. In many cases the toxin had affected their central nervous systems, causing them to stagger around as though drunk and blind. Their hair had fallen out in patches, exposing dry, grayish skin, wrinkled like elephant hide. The flesh hung on their bones.

One of these suffering animals was a cow named Flopsy, a special pet of Rick and Sandra Halbert's small daughters. The girls were heartbroken when told that she would have to be taken away, and plotted hiding her in the woods where they would nurse her back to health. Rick and his wife were unable to bring themselves to tell the children the truth, and for a while let them hope that Kalkaska was a place where Flopsy might be made better.

Between that summer of 1974 and 1977 William Hughston, a Michigan livestock dealer and trucker, took eighteen thousand doomed animals to the burial pit—cows, calves, lambs, and hogs. The smaller species were dealt with at the beginning, but soon it was only cattle. "They were awful hard to load and unload," he said. "You would have to drag them out of the barns and they would act like they were retarded. Normally you can walk a cow out, but these you would have to push. You would drive them one way and they would try to go another. They were not all sick and thin looking, but the worst of them looked awful bad. Stunted, lame, and a lot of them blind. Many had open sores on them. The young ones were the worst. They could not stand any stress and if you pushed them together they would just die. Some died on the truck before we reached Kalkaska."

Farmers followed in their cars to assure themselves that the slaughter was humane, which it was. Heaved off the huge double-deck trailers, the animals were led into pens, then one by one towards the burial pit. Those who wavered were often encouraged to go to their deaths by their owners—strong men who stood by with tears running down their faces. "C'mon Susie. Easy there, Blackie"—said in the soothing tones that were part of the ritual at milking time.

An MDA veterinarian would inject a powerful muscle relaxant into the cows' veins. Instantly they would drop unconscious, and be shot through the brain with a rifle. A crane would lift their bodies into the air, and for an instant they would hang stupidly suspended before being dropped on pallets and towed to the trenches. The animals' abdomens would be slit to prevent bloating, and the carcasses covered first with sand, then clay, then topsoil. The bizarre sight etched itself on the memories of men like Bill Hughston. "None of us ever stayed around to picnic in the woods afterwards," he said.

Much as the work depressed him, some of the cattle which he did not truck to Kalkaska bothered him even more. It was Bill Hughston's regular business to take livestock to the market. His operation, which was sizable, was based in the farming community of McBain in Missaukee County, immediately south of Kalkaska, and his six large cattle trucks traveled throughout Michigan and into neighboring states. For many months local farmers believed that the McBain area was safe from PBB contamination. Most of them bought cattle feed at the McBain, Falmouth, or Merritt co-ops, which were Farm Bureau affiliates and which had sent them written assurances that they had NEVER (the capitals were the co-ops') handled the contaminated feed.

Hughston, however, had been troubled for some time by the condition of some of the local cattle destined for the Detroit and Cleveland meat markets. They were not in the wretched state of the animals he took to Kalkaska, but they weren't right either.

"They were getting a tremendous lot of bruises on the journey," he recalled. "As they touched the door of the truck, their skin, instead of rippling and rolling, would slide and tear. The rib bruises were the worst and they were not caused from loading because we had been loading cattle all our lives and it had never happened before. Normally the hide protects a cow, but with these cows it was no protection at all."

All this was costing Bill Hughston money. He carried cargo insurance in case of fire or accident to a truck, but with the number of cattle his firm hauled every day, insurance on individual animals would have been prohibitive. "So," he said, "we took our chances. We had hardly ever had any claims before 1974, but suddenly I was having to pay three hundred to four hundred dollars a week off my freight bill for bruised cattle.

"I never thought about the problem being PBB, because we had been told there was no chance of it in this part of the state. But I talked to Doc about it. He didn't say much, but I could tell that something had registered with him."

"Doc" Alpha Clark, veterinarian, was Bill Hughston's next-door neighbor. Their houses were on a lonely stretch of highway on the outskirts of McBain, which in itself is little more than a rural crossroads. Both homes looked out on miles of flat fields, punctuated by large clumps of tall trees—remnants of dense forests which nineteenth-century settlers cleared first to make a living in the lumber business, then as dairy farmers. An artist capturing the essence of the landscape would paint the horizon near the bottom of his canvas, and a vast clear sky. A quality about the light would bring the scattered farmsteads, the groups of quietly grazing cows into sharp focus. The overall impression would be one of timelessness and tranquillity.

Alpha Clark, who was the most popular cow doctor in Missaukee, Wexford, and Osceola Counties, chose this spot for his home and animal clinic because it was in the center of a large circle of farmer clients which he had developed over the years. He was known to everyone as "Doc," even to his wife, Marlene. His father and both his brothers were Michigan farmers; he had lived all of his forty years close to the land, and, next to Marlene and his children, he felt that there was no creature on God's earth so satisfactory as a carefully bred dairy cow. He loved his work and could not conceive of doing any other. But when his neighbor Bill Hughston was fretting about his increased freight costs, Doc Clark had a more profound concern which he wasn't discussing with anyone. Professionally he feared he was slipping.

Starting in late 1973 and into the summer of 1974, many of his clients had been having herd problems which baffled him. Nothing specific, but nothing he had failed to treat successfully before. Mastitis, pneumonia, digestive difficulties, excessive calf losses—everyday complaints, but suddenly there were too many of them and the proven treatments were not working. Antibiotics did not act. Conditions which he had always been able to overcome seemed only to worsen.

Doc Clark was such an honest man that if he could not improve a cow's health in two or three visits he did not charge for subsequent calls. He explained, "I try to render a service that's going to be economical for the farmer, and if a cow don't get better or dies on me I may just write the whole thing off. I don't feel right about billing that farmer any more than I feel right after I pay an auto shop to fix my car and the dumb thing still don't work." In the spring and summer of 1974 Doc Clark made a lot of farm calls for which he did not charge.

From herd to herd the symptoms varied slightly—just as the cattle which had been more heavily contaminated had been different enough from one another to confuse Farm Bureau's veterinarian, Jim McKean. The toxin reacted differently with variations of farm management and feeding practices. In the summer of 1974, Doc Clark did not even suspect PBB. He had barely heard of it, no farms had been quarantined anywhere near his area, and there was the feed company's assurance that no contaminated products had reached local co-ops. Nevertheless, Doc Clark questioned the co-op managers and described his herd problems to experts in the veterinary college at Michigan State University, of which he was a graduate. They all looked at him blankly. "It made me feel like the lone ranger," he said.

Yet he still had this gut feeling that something was profoundly wrong with most of the herds under his care. "It bothered the hell out of me that they would not respond to treatment," he said. "It was a subtle situation and I couldn't get a hold on it. It was like trying to catch a snake in the sorghum. You grasp him by the head and then by the tail, and still he gets away from you."

Soon Doc Clark was blaming himself. "I had had plenty of experience and I still had plenty of energy, so I should have been at my peak professionally. But I couldn't seem to do anything right. Some of the farmers got so they quit calling me. They would just come to the clinic for medicine because they was having just as much luck doctoring their cows as I was. I had always had a good rapport with my clients but all of a sudden they was getting this feeling of uncertainty about me. They would question me on everything—why this was happening to their cows and why this wasn't.

"I didn't know, and this made me insecure. I thought I must be slipping, just plain fizzling. So I worked harder and spent more money buying drugs and traveling the countryside, going back and back to the farmers, but it was like dumping all my efforts down a rat hole—the most tired exasperated feeling I had in my whole life. In 1974 I spent ten thousand dollars more than I took in, mostly in medicines and gas. Yet all my instincts as a practitioner kept telling me that it wasn't the farmers' faults or mine, but something else which was making my treatments go haywire."

On August 15, 1974, Doc and his brother Clyde—both of them experts in cattle breeding—drove south to Fremont to judge cattle at the Newaygo County Fair. Doc did not want to go. The date conflicted with a local farm show where his children were exhibiting some of their favorite cows, but afterwards he felt that fate led him to accept the Fremont invitation.

The mass burials at Kalkaska had not yet begun and nothing about PBB had appeared in the newspapers which reached McBain. All Doc had heard about it (aside from the reassuring notice from the local co-op) was a brief item in a television news program, three months earlier, from which he gathered a vague impression that somewhere in the state a few cows had eaten poisoned feed. He was therefore amazed to find a dearth of decent cattle at the Newaygo fair, which was always a good one, and that several experienced breeders in the area were not exhibiting. He also caught undertones of arguments, not meant for his ears, about whether animals from some farms should have been allowed at the fair because of uncertainly about "it" being contagious.

Doc, who was good at picking up information and giving little away, soon learned from farmers at the fair that several herds in the county had been quarantined because of a toxic chemical which had got into feed mixed at the Fremont co-op. Over a midday meal one of them described how his contaminated cows looked and acted. Doc listened avidly, revealing nothing.

"Holy mackerel," he thought to himself. "I got a whole bunch of cows acting like that and now I know why!"

On the drive back to McBain his anger mounted.

"I got to thinking," he said, "and I realized that not only my clients' cattle was poisoned, but their land and their bodies was poisoned. I didn't know what to tell them. Since no one would admit there was PBB in their feed, I could have advised them to cover up and ship their cattle to market while they could still make a few bucks. Or I could have tackled the problem. Well, I tackled it. But once I found out what was wrong, how we had been lied to, it pissed me off so bad . . . but I still didn't know what to do.

"What would you do if you were working for a hundred farmers and at least sixty of their herds were poisoned with something you knew nothing about?"

14

Beyond Medical Experience

WITH knowledge which he lacked at the time of the Geagley Laboratory conference, Dr. George Fries estimated, almost three years afterwards, that by the time he had identified the toxin in Rick Halbert's feed 95 percent of the PBB which contaminated Michigan was already in the environment. Much of it was in farm animals. Some was still in the cattle feed. Some was in the state's food supplies, in home refrigerators, and on grocery shelves. Some was in the bodies of its citizens and in the cells of unborn children. Any action taken after May 1974 could not change this, according to Dr. Fries.

And yet, if expert help had been mobilized, plenty could have been done to prevent the situation from growing gloomier. It was known from the beginning that since PBB would accumulate in the tissues, even the smallest amount might unbalance the health of a person or animal already carrying more than a safe body burden of the chemical. Extraordinary efforts should have been made to prevent this persistent substance from being recycled—from cattle to people, from one foodstuff to another, from mothers to their young. Instead, all the state and federal regulatory agencies reacted, in May 1974, as if Michigan's PBB contamination were merely an overnight accident, not a slow poisoning which had been going on, statewide, for nine months; as if the food supply were safe; and as if the PBB content of any contaminated milk could be brought down to imperceptible levels in bulk milk tanks. Nobody worried publicly about meat—which can't be diluted—because the public was allowed to assume that the contamination had been discovered before any animals went to market. State officials did not lie about this. They just did not tell the whole story—the automatic bureaucratic reaction to a situation which does not look good and cannot be remedied. For many months Michigan newspapers saw no cause to question the line taken by the state agencies; neither did Governor Milliken.

It was much the same with the one federal agency which took action at the beginning. In May 1974 the Food and Drug Administration reacted to the PBB discovery by imposing a "tolerance" level of one part per million (1 ppm) in milk or meat, measured on a fat basis. Anything above this level must be destroyed. Anything below could be sold for human consumption. Again the public did not realize it, but these guidelines (which within six months were acknowledged to be far too lax) had no basis in toxicological experience. One part per million was merely the lowest level which could be detected with instruments generally available to laboratory technicians at the time. More refined testing was possible but much more difficult and costly. According to Dr. Whitehead at the state's Department of Agriculture, even the test which was being performed was "relatively difficult . . . requiring about eight hours of procedure time and the use of expensive instrumentation and chemical solvents." MDA professionals were already working seven days a week on PBB testing, with no federal help, and none of them was anxious to make the job more complex.

For its part, while the FDA was charged with protecting the public health, it also had to consider the cost of its actions. The lower the tolerance level, the more farms would have to be quarantined, the more food must be seized and destroyed, and the more farmers, retailers, and citizens, would have to be indemnified. By whom? There was neither the money nor the mechanism to provide an answer. The FDA had to tread a tightrope between how much toxin a human body can stand and how far a government agency can commit itself financially.

The FDA's associate director, Dr. Albert Kolbye, spelled out the dilemma frankly.

"Government and society are faced with difficult and critical decision-making responsibilities concerning the presence of environmental contaminants in food, especially those that are known or suspect carcinogens," he said. "In carrying out its regulatory responsibilities in this area, FDA's goal is to take every possible precaution to reduce potential hazards to human health that could result from dietary exposures to these contaminants. At the same time, the Agency cannot overlook other quite relevant factors, such as continued availability of food and the socio-economic considerations of unnecessary restrictive regulations."

All these factors shaped government decisions early in Michigan's PBB crisis—decisions which led to widespread public misunderstanding. The PBB tolerance level, designed by FDA and enforced by MDA, was never meant to be an infallible dividing line between contaminated and wholesome food, between sick and healthy animals. Nevertheless, this was soon the accepted belief. It went along with the philosophy, encouraged for many months by Michigan's Department of Public Health, that while some people living on quarantined farms might suffer ill health (Dr. Kolbye later revealed that he was surprised some of them did not die), nobody else in the state of Michigan was likely to be endangered by PBB.

None of these government departments could afford to give a different impression. The situation was unprecedented. The handling of the crisis was nobody's assigned task. The federal authorities had deemed PBB to be Michigan's problem—which left the state agencies to do the best they could with no outside support, no extra funding, no experience in chemical contamination, and no comprehension of the disaster's enormity.

"The root of the whole problem was lack of money," said Harry Iwasko, who had the major responsibility for handling the legalities of the PBB crisis. "There was no money from the Feds. The amount needed was more than any state could have afforded. To have handled this thing properly we should have tested every animal in every herd in Michigan, and the minute we found any detectable level of PBB we should have exterminated it. We should have imported all the milk and meat for nine million people until we were sure we were safe."

Iwasko sighed when he said this. He was sitting in his office in the state government's Law Building, a small room where a photograph of his own smiling, rosy-cheeked children was prominently displayed. The more he became involved with PBB, the more he worried about their well-being.

"We are talking about Utopia, though," he went on. "We are talking about enormous federal funds and mobilizing the whole country. Maybe I am over-stating the case, but something like that should have been done. There is no way the state could have done it. No way. It would have taken something like two hundred and fifty million dollars. It was not just the dairy animals which should have been tested, but the beef animals as well. And the chickens.

"The Feds gave us suggestions, but not the money. Without it there was no way we could have hired laboratories all over the place and mobilized the veterinarians. In order to get this thing off the ground as it should have got off the ground we would have had to convince an enormous number of people: the governor, the legislature, the Departments of Agriculture and Health, the federal agencies, Congress, the president. Keep in mind who became president that summer. Even with that man in the White House we could not do it."

The man occupying the White House through the worst years of the PBB contamination was Gerald Ford, who took office in August 1974 after Richard Nixon resigned in disgrace. Ford was raised and educated in Michigan, and served as a Michigan congressman for twenty-five years. Yet, so far as is known, no Michigan official ever appealed to him directly for help with the problem which engulfed his home state. They all knew, innately, that wherever his sympathies might have been, he was as limited by political considerations and bureaucratic boundaries as they were. His was also the burden of healing the wounds of American society which had festered under a corrupted presidency. A few thousand ailing cows were a small consideration compared with that.

The analogy of those times was soon picked up by the farmers of Michigan. "They thought what was happening to them was like Watergate, and they called it Cattlegate," Iwasko recalled. "But there was no big cover-up. It was not what was done that was wrong, but what was not done by a number of people in authority who did not realize the magnitude of the problem.

"If I go back over the whole thing day by day and think how we might have acted differently, I do not think we could have done better than we did. Within the power that we had, I believe we did a commendable job."

Dr. Maurice Reizen, director of Michigan's Department of Public Health, undertook a health study of some of the families living on quarantined farms as the contaminaton became known. But state health departments were not created for this kind of crisis. They could respond superbly to outbreaks of contagious or infectious disease, but chemical contamination on a massive scale was beyond their means and expertise in the early 1970s. Dr. Reizen had tried to develop an environmental epidemiology unit within his department, following the mercury pollution of the Great Lakes. It was a modest project which the state legislature funded for less than three years. In the spring of 1974, Dr. Reizen's department requested an appropriation of $89,800 to continue the work for another year, but the legislature refused and the mercury project came to an abrupt end.

As soon as the mercury project ended, the PBB contamination emerged. Dr. Reizen asked for funding to be restored and again was refused. To study a few families from quarantined farms, he had to reassign scientists and technicians working in other department sections. He realized that this limited short-term study was inadequate, but thought it was good enough to determine whether the state had a human health problem on its hands. Anyway, it was the best he could do. Reizen's staff went out to the quarantined farms, did physical examinations, took blood samples, and asked questions of 165 farm people.

How did they feel since the cows became ill? Had they noticed unusual new symptoms? They followed the same procedure with 133 persons from Michigan farms which had not been quarantined, and compared the two. It was a standard medical study conducted along accepted lines; a group of potentially affected people compared with a group of "controls," but it was meaningless because it lacked two essentials.

Firstly, most of the "controls" had been exposed to PBB. Many of the nonquarantined farms were contaminated, but their owners had yet to discover the fact. Many families living on them had been ingesting contaminated food, whether it was their own produce or food from local supermarkets. This possibility did not occur to the health department.

"We were mired in a swamp of ignorance. If we had had the money and known more, we would have gone outside the state for our controls," Dr. Reizen conceded. But that was long afterwards.

Secondly, Dr. Reizen's staff expected that if PBB were highly toxic to humans, some farm dwellers would become very ill very quickly. When none of them did, it was assumed that the worst danger was over. All the state researchers, including Reizen, had had a standard medical training which teaches that a classic toxicology works with two variables, dose and time. The larger the dose, the more seriously ill a person becomes. The more toxic the poison, the quicker it acts. Known as a dose-response relationship, this is not always true of chemical contaminants. For example, not until twenty years after the exposure of thousands of workers was it realized that asbestos fibers can cause a rare form of lung cancer; it took that long for the disease to develop. The U.S. airmen who inhaled vapor from some of the defoliants which they sprayed over Vietnam did not become seriously ill with related diseases until a decade later. It was a quirk of chemistry, a reversal of the rules of toxicology which few doctors had experienced, let alone studied, when PBB began to filter through Michigan. Furthermore, PBB turned out to be more perverse than most chemicals. Some people who received low doses became very ill after a lapse of time. Others with higher levels of PBB in their blood and fat appeared to remain healthy. This, too, confused the health department team, which concluded that some quarantined farmers' symptoms might be psychosomatic—that they felt wretched because they were overwhelmed by the disaster striking their cattle.

The health department's report was not only sketchy and unscientific; it came out too soon to make an informed assessment. It was published within a year of the contamination being discovered, and its most significant conclusion was that no case of human illness could be attributed to the acute effects of PBB. This made comforting headlines.

"We could not find any cluster of symptoms which we could call

PBB-itis," Reizen remarked. "We did find symptoms, but of the kind one might find in the general population, and we could not link them with PBB in a dose-response relationship. This was frustrating, because physicians are used to seeing the most symptoms in the people with the highest toxicity. What we saw did not follow what medical school had taught us. It opened a vista of a field of epidemiology new to us. I liken what we saw to a small ripple on a pond. While we were trying to understand why it was there, a much bigger ripple had already started."

The second ripple had become so large that before the study was finished more than three hundred farms had been quarantined. Finally realizing that there might be delayed toxic effects upon people, Dr. Reizen went back to the legislature and asked for money to start a long-term epidemiological study. Once more he was refused, and told to seek federal funding.

"This is a problem for cows, not people," a senator told him curtly.

When the cattle contamination story broke, at least one Michigan physician quickly grasped the potential for human disaster. He was among the very few to do so, and the only one to translate his fears into immediate action. Thomas H. Corbett, chief of anesthesiology at the Veterans Administration Hospital at Ann Arbor and a faculty member of the University of Michigan's medical school, was a bouncy, energetic man in his mid-thirties with the same persistent scientific curiosity as Rick Halbert. This had led him to do some fascinating research concerning the effects of anesthetic gases upon those who administered them, and some of his experiments with mice had led him to suspect that long-term, low-level exposure to such gases might cause cancer.

It was a frightening theory, which grew in Tom Corbett's mind soon after he began to work with anesthetics. At the end of a day in an operating room, his wife, who had been a nurse, could tell which anesthetic he had administered by the smell which hung about him, even after he had showered and changed clothes. This made him realize that his body must be absorbing and storing anesthetic gases. Troubled about the consequences, he conducted experiments and found some gases had the potential to cause cancer and birth defects among people subject to prolonged low-level exposure, notably operating-room staff.

Tom Corbett presented this thesis at a meeting of the International Anesthesia Research Society, held in Florida in March 1974. It attracted the attention of Dr. Irving J. Selikoff, a pioneer in the field of chemical contamination. Selikoff felt that Corbett was on the right track but that he needed to make more carefully detailed studies, and encouraged him to do so.

Irving Selikoff was an extraordinary man, occupying a unique place in American medicine. By drawing upon a number of medical disciplines and developing others over many years, he had done much to create an understanding of man-made diseases. Selikoff was one of the first to sound the alarm about the danger of cancer to asbestos workers, and to comprehend the time-lag between exposure to an industrial carcinogen and development of disease. He was a generation older than Tom Corbett, and it was natural that the younger man should consult him. Lots of people did. As director of the Environmental Sciences Laboratory at Mount Sinai Hospital, New York, he headed the country's national resource center of environmental medicine and was its most distinguished expert.

Corbett kept in touch with him after the Florida conference, discussing the shape that his future anesthesia studies might take. But during May, Tom Corbett's fascination with this subject was suddenly diverted. In the Detroit newspapers he began to read about PBB.

The first story was brief and told of a few herds quarantined. Within a few days there were more, then more still. His curiosity aroused, Corbett telephoned the headquarters of Michigan Chemical and asked for a sample of PBB. He also asked some awkward questions about what kind of studies had been done concerning the long-term effects of the chemical. The answers were vague and left him apprehensive. His concern increased the following morning, May 29, when he read in the *Detroit Free Press* that the number of quarantined herds had increased to thirty. One sentence in the news story disturbed him deeply:

"One of the dairymen sold his animals for slaughter last January."

Corbett read it over and over. "I realized," he said, "that while Michigan Department of Agriculture and Farm Bureau were thinking in economic terms of how soon they could get quarantined farmers back on the market, many people had been eating this chemical."

He telephoned MDA to enquire whether its Bureau of Consumer Protection was planning toxicology studies on PBB to evaluate the human health hazard. The answer was no, MDA was not a research agency and it was already working overtime on the farm problem. Since his hospital contract allowed him to do some research, Corbett decided to tackle the job himself.

He ordered some laboratory samples—pregnant mice and rats—and while he was waiting for these he drove his family to northern Michigan to spend a weekend with his brother, who had rented a house on a farm near Traverse City. During the visit his sister-in-law proudly served eggs which had been hatched on the farm. Corbett looked at the chickens and commented that they were a miserable, scrawny bunch which looked as though they had been half plucked already. When he learned that the previous owner of the chickens had given them Farm Bureau feed he took some eggs back to Ann Arbor for analysis and, sure enough, they contained PBB—despite the fact that his sister-in-law had been giving the hens another brand of feed (presumably uncontaminated) for the past six weeks. The PBB in the eggs was below FDA's tolerance level, which meant they were officially safe to eat—a fact which alarmed Tom Corbett even more, because he felt sure they weren't.

His experiments with mice were even more disquieting. Those given PBB-spiked feed produced deformed fetuses and developed abnormal liver enlargement. The deformities included cleft palate and exencephaly (brain outside the skull). Scientifically, Tom Corbett was excited by the latter because an exencephaly is rare and he had never seen one before, but personally he was most deeply concerned for the public health of his home state.

At his hospital he tried another experiment. He took tissue samples from three people who had just died, and found traces of PBB in all of them. All three were urban dwellers who would have bought food commercially, and their deaths were unrelated to one another and to the farm contamination. Corbett concluded that contaminated food had been sold indiscriminately throughout the state.

During the summer he kept in touch with Dr. Irving Selikoff and told him about the PBB. In September he went to a meeting at Farm Bureau headquarters in Lansing, attended by dairy farmers, scientists, attorneys, and officials from MDA and Michigan Chemical. Dr. George Fries spent some time detailing his work with PBB and answering questions. Discussion centered on the economic loss to farmers and on how long it would take the animals to reduce PBB in their body fat to below the FDA tolerance level. Or, as Corbett put it, "how to get out of this mess as cheaply as possible."

After listening for several hours, Corbett took the floor. He showed slides of his mice experiments and expressed concern for the health of the Michigan populace. He explained that PBB had yet to be thoroughly analyzed for chemical impurities, and that these unknown contaminants might be more toxic than the chemical itself. Firemaster, he said, could be a material like some versions of the herbicide 2, 4, 5-T—a substance which could become contaminated with dioxin in the manufacturing process, and in which the contaminant was more dangerous to humans than the herbicide.

Tom Corbett got the impression that his audience was unimpressed. "I was not able to cut any ice with them," he said.

A few weeks later, on October 9, MDA held a fact-finding meeting on PBB and Corbett was asked to present his data. Delegates included researchers from Michigan State University, FDA officials, a few veterinarians, and several dairy farmers—Rick Halbert among them. Discussion took place about the unquarantined farms whose cattle had symptoms of PBB toxicity—the low-level syndrome which Doc Clark was seeing in herds under his care. It was becoming more and more apparent that the official tolerance level of one part per million was too high, and that some animals with much less PBB in their body fat were sick enough to be quarantined. Several delegates recommended lowering the tolerance level to prevent milk and meat from these herds going to market.

On the other hand, there were attempts to persuade Corbett that his experiments were irrelevant because even farm families had not received such comparatively high doses of PBB as he had fed to rodents. This strengthened Corbett's conviction that nobody in authority seemed to appreciate the risk to public health. He suggested a study of Michigan's general population by the state health department, but quickly discovered that the department felt it was doing as much as it could by examining a few families on quarantined farms.

"It was obvious to me that none of the state agencies knew what they were doing," Corbett summed up. "PBB was being recycled many times over in all kinds of ways, nothing was being done to prevent it, and nobody really knew what was in the chemical.

"After that meeting I telephoned Selikoff and said, 'Irving, you have got to come here . . .' "

Dr. Henry A. Anderson, assistant professor at the Environmental Sciences Laboratory in New York, was due to fly to his home state of Wisconsin on a family visit towards the end of October.

"Andy," Selikoff told him, "I'd like you to stop off in Michigan and have an informal talk with Tom Corbett about the cattle contamination there. See what needs to be done and whether it's a job for us."

Since its foundation in 1963, the Environmental Sciences Laboratory had assembled a remarkable group of doctors and medical technicians whose expertise covered every known aspect of chemical disease: neurological, gastro-intestinal, orthopedic, immunological, pediatric—any specialty which might be involved. Backed by laboratories with sophisticated testing apparatus, they functioned as a detective team which could go into an area of contamination, investigate the risks to human health, and make informed recommendations. Even when there was no known cure for the ensuing toxicosis, they could do much to minimize and contain an environmental disaster. They were able to function independently of industry or local government, being funded by the American Cancer Society and the National Institutes of Health.

Henry Anderson and Tom Corbett spent a day together in Ann Arbor. They began by discussing the usefulness of expanding Corbett's own studies with help from the Environmental Sciences Laboratory, but it soon became clear to both men that the job was too big for this. Corbett was concerned that the birth defects he had seen in rodents might show up in humans; Anderson worried that the contamination could reach epidemic proportions. Both felt that the medical work should be tackled promptly, since no one knew how far the contamination had already spread.

Back in New York, Selikoff heard Anderson's report and quickly agreed that his team had a responsibility to investigate the PBB problem. Selikoff's scientists were already working on PCB contamination of the Hudson River and had amassed a lot of knowledge of this closely related chemical. There was enough funding to pay for the team to travel to Michigan, a permanent faculty which could be organized to start a prompt investigation, and an unmatched experience in chemical contamination. No state health department had anything like this expertise and it was available at no cost to the state of Michigan.

Selikoff explained this in a telephone call to Corbett.

"I would be glad to come," he said. "But I must have a formal invitation from the state. I can't just barge in."

Thinking that Michigan's Department of Agriculture would be a suitable agency to issue the invitation, Tom Corbett called its Bureau of Consumer Protection and told the story. Again he felt he had failed to arouse interest, so he called the state Capitol and asked to speak to Governor Milliken. He got no further than an aide who promised to deliver the message. Corbett was persistent. He explained his fear for the health of the people, the importance of knowing precisely the nature of PBB, and he pleaded for a few minutes of direct conversation with the governor. The aide politely refused, but promised that the information would be "taken under advisement."

It was not followed up. The Selikoff team was left to assume that Michigan's Department of Public Health felt it had no need of outside assistance.

Several years later, Governor Milliken stated that he had never received Tom Corbett's message. By then the aide who answered the telephone was no longer working for the state and the log in which he had noted incoming calls had been destroyed. Unknown to Corbett, a memorandum of Selikoff's offer did exist in the files of FDA, to whom it was communicated by MDA, but neither agency acted upon it. Corbett was mortified that he had not put the request in writing, but at the time he did not feel this was necessary. He was not used to thinking like a bureaucrat. He had no secretary, he did not type and, like Rick Halbert, was a man of immediate action. When he had something important to say and wanted a quick response, it was automatic for him to pick up the telephone and expect an answer.

15

Farmer Against Farmer

HAVING failed to get money from the legislature for the quarantined farmers, Dale Ball assured them that their losses would be covered by the insurance of Farm Bureau Services and Michigan Chemical Corporation. "That's why we have insurance," he reminded them cheerfully.

It was not so simple. Protracted arguments went on about liability, complicated because the feed company's insurance was spread among several companies—and one of the policies did not take effect until *after* the contamination had begun. The bickering went on for weeks; it was late summer before the first settlements were paid to farmers.

"We had no income for a few months after the quarantine," Rick Halbert recalled, "at the same time, we had the expense of maintaining the farm and paying our employees. Our attorneys were advising us to bank our fires, to find a way of getting through the next two years, because they thought it might take that long to resolve the dispute in the courts. But at that point we determined our survival would best be served if we restocked with animals, even before we got rid of the most contaminated. Farm Bureau came in and tested all our herd and found that some of our young stock was below the one part per million level, so we decided to keep them. It would have been better if we had got rid of them all, but we had no idea what, if anything, the insurance would pay.

"We borrowed one hundred thousand dollars from the bank to buy new animals and managed to keep going until October, when we received a settlement. At the time we felt it was satisfactory. It was actually more than we had hoped for. After all that had happened, we did not expect very much."

Owners of some other heavily contaminated farms, lacking the Halberts' financial resources, were less fortunate. Many who were already making heavy mortgage payments could not raise the money to put their farms back into business. Jerome Petroshus had no livestock on his farm for thirteen months after his animals were slaughtered, and had to take odd jobs to make a living. Art Laupichler had to wait more than a year for a satisfactory settlement because, through no faulty of his own, he bought 402 pellets *after* the feed company issued its recall—and there were lengthy arguments about who was liable.

"I became pretty discouraged at the way public officials handle a disaster," Laupichler commented.

"We have all these people in high places and we pay them good wages to take responsibility. But when an emergency happens, nobody will take the initiative of deciding what needs to be done: not the state officials, Farm Bureau, or any of them. They all pass the buck."

Bill Coomer was similarly disenchanted. "We lost everything," he said. "The only animals we kept were the children's pony and the dog. Eventually we restocked with cattle from five states, and hogs from three states. They were all clean when they came in but the contamination on our farm was so bad they couldn't stay that way. We got so damn much of the stuff."

On farms that had received very heavy doses of PBB, recontamination was almost inevitable—although this possibility was not officially recognized for about four years. Although barns were steam-cleaned and new concrete laid, the heavily contaminated farms still held traces of the chemical. It was in the dust. It was in soil which had been fertilized by manure from contaminated cows—soil which was used to grow grain for the untainted cattle which farmers had taken the trouble to go outside Michigan to buy. Like the Halberts, many of these farmers had decided to keep their least affected animals, not wanting to lose irreplaceable genetic strains which they had bred over years, not realizing that the slow process of detoxification would endanger the new stock.

On some farms it did worse than that. The toxin was picked up by wildlife. Five years after Rick Halbert noticed the first symptoms in his cows, he accidentally caught a fawn in a mowing machine, on a section of the farm where no cattle had ever been kept. Although the fawn must have lived off wild vegetation (the woods, perhaps, where the Halberts had thrown their unsaleable milk) a biopsy of its tissues showed traces of PBB.

In various parts of the state, Michigan's Department of Natural Resources found measurable amounts in the bodies of rabbits, coyotes, ravens, starlings, gulls, pheasants, rats, raccoons, bears, deer—long after the worst of the cattle contamination was over. A dramatic tale was told by Gerald Woltjer, the dairyman who bought Peter Crum's highly contaminated farm. Crum was the Dutch immigrant farmer so impoverished by the effects of the undiagnosed cattle plague that he sent his herd to the slaughterhouse and sold his farm, only a few weeks before the cause of the plague was identified.

Crum was frank about his problems to Woltjer who, not suspecting anything contagious, moved to the farm with his own herd. He had been told that Crum's cattle sickened because they ingested pesticide, and he had thoroughly cleaned out the barns to the satisfaction of a Michigan Department of Agriculture inspector. A few days later, another state inspector called to tell Woltjer that the problem was not a pesticide, but PBB.

"What's PBB?" he asked.

"We don't know ourselves yet," the inspector replied.

In any event, it was too late. Along with the farm, Woltjer had taken over Crum's remaining stock of feed and had been using it for his own cows. Two or three times a year Crum had also spread the 180-acre farm with manure from his herd.

When PBB was discovered, Woltjer said, "Farm Bureau wanted me to skim off a foot of dirt where Peter Crum had spread that manure, but it was a low-lying farm to start with. We did it as best we could, except for one little piece of land where there were trees, a nice place for the cows to lay."

Within two years Woltjer's herd—which was never quarantined—was so sick and useless that he was on the verge of bankruptcy. He told of scrawny cows with perpetually bloody noses "who acted like they were blind"; cows so weak that they could not get up to be milked; cows which had bodily infections but passed inspection to be butchered for human consumption. Yet Gerald Woltjer's description of the devastation of his land was even more terrible than the contamination of his herd.

"The longer I lived on that farm the worse it became," he said. "After a time there were no worms in the soil. There were no field mice, no rats, no rabbits, no grasshoppers. As the cattle were dying, the cats and dogs were dying too. A fully grown cat would live only six weeks on that farm. Our three dogs went crazy. Our neighbors had bees that were dead in the hives. The frogs were dead in the streams. There was a five-acre swamp that used to croak at night so you could hardly sleep. Then it was silent. And it was a long time before I knew why."

None of these consequences was foreseen when the first damage settlements were paid to quarantined farmers. Farm Bureau believed that there were only a few of them, and that once they had cleaned up their premises and replenished their herds their troubles would be over for good. Most of the farmers believed this themselves. "This is better than having a disease like TB," Myron Kokx remarked. "At least when we get new cows, we'll have clean milk again. The premises are not contaminated." Rick Halbert was probably the only farmer in Michigan to recognize, in the summer of 1974, that this was not true. He knew that now PBB had entered the environment, traces of it would remain on farms even after the most thorough cleanup, just as they hung about the atmosphere of Michigan Chemical's laboratory, contaminating the testing equipment and making it impossible for technicians to get accurate readings.

While he was aware of all this, Rick did not think it a serious concern. Coupled with his scientific knowledge, he had a modern farmer's attitude towards chemicals. Every year his family spread between three and four tons of pesticides on their land, and if they had been growing cash crops instead of rearing dairy cattle the amount would have been far greater. Chemicals had become essential in many ways to America's agricultural economy, and the risks which accompanied them were part of the trade-off for cheap food.

"If we don't want to pay the price of having people out in the fields doing the work, we have to accept chemicals on farms," Rick philosophized. "The general consumer may think this is bad, but he does not understand that virtually all food is contaminated with something. Even before the chemical age, purity was elusive because food was contaminated with fungi and bacteria, and animals were contaminated with brucellosis and tuberculosis. So to those nostalgia buffs who think that food was better or safer in earlier times, I would say, 'Prove it.' For us, the alternative to using chemicals is to cease farming, and I think that is too much to ask."

Rick even had reservations about the wisdom of research to find "safer" chemicals for agricultural use. "I think we can work our butts off trying to develop chemicals which don't hurt us," he said. "But even if we come up with a new set which are not now deleterious to our being, we cannot be sure that they will not be so in the future. If they enter the food chain in quantity, they may cause things other than cancer or genetic damage, things we have not envisaged."

Rick's pragmatism was shared by a growing elite of successful, high-producing farmers to which the Halbert family belonged. Operating with huge investment, efficient management, computerized and scientific farming methods, they welded together a diversity of technical skills to create an industry for which a new word had been coined: agribusiness. It was farming of the future, and much of the present. In agribusiness, machinery replaced manpower, production was paramount, and chemicals were as essential as the rains. The very low trace levels which inevitably entered the food chain had not been proven to do harm. By 1974 it was widely acknowledged that every American had measurable levels of DDT and PCB in his body, apparently without ill effect. An even smaller exposure to PBB could not hurt, so the agribusiness farmers rationalized, provided the farm animals remained healthy and productive.

This was where Rick and his family stood in November 1974. They had received a handsome settlement. They were back on the milk market. Their most damaged animals had gone to Kalkaska and had been replaced with new ones. Their barns had been thoroughly cleaned and they were nursing the slightly contaminated cattle back to health. They had been through a terrible year but the future looked hopeful. Rick knew that PBB might remain in the dust of his farm indefinitely, but probably to a lesser extent than the residues of all the other chemicals which were regularly used on the land, and he was not seriously worried.

Conditioned by this thinking, some agribusiness farmers resisted sending their animals to Kalkaska. Paul Greer's neighbor, Blaine Johnson, was one. He said he would rather "work with" the sick animals in his pure-bred herd than lose them, and for months he tried. But by November it had become obvious to the Food and Drug Administration that the "tolerance level" of one part per million, measured in milk or in a sample of the animal's body fat, was too high. Far too many animals below this level were sickly, with the same kind of symptoms that had been noted on the highly contaminated farms. By this time the manifestations of PBB toxicosis in cattle had been chronicled in an article, jointly written by Ted Jackson and Rick Halbert, and published in the *Journal of the American Veterinary Medical Association* in September. The publication of this article in a leading professional journal was a most prestigious event for a country practitioner like Ted Jackson, and he labored over it with such care and precision that it has stood ever since as the classic text on the subject. In it he and Rick described all the clinical signs of "PBB-itis": the loss of hair, the wrinkled and thickened skin, the abnormal hoof growths, the damaged livers, the calves born late and dead.

When it realized how many of these things were happening to so-called low-level herds, the FDA lowered its PBB tolerance level from one part to .3 parts per million on November 4, 1974. Immediately at least five thousand more Michigan cattle were doomed. They included herds like Blaine Johnson's, which had eaten what Rick called "hot" feed, and herds like those of many of Doc Clark's clients, which had been contaminated at second hand from other Farm Bureau products and from equipment at the mills. FDA officials explained that they were able to lower the tolerance level because development in testing equipment now made it possible to measure down to .3 ppm with reasonable accuracy. Some scientists insisted that this had been feasible all along—that it was economics which dictated the initial tolerance level and only political pressure which reduced it six months later.

Another factor was considered in arriving at the figure of .3 ppm. The tolerance level for polychlorinated biphenyl already stood at 2.5 ppm in milk fat, a figure arrived at on the basis of the *yusho* experience. Tests indicated that polybrominated biphenyl was in some respects five times as toxic as PCB; so the FDA made a calculation which went like this: Divide 2.5 by five, halve that figure for safety, take it to the nearest decimal point and .3 is the answer. Nobody could be sure whether meat and milk below this level was safe. The figure was only a piece of educated guesswork, influenced by the capability of gas chromatographs which were standard at the time, influenced also by the condition of cattle which were above that level. On live animals the test was usually performed by a veterinarian who would make a small incision near the tail, remove a piece of fat, close the wound with sutures, and send the sample to MDA for analysis. The cost, about fifty dollars for every animal tested, had to be borne by the farmer, who had only one incentive: if any of his animals were above tolerance level he could get his entire herd quarantined and claim compensation from Farm Bureau Services. There was one major difficulty. After the level was lowered in November, the feed company was inundated with a new batch of farmers' claims—at exactly the time when it thought it had taken care of most of its liabilities, a time when its insurance coverage was beginning to run out.

Added to this, Farm Bureau Services still had to pay for the extermination at Kalkaska, which became daily more dreadful. The state's original plan had been to seal the pit in October after burying 5,000 cattle, but eventually about 30,000 farm animals were buried there (approximately 25,000 cattle, 4,000 swine, and 1,000 sheep, goats, and horses). About 1,500,000 contaminated chickens were buried in landfills in various parts of the state. Michigan's Department of Natural Resources was not happy about this, but "we had no better alternative," the department's director, Dr. Howard A. Tanner, said. "And so we continued as long as the flow of cattle continued. . . ."

Some days the flow was more like a flood. Bill Hughston remembered one Sunday afternoon, towards the end of 1974, when he took a load of condemned cattle to the pit and found eleven trucks ahead of him. "They were killing two or three thousand head of cattle a week at that time. They would pile the dead animals on wagons like cordwood, and dump them."

When the tolerance level was lowered to .3 ppm, the public press again missed the essence of the story. For six months state officials responsible for public health and consumer protection had insisted that there was no danger to food supplies under the old level. If the FDA now felt it necessary to make the guidelines more stringent, had Michigan's meat and milk really been safe all this time? The general assumption, oddly enough, was yes—and that with the lowering of the tolerance level it was being made even safer; that the FDA was simply taking the precaution of keeping PBB at the lowest measurable amount. The state's newspapers, for the most part, accepted this explanation, and the general public continued to regard the story (if it made an impression upon them at all) as a rural event outside their interest. In his role as consumer protector, Dale Ball had assured and reassured them that Michigan's food continued to be wholesome and that "scare stories" about PBB contamination were unwarranted.

"I think the public should realize that almost any compound, eaten in sufficient quantity, can poison both people and animals," he stated. "For example, common table salt can be lethal if consumed in large amounts. It is not unusual for animals to die of salt poisoning."

This quote of Ball's was long remembered. It was turned into a misquote ("Dale Ball said PBB was no more harmful than table salt") and the misquote was so often requoted that it stuck. It was used both for and against Dale Ball. Some farmers scoffed at it and some of them believed it. Later, after a long period of indifference, the public became similarly divided into two camps, those who were concerned about public health and those who thought the "PBB scare" was nonsense. As time passed, attitudes hardened—and that was the way it stayed for months, even years, after the event.

Reduction of the PBB tolerance level caused the first battle lines to be drawn between farmers. On one side were those who had been severely contaminated, reimbursed by the feed company, and were back in business. Most early settlements were made on the basis that cattle with PBB levels below one part per million were retrievable, which most of their owners believed to be true. In general, these settlements precluded the possibility of future suits from the farmers—certainly after a period of three years. "At the time that seemed pretty good," commented Bob Demaray's wife, Violet. "If no problems showed up in three years, we couldn't imagine PBB bothering us after that."

Many of the agribusiness farmers received relatively quick and generous payments from the feed company. Because they had such well equipped, high-producing farms, they were able to absorb the problems of the less contaminated animals which they decided to keep. With their shared attitude about chemicals they truly believed their lightly contaminated cattle to be harmless and redeemable. Even if the PBB in some of their milk was above the new level of .3 ppm it could go to market, if diluted by the clean milk of the new cows. But the more stringent guidelines of November 1974 posed a threat for the future. Suppose public pressure were to cause the tolerance level to be reduced still more: then these farmers who had successfully come through one quarantine might have another imposed on them—one for which they saw no need, nor sense, one which could cause ruin because they had signed away their rights to further claims. While there were few quarrels about reduction of the tolerance level to .3 ppm, some farmers whose herds had been heavily contaminated fought with a vengeance to deter federal and state authorities from making it even lower.

They had two powerful allies: Farm Bureau and Michigan Department of Agriculture. Farm Bureau did not want its subsidiary feed company paying out any more claims, and the Department of Agriculture was concerned that if any more farmers had to go out of business, Michigan agriculture (which, next to automobiles and tourism, was a leading industry in the state) might be irreparably harmed. Also, its staff was already being worked beyond capacity, with no help from the federal government, in coping with the PBB problem which it already had on its hands. If the job were any more complicated it would be beyond the capacity of MDA, and no other branch of the state government was willing or able to assume responsibility.

The agribusiness farmers felt that none had suffered as severely as they, and they had been able to recover. Their families appeared to be unharmed. How was it possible, they reasoned, for a farmer who had received only a tiny amount of PBB in a load of feed—the amount left in the bottom of a mixing machine, already diluted by other feed—to be having problems? It was like a man recovering from a badly broken leg, getting back on his feet, finding how good it was to walk again, how well he could move, wondering impatiently why a person with a sprained ankle should be as incapacitated as he was—or should expect to be paid the same compensation.

On the other hand, owners of slightly contaminated farms had a different set of problems, not understood and barely recognized. Some of their cattle had been sickening for months, were as unproductive as animals which had been buried at Kalkaska, yet fat biopsies showed them to be under, or only slightly over, the new .3 ppm guideline. It was impossible to determine when or how these herds had ingested the toxin, but they must have been taking in low levels of PBB over many months. Rick Halbert's cows, along with others, had swallowed very high quantities for a much shorter period. Their symptoms appeared quickly and were so clearly related to the feed that the cause was soon removed. Symptoms in the low-level herds were slow to appear and more subtle, making diagnosis difficult.

For a long time, this apparent lack of a dose-response relationship did not make sense to most doctors and veterinarians. If an animal had been badly poisoned they expected that it would become ill quickly; if it didn't, symptoms which appeared weeks, even months, later could not be blamed on the toxin. They did not appreciate the fact that PBB given in low doses over a long period could build up in the tissues and eventually produce a toxicosis just as devastating. Ultimately the amount of PBB ingested in this way was often greater than that from a single heavy contamination. The really "hot" feed tasted so terrible that the animals soon rejected it, but feed with residual amounts of PBB apparently did not have as much unpleasant flavor, and the cows went on eating it.

Doc Clark took a while to grasp this concept. So did the farmers' attorneys—for in the summer and fall of 1974 several of the quarantined farmers had engaged lawyers to process their claims. The two who gathered the most clients were Paul Greer of Fremont and Gary Schenk of Grand Rapids who, at Greer's suggestion, had joined forces because there were more quarantined farmers in the Fremont area than he could handle alone. It was an odd alliance. Greer, a handsome, middle-aged man who had built up a successful practice in the small town where he was born, and who enjoyed the life of a country gentleman; and Schenk, a generation younger, bright, aggressive, and barely five years out of law school—a city boy from a large, working-class family who had worked his way into a junior partnership of a prestigious Grand Rapids law firm by his own determination and ability. Where Paul Greer was astute, shrewd, and practiced in legal and political strategy, Gary Schenk was forceful, confident, and ambitious. Greer had a natural grace and flamboyance which Schenk might never acquire; Schenk had an adroitness of mind, sharpened by a boyhood in mean back streets, which Greer privately envied. They were not each other's kind of person yet each recognized in the other qualities which he would like to have added to his own.

They began to cooperate on PBB cases shortly after the first quarantines were issued in May 1974, Greer's farm among them. Schenk had recently filed suit against Farm Bureau Services, on behalf of his client, John Williams of Westmac chicken farm. When Schenk read about the PBB contamination he felt sure this was the "unknown toxin" which he had alleged to be in the chicken feed, but at first the feed company insisted that only dairy cattle were affected. Almost a month passed before he was able to persuade the state Department of Agriculture to test Williams's chickens for PBB—a test which resulted in the farm being quarantined. Meantime, Williams had innocently sold as "spent hens" 69,000 birds which had mysteriously stopped laying. Spent hens are ones which have gone through their breeding cycles, whose flesh has become too tough to be sold on butchers' counters, and who traditionally end up in cans of chicken soup. It was Gary Schenk's belief that this is where Williams's 69,000 birds went.

It made him angry to think that his client's contaminated chickens had probably gone into the food chain instead of the pit at Kalkaska. He was also concerned because he, his wife, and his two small children had been eating Westmac eggs for months past—eggs which had tasted so good and fresh but which he now believed to have been severely contaminated.

About this time Greer telephoned him. He had read of Schenk's involvement in the Westmac case and was inundated with requests from neighbors whose farms were quarantined, as his was.

"There's a bunch of quarantined farmers coming to my office tonight to discuss their cases," Greer told him. "How would you like to drive over from Grand Rapids and join us?" He suggested that Schenk bring a copy of the complaint he had filed on behalf of John Williams to serve as a model for his own clients.

The group met in Paul Greer's library. The farmers argued long and late about their claims against Farm Bureau Services: how to calculate the actual milk loss, how much the normal cull rate in a herd was increased by a quarantine. Gary Schenk listened fascinated and nonplussed. He knew nothing about dairy cattle. He didn't know what a cull rate was (the process of weeding out the less productive animals in a herd) or what the farmers meant when they talked about their cows freshening (giving birth).

Afterwards he told Greer, "I won't be any good to you on this case, Paul. I don't know the lingo."

Nevertheless, the farmers' stories intrigued him. He was impressed by their descriptions of how the contaminated cattle were rough-looking and developed tremors before they died, just as he had noted with John Williams's chickens. He had never concerned himself with rural affairs or environmental matters before; he remembered hooting with derision when he read that American eagles were not reproducing because DDT had caused the shells of their eggs to be too thin for hatching (it had appealed to his college-boy sense of humor that the symbol of American supremacy was being threatened by a bug spray), but suddenly he realized that this was the same family of chemicals which had poisoned John Williams's chickens, and Paul Greer's cows and—who knows?—his own wife and children. They were all, animals and people, helpless victims of profit-motivated businessmen and bungling bureaucrats: that was how the situation crystallized in Gary Schenk's mind. Paul Greer's attitude was more simplistic. He was fighting for himself and other farmers in his community. Farm Bureau Services had misled them all; he was a sore loser, and he was not about to let the feed company get away with it.

Together, Greer and Schenk sued the feed company on behalf of a number of quarantined farmers. One stricken farmer told them, distractedly, "I can't bear to hear those cattle out there bawling any longer"—a remark which haunted Schenk for months. Most of these farmers had such severely contaminated herds that their cases, like Rick Halbert's, were not difficult to settle. The feed company's insurance paid out sums satisfactory to most of the farmers, and most of the settlements which Greer and Schenk handled in the summer and fall of 1974 were paid without much delay. They ranged from several thousand dollars to settlements close to a million. Working on a percentage basis, the attorneys made handsome incomes for several months.

"Then," said Schenk, "we did something that will rank as one of the greatest bungles by plaintiff's attorneys. We had a farmer in Fremont whose milk was below one part per million—the tolerance level had not yet been lowered—but whose herd was obviously damaged by PBB. He had not been quarantined, and the insurance company said they would like to put his case on the back burner because they didn't know what to do with it. We agreed to that and so did the farmer, because the quarantined people seemed to be having worse problems than he was.

"We were wrong. We should have held out and said, 'The problems are the same for all kinds of farms, and you are going to take care of all or none.' Then we would never have run into the artificial distinctions which existed for ever after, between quarantined and nonquarantined herds, between high-level and low-level contamination, between the cows which could go to Kalkaska and those which could not. From then it became established practice that the insurance companies could get away without paying the nonquarantined farmers, regardless of how badly their herds might be afflicted.

"At the time we thought we were acting in the best interests of our clients, but at the time we did not understand low-level contamination."

16

Only a Litmus Test

In September 1974, less than a month after his revealing conversation with farmers at the Newaygo County Fair, Doc Clark received his copy of the professional journal with Halbert's and Jackson's article, describing symptoms of PBB toxicity in dairy cows. "I must have read it a hundred times, and the more I read it the more I got out of it," Clark said. He returned to the rounds of his farmer clients, wiser and heavy-hearted. He wasn't sure how to suggest to them that perhaps their cattle problems were due to PBB.

He knew he might not be believed. Most of his clients were staunch members of Michigan Farm Bureau who would not question the organization's word, and its word had been that no contaminated feed reached their area. There was also a stigma about "having PBB." Farm families talked about it in whispers as though it were some social disease. Neighbors avoided quarantined farms in case of contagion; children who lived on them were sometimes shunned in school. For Doc Clark to call on a farmer he had served for years, and tell him, no matter how tactfully, "I think you've got PBB," would be devastating.

Doc Clark had shrewd understanding of the idiosyncracies of his clients. Most dairy farmers in his area were of Dutch Protestant origin, only a generation or two out of "the old country," who maintained clean herds of Holsteins on immaculate farms. They were profoundly religious, political conservatives who secured their heritage by marrying into one another's families, perpetuating a thrifty, proud community in which every member of a household had some share in running the family farm—to the end that one piece of land could comfortably support three generations: the original owners, a son and his wife, who were now managers, and their children, who would ultimately take over. Since the late nineteenth century, when this Dutch settlement of parts of the midwest began, the pattern had changed in only one particular. Farms had become bigger and fewer. Automation had brought about the economic necessity for expansion, and those unable to make the change had sold their land to the others and moved to the cities.

Doc Clark was of Canadian and Northern Irish Protestant origins. His mother's maiden name was McNeilly, and he had a commercial copy of the McNeilly coat of arms framed in his living room. *Vincere vel mori,* conquer or die, the motto read, and he took it personally. But while his approach was fearless, it was rarely direct. It was part of his country wisdom to assess a problem warily from all angles, and then, most often, to tackle it circuitously. In some ways he knew "them Dutchmen," as he called them, better than they knew themselves. He was a frequent guest in their generous farm kitchens: homes where the Bible was read aloud at a well-spread dinner table.

One day towards the end of the summer, he called at Garry Zuiderveen's farm at Falmouth, a rural community adjoining McBain, and dropped one of his first hints about PBB. Some of the Zuiderveens' young animals were having respiratory problems, which was most unusual in warm weather.

After a tour of the barn, Doc remarked quietly, "Garry, looks like a dozen of those calves are going to die."

Garry Zuiderveen was stunned. In the late winter and early spring of 1974 he had lost an unusual number of calves, and some of his cows had developed inexplicable problems: lameness, emaciation, intermittent diarrhea, tearing of the eyes, reduced milk production, and abnormal behavior—animals becoming neurotic and hyperactive, then unaccountably lackadaisical. Sometimes they acted like demons, sometimes as though they were blind and stupid. He had called Doc to the farm much more often than usual. Neither of them had known what to make of it. Garry and his wife, Lois (a farmer's daughter, also of Dutch stock), had become increasingly worried. They had bought the farm from Garry's parents in 1963 and had built it into a thriving enterprise which one day their son, Garry Junior, would take over. Garry's Dutch-born father had built his retirement home near the main farmhouse and still kept his hand in by taking care of the calves. He had been a conscientious farmer all his life, and now he blamed himself because so many of the young animals were deathly ill.

"You had better take them over," he told Garry with tears in his eyes. "It seems I can't farm any more."

The remark saddened his son dreadfully, but Doc Clark's observation was the worst blow of all. A dairy farm has to be steadily producing calves to stay in business, and this year the Zuiderveens had already suffered too many losses. Garry Zuiderveen, a strong, stockily built man, was an expert farmer, one of the best in Missaukee County. Like Doc Clark, he should have been in his professional prime, and like Doc he was beginning to wonder if he was slipping. When he began in partnership with his father they farmed two hundred acres. Now it was eight hundred acres; Garry Junior was at agricultural college preparing to join the family enterprise—and suddenly the business was tumbling downhill.

The trouble began, he remembered clearly, with his "Number Eight" cow. Some farmers give their cattle names. Some call them by the numbers on their ear tags. Number Eight was the best cow on the Zuiderveen farm. She produced a calf in February 1974, and a few days later it died. She continued to lie in her pen, sickly. Given the standard treatment of glucose and calcium for milk fever, she failed to respond. She went off her feed, was treated with Vitamin A (which Ted Jackson had found helpful with the Halbert cows—a common treatment for digestive problems in cattle) but became terribly emaciated. Six weeks later she died, with blood running from her nostrils. Since then several other cows on the Zuiderveen farm had gone the same way as Number Eight.

All this time Garry Zuiderveen had been buying feed from the Falmouth co-op. He had used only a little 402 but had also bought soybean meal and a high protein ration for young calves, known as Mannamate—all of which, the co-op promised, were free from contamination. Now Doc Clark's prediction about the calves and the few hints he dropped about PBB made Garry Zuiderveen fearful. If he were to lose a dozen of the latest batch of calves, if the cows due to freshen in the fall were to die like Number Eight, next year would be catastrophic. After pondering for a while, Garry confided his fears to his family.

The reaction of Garry Junior, who had been helping on the farm during his summer vacation, typified the mutual support which existed in that household. "Dad, I can't go back to college for another year and leave you in this mess."

"Son, the way things are you had better finish your education and learn to do something other than farming," his father replied.

A few weeks after this exchange Garry Zuiderveen was one of many farmers in the area to receive a letter from the joint management of the Falmouth, McBain and Merritt Co-op.

Dear Customer,

Last spring we wrote you a letter stating that the feed we sold that came from the Battle Creek plant was, as far as we knew, free from PBB contamination and it still is. The feed is being tested for PBB on a regular basis and is O.K.

> But in the year 1973, there could have been contamination and at that time no one knew about it. The Health Department requested your names and are drawing milk samples to check possible contamination. If your test shows positive, please call on us and we will help you. Farm Bureau also has help available on call.
>
> We are very sorry about this but did not know what was going on at the time.

By now it was early November 1974. The PBB tolerance level in meat and milk had just been lowered from one part per million to .3 parts per million. Doc Clark suggested to Garry Zuiderveen and to a number of his clients in similar straits that they permit him to take fat samples from their worst afflicted animals. He did so, and sent them to Michigan Department of Agriculture for analysis. The results showed that a few of Doc Clark's clients had animals above the .3 level. Their farms were quarantined. A surprising number of farms with herds which, in Doc's judgment, were just as ailing passed the MDA test. Garry Zuiderveen's was among them. He therefore had no chance of financial help from Farm Bureau, as indicated in the co-op letter. He was left with no alternative than to arrange for a mortgage on the farm —he who proudly owned his land free and clear—and try to struggle on.

Garry Zuiderveen realized months later that he had chosen the wrong animals to be tested. After the first round of highly contaminated farms had been dealt with, testing became a farmer's responsibility. The Department of Agriculture made spot checks of bulk milk tanks and of cattle going to market, but it did not have the resources to check every animal. Farmers had to make their own decisions about testing and no dairy farmer who was losing milk production could afford to have fat samples drawn from his entire herd—about 185 animals in Garry Zuiderveen's case, 65 of them dairy cows. Like Garry, he would choose a few which looked the roughest.

"The first animals we picked for testing were the skinny ones which gave practically no milk and were so lame they could hardly walk," Lois Zuiderveen said. "It was months before we realized that they were the ones who had already got rid of most of the PBB through their milk, or through calving, and what we were seeing in them was the damage which the chemical had already done to their systems. It was the better looking animals which still had it in their tissues."

The Zuiderveens had to develop this understanding from Doc Clark, who took months to find it out for himself: that so long as the toxin was quiescent in an animal's body fat it appeared to do little harm, but as soon as the animal mobilized its fat to cope with stress, then the PBB began to devastate its system. The most common stress for dairy cows is pregnancy and calving, and this—in the experience of most farmers—was when the worst toxic problems were manifested. Animals which the Zuiderveens picked out as their weakest were probably the best survivors.

Farmers such as the Zuiderveens were faced with a dilemma. If they suspected their animals to be lightly contaminated, should they try to get quarantined in the hope of a substantial enough insurance payment to re-establish the herd? Or should they sell the least profitable cattle and try to nurse the others back to full production with a high protein diet and vitamin supplements? Depending upon their upbringing, their personal philosophy, their financial need, it became an economic decision for some, a moral judgment for others—and a deeply troubled area of uncertainly for most.

Sending contaminated cattle to market was easy enough to do at the time. Of the few that were spot-checked for PBB some of the worst-looking, the most worthless to a farmer, would pass the test for the same reasons that Garry Zuiderveen's worst looking cows were found to be below the PBB tolerance level. Pitiful as they looked in the slaughterhouse, once their hides were stripped and internal organs removed, the flesh that was sold to butchers resembled the flesh of healthy animals. A farmer might have to take a lower price, but at least he was selling his problem instead of spending money on it. To some, it seemed the only way to survive.

The thought of selling his unthrifty animals for slaughter crossed Garry Zuiderveen's mind, but he dismissed it. In ignorance, his family had eaten meat from the farm which they now believed to be contaminated. They would not risk eating more and neither did Garry feel he could let anyone else to do so.

He also wondered whether the FDA tolerance level was an accurate way to determine if an animal was fit for human consumption. A recent experience, in October, had caused him to doubt. He had a farmer neighbor, Peter De Ruiter, who had given his cattle similar feed and was experiencing similar problems. Doc Clark suggested that they both take some of their worst animals to the pathology laboratory at Michigan State University "to see what they can find."

The Zuiderveens' Number Five cow was wasting away like Number Eight, so Garry took it and a ten-week-old calf which was eating voraciously, yet too weak to stand. Pete De Ruiter took a steer and a cow. The two farmers had compared notes only a few days previously; otherwise, Garry, like most of the other afflicted farmers, was keeping his problems to himself. He gave his two sickly animals feed and water before leading them on to a truck for the 125-mile drive to Lansing. When he arrived there the calf had died. Veterinarians at the laboratory performed a gross pathology examination while the farmers waited. "We saw nothing about that calf to alarm us," Garry reported. "A year later, after we had opened enough cattle and knew what we were looking for, we recognized the enlarged livers."

He and Pete De Ruiter asked for fat samples to be taken from the living animals and sent to the state laboratory for PBB analysis. Two months passed before they received the results, and by then the official PBB tolerance level had been lowered. His dead calf showed no detectable amount of the toxin. The two cows, his and De Ruiter's, contained a small but permissible amount. De Ruiter's steer contained slightly more PBB, just above the new guideline, and as a direct result his entire herd was quarantined.

It was Garry's first insight into the chanciness of the testing procedure. "In one of our animals we eventually took at least thirteen tests and every one of them was different," he related. "All showed traces of PBB but the variation was enormous. It proved to us this was only a litmus test which showed whether the compound was present in the animal, but it gave no insight as to the damage done. There were so many variables. The concentrations of PBB must have varied in different batches of feed, and the chemical itself must have varied from one batch to another. What you found in an animal when you tested it wasn't all that relevant. There were so many things you couldn't measure. Every animal had a different metabolism, a different threshold of endurance. There may have been micronutrients in my soil that were quite different from another farmer's, and these could have been a catalyst for the PBB. It was all so elusive."

At Farm Bureau and at the Michigan Department of Agriculture none of these factors was acknowledged. Animal testing continued on the presumption that any animal below the new tolerance level was fit to provide milk and meat for human consumption, no matter what symptoms of PBB toxicosis it exhibited. Experts in both organizations were beginning to hint that sick cattle below the tolerance level were suffering from poor management, not PBB. There was no such problem as low-level contamination, only incompetent farmers. The charge sounded plausible. There is always a minority of poor farmers, as there are inefficient secretaries, inept attorneys, bungling surgeons. But men like Garry Zuiderveen had spent all their lives on the land, with never a herd problem before this one which they had been unable to overcome.

"My clients were all good farmers," Doc Clark insisted. "Those guys got more instinct about taking care of a cow than the experts will know in a lifetime. They've been pulling teats ever since they was knee-high to a grasshopper."

Three years later, in 1977, an astonishing piece of information came to light. Farm Bureau Services had known about the testing variables all the time. There was a Michigan farmer named Jim Cronin whose herd was among the first to be contaminated, and in June 1974 the feed company asked the Environmental Research Group of Ann Arbor, which was doing some of its early PBB analysis, to make a special study of two of Cronin's cows. The study involved measuring the PBB level in the cows' butterfat six times a day (before, during, and after each milking) over a five-day period.

The results were remarkable. PBB levels varied enormously from milking to milking, even during a milking, and the variation was greatest in the animal with least contamination. One cow showed less than .3 ppm on six of the thirty occasions when she was tested—and therefore her milk could have escaped quarantine if the laboratory had depended upon a single test. But the other twenty-four tests were above .3 and one was as high as 3.5 ppm. The variation in Jim Cronin's other cow, much more highly contaminated, was from 420 ppm to 950 ppm.

In a letter to Dr. James McKean, Dr. Richard A. Copeland of the Environmental Research Group commented on his laboratory's tests. He confessed to being baffled by the discrepancies. Then he added, "The wide variation within a cow from day to day, *especially at the lower PBB levels,* helps to explain why the Department of Agriculture numbers for a composite sample of a herd could vary significantly from week to week."

The italics were not in Dr. Copeland's letter. But those words should have made some impression upon the officers of Farm Bureau. They explained a great deal which it took a long time for farmers, veterinarians, and state officials to understand: that the toxin was moving around in an animal's body and no single fat biopsy could provide a true test of the extent of contamination. Nevertheless, the state was depending upon isolated tests, arbitrarily taken, to determine whether an animal or its milk should be quarantined.

The variation of a few parts per million had made no difference to severely contaminated farms like the Halberts'—their animals were so much in excess of the tolerance level that there was no doubt about quarantine. But once the level was lowered, the smallest variation in tenths of parts per million could make the difference between whether or not an animal or its milk were to be kept off the market.

Unrelated economic factors also shaped the fate of "the low-level farmers," as they became known. Farm Bureau Services was running out of insurance coverage, and the reduced tolerance level caused a flood of new claims. The feed company could no longer afford generous settlements, nor did it have the same motivation to settle quickly. The scene had changed. Many low-level farmers had animals of quarantine level on the basis of a body fat test, but whose milk was below .3 ppm. So MDA quarantined the herd only—which meant that none of the animals could be sent to market yet their milk could be sold legally. This left a farmer to decide whether to keep the unthrifty animals or send them to Kalkaska, in the knowledge that he might have to wait a long time for a settlement. Either way, he could not run a profitable farm. If his animals, like Garry Zuiderveen's, showed traces of PBB but were below quarantine level, he was faced with similar financial loss with even less prospect of recovery. Both kinds of farmers had no economic choice but to send their milk to market.

These financial considerations made veterinarians reluctant to suggest to farmers that they test for PBB, especially after the insurance money ran out. In hopes of increasing the fund and covering its own costs, Farm Bureau Services sued Michigan Chemical in November 1974 for about $270 million, but there did not seem much likelihood of such a huge claim being settled before some farmers went broke. So Doc Clark had his temptations too. It might, as he remarked, have best served his clients to advise them to send their problem cows to the slaughterhouse while they could still exchange them for cash.

Veterinarians were under other subtle pressures. They depended upon Michigan Department of Agriculture for annual renewal of their accreditation. Without this, they could not continue to practice in the state—and MDA and the veterinary school at Michigan State University kept warning them not to jump to conclusions in making a PBB diagnosis, but first to eliminate all other possibilities. Pondering on why so many afflicted farmers should have felt for so long that they were alone, Rick Halbert once complained about lack of communication in the veterinary profession: "All these practitioners are licensed by the state and if there is an outbreak of brucellosis or tuberculosis they are all sent all the relevant information. But with PBB I doubt if the state told them much. There was no feedback mechanism, no way that all the symptoms, statewide, could be catalogued on a daily basis. If there had been, someone would have recognized early how many cows were refusing feed; not breeding; dropping dead."

Doc Clark put it more bluntly, "If you felt like speaking out, you had to be prepared to buck the system. There was unspoken pressure from MDA, and in some areas there was client pressure. If a lot of your farmers were devoted to Farm Bureau and you criticized it too loudly, you could lose clients—and your living."

Michigan's veterinarians had no authority to insist that a farmer test his herd for PBB, no matter how strongly they suspected it. They could only recommend. Many practitioners genuinely believed that the PBB scare was exaggerated, and went on treating contaminated cattle for everything but the contamination (which no one really knew how to treat). When the animals failed to recover, they hinted that the farmer must be doing something wrong. This was the least controversial course, and veterinarians, too, needed to make a living.

No prospect ever existed of MDA taking over the testing which many of these veterinarians by-passed. It lacked the means and the money. One state official estimated that on its existing budget MDA would have taken a hundred years to test every animal on every farm.

From the beginning, Doc Clark did not doubt the course he must follow. He had some idea how difficult it would be, so he began by making discreet inquiries—not the kind a country practitioner would be expected to make in such circumstances—about the nature of the chemical and the company which made it.

"The co-op sent a letter in May 'seventy-four saying the contaminated feed did not come here, and that fall they sent another letter saying it did not come in 'seventy-four but may have come in 'seventy-three," he said, "I thought we were being lied to, so I decided to find out all I could about this PBB, what it was and how it got into the feed. I figured I had better be one step ahead of the authorities if I was going to take on this job, so I hired a professional to make some inquiries. He did not come cheap, but he told me a lot of things that turned out to be right. I discovered that Farm Bureau was not just a farmers' co-op but a political power structure, and that Michigan Chemical was not the small outfit it seemed, but part of Northwest Industries, which is a huge corporation.

"When you decide to get involved with people like this, it helps to know who you are taking on."

A hundred and fifty miles south, in Battle Creek, Ted Jackson was fighting the same battle with different tools. Although the Halberts' problem was behind him, Jackson could not leave the PBB issue alone. For months he thought he was dealing with a unique case of poisoning in a single herd; now he had an inkling of how widespread it was and how poorly farmers were protected against its spreading further. His clients had been affected by high-level contamination, but he was beginning to understand how lower levels of PBB could build up in a herd. In his spare time, which, with his heart condition, he should have used for resting, he was writing letters to state legislators, urging firmer controls over the manufacture, distribution, and testing of animal feed; and stressing the importance of more sophisticated research facilities for the state's veterinarians. Some of his letters were long and emotional. Writing them, he would become so carried away with the urgency of his appeals that he would forget that most legislators are unfamiliar with the jargon of the agricultural world. Sometimes he received interested responses, and these spurred him to write back in even more detail. Often he received brief formal acknowledgments which indicated that his letters were lodged on some secretary's desk in the Capitol.

At the same time, letters of a different kind were pouring into Ted Jackson's office. Veterinarians as far away as England and Canada had heard of his article in the *Journal of the American Veterinary Medical Association* and wanted reprints. Among the enlightened of his own profession, Ted Jackson had made an overnight impression as a meticulous and perservering diagnostician. State officials who received his letters and practitioners who were his immediate colleagues saw him differently: Ted Jackson, they thought, was a man with a one-track mind who had already expended too much effort on the Halberts and who should not waste his time meddling with politics.

17

A Doctor's Dilemma

In a phrase they are fond of using, Michigan dairy farmers don't go in for doctoring. They are strong, firm-muscled men conditioned by outdoor life in a clear, cool climate which strengthens the sinews and satisfies the soul. Michigan winters are long and cruel, with icy winds sweeping in from the Great Lakes, mile upon mile of them frozen solid, and deep snowfalls which bury one another. Spring, summer, and early fall merge into one tenuous season when the land is green and beautiful, and night temperatures are tinged with frost. Whatever the weather and how ever exhausted he may feel, a dairy farmer must get up early for milking and repeat the process every evening. If there is no evening dew in the haying season, he knows this to be a sign that it will rain next morning and he must work through the night, if need be, gathering hay to save it from a ruinous soaking. If a cow has a difficult calving he must help her to deliver—a muscle-wrenching job when the beast in panic can kick him sore. If he aches from a long day in the fields, or if he has the flu, he must still force himself to fulfill these relentless demands of the farm, accepting that his indispositions must be overcome because the demands of animals and land are inflexible. There are no days off for dairy farmers. The pattern of life is circumscribed, repetitive, and unalterable.

Since illness has no place in a life like this, it is seen as a shortcoming. Farmers who never hesitate to call in a veterinarian for herd problems are reluctant to see a physician for themselves. Most recognize only two types of human sickness: the kind a man can work off, and the kind which kills him. Since each eventually resolves itself, neither has need for doctoring.

During the winter of 1974–5 some Michigan farmers and their families began to be bothered by symptoms which did not fit these categories. They didn't go away, they didn't get better, and they couldn't be ignored, because they affected a person's ability to work. The symptoms were diverse, apparently unrelated and inexplicable. Joints ached. Body sores appeared and would not heal. Strong men and women felt debilitated, and after falling into bed exhausted would sleep around the clock, then wake up tired. Some suffered stomach cramps and diarrhea, swollen limbs, unaccountable bleeding from the gums and nostrils, fainting spells, and visual disturbances. Children were tired, fretful, with aching limbs and colds which lasted for weeks. Other health problems were less tangible. Some farm people were troubled by a lack of coordination, an erroneous perception of distances. Garry Zuiderveen's father complained that he was becoming "as clumsy as those dumb cows"—the contaminated animals which stumbled around with "the blind staggers."

Garry had different symptoms, which he attributed to personal stress. In addition to his own farm he felt morally responsible for the well-being of one which had belonged to his brother, who had recently died. His brother's son, Kenneth, had left college to take over the operation for his mother. Ken Zuiderveen had tackled the job with pluck and determination and Garry was giving all the help he could; now Ken's herd also seemed to be contaminated by PBB.

"There were nights when I could not sleep," Garry recalled. "I would walk down the lane and sit on a stone pile between the two farms, Kenny's and mine, and just meditate."

At times Garry's memory played tricks with him. These were passing aberrations, but they troubled him more than he would admit, conscious as he was of all the extra responsibility he had taken on. "Sometimes I felt like my head was disconnected from my body—and long afterwards I heard other farmers say the same thing. I can remember times when known distances would seem like half a mile off. I can remember sitting in church and the minister getting farther and farther away. I can remember driving up to the traffic light at the junction of Route 131, and now knowing where I was, although I had been there a hundred times before." At the same time his wife, Lois, suffered uncharacteristic bouts of depression, bursting into tears for no apparent reason.

These were not ailments which they would take to a doctor. Eventually they passed and the Zuiderveens tried to forget them. In other families the symptoms of toxicity were too pervasive to pass off—and most of them were happening on the low-level farms, not all of which had been quarantined, but where families consistently ate their own produce.

Some of the earliest symptoms appeared in a family which reared chickens as a side-line and used some of its cattle grain to feed them, then ate the eggs as part of the daily diet. At first the birds seemed to be such good layers that the farmers, Floyd and Yvonne Yarnell of Merritt, gave several to Floyd's parents as a present. Soon afterwards the entire family—three generations—suffered severe stomach aches and diarrhea. Floyd had blackouts, and once, when he passed out, badly injured his head. Another time he developed an abscess on his nose, his head became grossly swollen, and he had to be rushed to a hospital for emergency treatment because the doctor feared brain damage. One teen-age son had severe bleeding from the nose and gums, another developed an abscess on his tonsils, a five-year-old son had skin lesions all over his body, a daughter suffered agonizing stomach cramps and the oldest daughter—the only one who was married at the time—had a miscarriage. She soon became pregnant again and delivered a baby which was weak and jaundiced. The infant lost weight and became even more sickly until Yvonne persuaded her daughter to stop breast feeding; then the baby's health improved dramatically.

Floyd's father had such devastating diarrhea that he lost twenty pounds in fifteen days. Yvonne became "deathly sick" with back and stomach pains and she reported, "they opened me up and found me full of pus and infection in the gall bladder, the appendix and the liver function." She added, "Four out of fourteen in our family was doctoring for liver trouble at that time."

Her daughter's miscarriage distressed Yvonne more than any of the family ailments: it made her fear for the future. In her rural community there appeared to be an epidemic of still births and spontaneous abortions, just as with the cows. Other symptoms also related to the cattle plague: the nosebleeds, the abdominal pains and diarrhea, the persistent abscesses and sores, even the fact that different members of the family were affected differently. It was tempting for a physician to dismiss these discrepancies as unrelated—even psychosomatic—episodes, but like most farm families, the Yarnells had no history of chronic sickness.

"The first fifteen years that we were married, I don't think my husband was ever in a doctor's office," Yvonne Yarnell recalled. "The only time I ever remember him being sick was after he had a flu shot.

"Right after we started noticing our own sickness we sold the rest of our cows. We were going broke on the farm. Our milk production had dropped, and the cattle were not even making their own feed. We didn't associate the two things—our own sickness and the cows'. We owed money at the bank and needed the cash, and the Department of Agriculture was saying that cows below the tolerance were all right to sell. But if we had known how sick we would become afterwards, there is no way we would have sold those cattle. We would have gone to jail first."

Their innocence was widely shared. Farmers, unaware that their animals were unsafe to eat, butchered unproductive cows for their own freezers and sold some of the meat to friends. Louis Trombley, a farmer in the rural community of Hersey, just south of Doc Clark's area, was at first advised by his veterinarian that the trouble with his cows seemed to be calcium deficiency—and by his feed supplier that it could not possibly be PBB.

"I had five cows with twisted stomachs and the vet advised me to sell them for hamburger," Trombley recalled. "So we slaughtered them and ground them up and sold them to all our friends and relations. Their insides were mushy but the flesh looked good. About that time we had just built a new barn and we had a barn dance to celebrate. About three hundred people came to that dance. We had some beer, hired a band, served Sloppy Joes with this PBB hamburger meat, along with tomato sauce, chili sauce and onion. It tasted so good a lot of people asked us for the recipe." Soon after this Louis Trombley's herd problems worsened.

"In late 1974 the cows started aborting. There was days I could walk through the barn with a scoop shovel and wheelbarrow cleaning up aborted calves. I had calves born with holes in their head. You could look right through to the brain. I questioned the vet and he said it was a calcium deficiency. When I would put the newborn calves in a pen they would beat themselves to death against the wall. Yet when I finally suspected PBB and had fat biopsies taken of the cattle, seven out of seven samples came back from MDA marked nondetectable."

A year passed before Lou Trombley knew that his herd was contaminated—another year in which his family and a great many others were eating what came to be known as PBB-burgers from his farm. During this time Trombley, his wife Carol, and their six children developed symptoms of PBB toxicosis: blinding headaches, digestive problems, skin eruptions, inexplicable weight loss. One son lost fifteen pounds in six weeks. Unknown to the Trombleys, the Green family—who lived ten miles east, in the village of Chase—was in a similar situation. The same veterinarian served the two farms—a man who like most of his colleagues in the state, took a while to become familiar with the symptoms of PBB toxicosis. The Greens also had low-level contamination, and as early as October 1974 Alvin Green was worrying about lost milk production. About that time one of his cows, Nettie, caught her head in the feeder, damaging a nerve, and became paralyzed. Most farmers deal with this type of accident by slaughtering the animal for family consumption.

"We sent her to a butcher, and when we went to get the meat he said the liver was bad in that cow," Al Green recalled. "We never tumbled to what it was and we ate pretty near all of her."

Within months the entire family was ill—Al and his wife Hilda; their son Douglas and his wife, Donna, who farmed with them; and Doug and Donna's children. Al had blackouts and dreadful sores on his legs which would not heal; Hilda and Donna had agonizing abdominal pains; the children suffered boils and abscesses. Al was the sickest. A formerly healthy man, then in his fifties, he had two minor strokes which he compared with the convulsions he saw in his cows, and which aged him considerably.

"The vet kept asking us if we had bought Farm Bureau feed," Al said. "We kept saying we didn't, because our supplier handled feeds from an Indiana corporation. It was a long time before we found out that he was adding a supplement which he bought from another supplier, who got it from Farm Bureau.

"At this time I had a neighbor who had a PBB problem. I used to go by his farm, get a sick feeling in my stomach. I thought it couldn't happen to me. Then, by God, it did happen to me."

Before the Greens realized this, Doug Green—then aged thirty-five—became so fatigued that he could not do his job properly.

"I just wanted to sleep all the time," he said. "I went to the doctor, and for me to admit I was sick—well, I am the strongest person on earth—but I could hardly manipulate myself. I was exhausted, mentally and physically.

"On October 27, 1974, I got caught in the power shaft of the tractor when I was unloading corn to put in the silo. I went round and round in the shaft and broke both legs. I did not care if I lived or died. I knew there was something wrong with me for this to have happened, but I did not know what it was.

"Lying in the hospital with two broken legs, I was getting rest. I was in pain but I was getting rest. I was relieved to be in bed because I felt so exhausted. If you can imagine a doctor saying it will be over a year before you walk, and you having those feelings—almost of relief—when he says it, well, you have to be pretty bad. They operated and put plates in my legs.

"Before my accident, when the herd started going downhill, we had tried everything the vet suggested: more calcium, more phosphorus, more protein, and in June or July he had said he would test for PBB. I got mad and said there was no way we could have PBB. We had good crops, the best in the area, and we ground our own grain. I didn't see how the farm could be contaminated and I fought my dad and mother and Donna about having the cows tested. The meat from the cow we slaughtered looked perfectly good to me, and I told the others that the whole idea of PBB was a bunch of B.S.

"While I was in hospital the problem was getting worse, and since I was not around to argue the other three got the vet to check for PBB."

Several tests of milk from the Green farm all showed barely perceptible levels of PBB, well below .3 ppm. Eventually the family sent some hamburger meat from their freezer to a private laboratory—the little bit that was left of Nettie—and the analysis showed 1.47 ppm of PBB.

"We had consumed seven-eighths of the cow by then," Doug said. Yet the Green farm was never quarantined.

Confronted with such a variety of baffling complaints, most Michigan doctors simply treated the symptoms. The aching joints were taken as signs of arthritis, the skin eruptions appeared to be localized infections, the respiratory conditions were often diagnosed as bronchitis. The exhaustion and the altered perception of distances which cased Garry Zuiderveen to forget a familiar landmark and Doug Green to have such an unlikely accident were explained as "nerves" or debility. The miscarriages and the sickly babies seemed unremarkable to general practitioners. Doctors were even less disturbed by the assorted female problems of which many women complained. Every doctor knows that missed menstrual periods are a common symptom of stress —and since these farm women were having financial difficulties, what could they expect?

Unused to "doctoring," these people could have been chronically ill for years, so far as any medical man could tell. But the farm women equated what was happening to their children, and to their own reproductive systems, with the symptoms in their cows. Among themselves they worried about these things, reluctant to burden husbands who had enough concerns already. Meanwhile, men noted changes in their own sexual responses which they felt embarrassed to discuss with their wives. As one middle-aged farmer, happily married, explained it, "We would have a cow in heat and the bull would not pay any attention to her. He would go right past that cow to the other side of the pasture. I never saw anything like it." The farmer paused, shyly. "It has affected me that way too." For some time many farm families were prepared to believe what their doctors genuinely thought to be the case: that anxiety over their sick herds and their financial reversals had caused physical conditions of emotional origin.

One Michigan general practitioner came up with a correct diagnosis as early as 1974, and he seems to have been the first to do so; certainly he was the first to speak out. He was David Salvati, a doctor and osteopath, living in Big Rapids—a town of twelve thousand in west central Michigan, surrounded by farm country. Dr. Salvati's acquaintance with PBB was an early, personal one like Paul Greer's; as a second profession he, too, operated a small farm.

In the spring of 1974 he became concerned that three of his seven children had vague complaints—muscle aches and joint problems—and, suspecting the local water supply, ran laboratory tests for typhoid and paratyphoid. They were negative. Shortly after this his chickens and hogs were quarantined for PBB—and his patient load increased.

"A lot of them were what I would call 'clean patients,' in the sense that they were very rarely sick—men who worked two jobs, sixteen hours a day, with no problem. Suddenly they could not do it any more. I was starting to suspect PBB and would tell these people to get their hogs and cattle checked, and almost invariably they would come back when they had done so and tell me that they were of quarantine level.

"The biggest complaint at the beginning was extreme fatigue. Then aching joints, muscle weakness, gastric complaints, chest pains—but nothing you could pin to anything medical by the standard tests. It affected men more than women, but this was probably because men were used to working in the fields, or because the fat-muscle ratio was different. Usually I was used to seeing more females than males, but suddenly the pattern was reversed.

"We took tests which showed that the patients' livers had been affected due to the ingestion of toxins. I called the state health department several times in late 1974 and described the symptoms I was seeing. I told them it couldn't be anything but PBB. At that time the health department did not think the cattle contamination could affect humans."

Dr. Salvati related that one of the department's physicians dismissed his concerns with the comment, "Don't worry, if it goes in one end it will come out the other."

"The fact that we found PBB in their fat and blood serum at least helped people to realize what was wrong," continued Salvati, "and that they weren't imagining their symptoms.

"I was also treating more infections than usual. Resistance was lowered. I had an unusually good opportunity to watch people who were not ill before the problem, as well as people who had had symptoms before which were now being aggravated. Much later I did get patients who had read about PBB toxicosis in the newspapers and thought they had the symptoms. But that was not true of any of the early patients. Their health problems were real enough. To an extent these were compounded by the fact that they were losing their animals, their farms, their livelihoods. But they weren't just suffering from psychological stress."

Among Salvati's patients, Ronald Creighton and Lyle Ringler were both small-time farmers who raised a little livestock and got their basic income from outside jobs. They were strong young men, used to working long hours, raising enough animals to satisfy their family needs. It was people like these, modest tenant farmers whose health was hit the hardest; personally, they were exposed to far more of the toxin than were large commercial farmers because they ate all their own produce.

Ringler worked in a sporting goods factory in Grand Rapids and raised a few pigs for home use. In 1974 he butchered three for family consumption. Soon he, his wife, and their two young children were chronically ill.

"Our boy used to wake up in the middle of the night with severe pains in his legs and feet. My wife would be up all night with him. He was four then. I was having the same pain but was a lot more tired than he was. My wife and daughter were tired too. We went to Dr. Salvati, who ran some blood tests on us and said our livers were wrong. He asked if we had done any butchering, and told us to send some of the meat to the MDA lab for analysis. We did, and it came back 1.2 parts per million. They sent a letter back and asked if we had any other animals. I said just horses, and we heard no more.

"I kept working, but felt worse. I had a lot of stomach trouble and would fall down, black out with no warning. Eventually it got so bad that I was put on disability. They did not have a name for what was wrong with me but they put in a lot of medical terms which means arthritis, some kind of stomach disorder, some kind of balance problems and something wrong with my glands. I was checked over by one of the state doctors and he said I couldn't lift anything more than ten pounds, I couldn't grasp things, I couldn't walk up steps and ladders."

Ron Creighton, a neighbor of Ringler's in the rural community of Stanwood, worked for the Mecosta County Road Commission and farmed forty acres as a sideline. He produced all the milk, eggs, and chickens which he, his wife, and three small children consumed. The Creightons also made their own lard and butter. Their poor health began as early as March 1974—a month before PBB was identified as the contaminant in Farm Bureau's feed—and this time it was the woman of the family who was the first to become severely ill.

For no apparent reason, Jeannette Creighton's legs and feet became so swollen that she had to wear her husband's felt hunting boots to go into town. It was the only footwear she could get on. She was admitted to hospital for tests on a suspicion of cancer. These were negative, but doctors thought there might be a possibility of sarcoidosis or Hodgkin's disease. After her release from hospital other curious symptoms appeared. As Ron Creighton described it, "She had no saliva in her mouth; had to drink, six, eight glasses of water to try to eat a sandwich. And then a scum come over her eyes and everything was blurry. She went to an eye specialist in Grand Rapids and he prescribed eye drops, and the trouble did clear up, mostly. Then she went through a period when she lost her hair and she went to another doctor who said it was just nerves, and not to worry."

Typically, every specialist the Creightons consulted saw the problem in terms of his own specialty. Several months after Jeannette's legs became painfully swollen, Ron noticed the same thing happening to his own. Another doctor diagnosed congenital arthritis.

"He told me to take aspirin and learn to live with it," he related. "I was taking from eighteen to twenty-two aspirins a day, and I was tired and slept a lot. We saw the same thing in the children. It was nothing for them to go to bed on their own at seven o'clock and sleep until it was time for the school bus."

Finally the family consulted Dr. Salvati.

"He asked me if I had been around PBB. I told him I did not think so. He asked me where I bought feed, and when I said Farm Bureau he said I should have the animals checked. We had them checked and the chickens were 1.7, the hogs were 1.7, and some of the cattle were over the .3 level and some below, so MDA put a quarantine on us, and three months later my animals were picked up and taken to Kalkaska."

In February 1975 Ron Creighton had to give up working. His workmates had covered for him for months, leaving him on the truck when a group of them went out to do some road construction, then pretending he had been with them on the job. The time came when they could pretend no longer, and after two years on disability pay, with his wife and children still ill, Ron Creighton remarked, "This thing makes you feel you are ninety years old, that you ain't worth a shit nor nothing, that life ain't worth living."

The Creightons received a $10,000 settlement from Farm Bureau Services for the damage to their farm. After legal fees, they were left with $8,800. Neither the feed company nor Michigan Chemical made any payments for the human health damage.

David Salvati was one of the few Michigan physicians who argued with state authorities that his incapacitated PBB patients were entitled to disability pay. It was a battle, because none of the bureaucracies wanted to admit that PBB caused human health damage. "I got disability for one patient on the basis of my diagnosis of hepatotoxicity due to the ingestion of a toxic substance. Otherwise we would describe these cases as rheumatoid arthritis. But a lot of them still were not awarded compensation."

As the only doctor in a wide area who was prepared to try and treat PBB toxicosis, David Salvati's reputation spread by word of mouth from farm to farm. Soon his patient load was enormous, and he and his nurse, Mrs. Kathy Horonzy, were working night and day.

"We would try to schedule the PBB patients in the evenings, and we were running through the night, often until five, six or seven A.M.

Some would come from out of town and stay in motels and we would call them when we were ready," Salvati said. "Many did not have much money and I was paying for the lab work until I could not do so any longer. Some of these people were losing their livelihoods, were not on Medicaid and could not get welfare. And all we could tell them was that we would take their histories, take fat biopsies and do our best to treat their symptoms, but that we knew of no cure.

"I do not think the state health authorities realized how toxic PBB was. When they did, they started dragging their feet. They kept stressing that no human health effects had been demonstrated."

Mrs. Horonzy had one telephone conversation with an official of Michigan Department of Public Health who, she said, "told us not to pursue it, that we would panic the public."

Salvati decided it was the better part of wisdom to treat his PBB patients, get what help for them he could, and make no public statements. But an incident in March 1975 changed his mind. He had an acquaintance, Jerry Lane, who was a sheep farmer in Stanwood—the same neighborhood where Lyle Ringler and Ron Creighton lived. Lane was convinced that his flock was badly contaminated, but forty-two tissue samples from the sheep all showed PBB levels below .3 ppm. Consequently, the animals could not be quarantined. Lane insisted they were worthless and tried to force the issue by sending two hideously deformed lambs to auction in November 1974. The animals were born without hind legs and Lane expected their sale to be stopped by MDA inspectors. It was not, and they were purchased by a meat packer. Lane was appalled.

"He wanted Dr. Salvati to make a statement that meat like this was not fit for human consumption," Mrs. Horonzy said. "We both went to his farm to see what he was talking about. It was grotesque. The little animals had multiple deformities. Several had no hind legs and deformed faces. Their coats were poor, they were emaciated and really small. Yet there was a meat packer ready to buy them. The farmer was outraged that they were okay for market. He had had them tested and they had all fallen below the limit, except for two.

"At that time we had said nothing publicly, but this changed our minds. I spoke to the *Detroit News* about possible human health implications, and they came out with a story on it. Later Dr. Salvati went to Washington to testify."

Meantime, Jerry Lane publicly announced that he was prepared to sell his herd of five hundred for human consumption. This caught the attention of some Michigan newspapers, and resulted in further tests by MDA inspectors. This time one animal was found to have .8 ppm of PBB, and so, finally, the herd was quarantined.

Repercussions were felt by Dr. Salvati and his nurse for many months. "We started getting threatening phone calls," Kathy Horonzy stated. "It was always a man, and I am pretty sure it was the same person. He would say, 'If you don't want anything to happen you should get Dr. Salvati to retract his statements.' This person threatened to kill the doctor and me and then kill the children of both of us. It was summer, so we sent our children away. Soon after that there was this anonymous call saying, 'Don't think your son is any safer in Bay City with your parents or his daughter is safer in New York with his parents. We know where they are.'

"Then Dr. Salvati's dog disappeared. He was an exceptionally friendly German shepherd. Next day there was a phone call saying, 'The dog is just an example. Soon it will be your children.' We were scared, but there never was any sign of the children being attacked."

The doctor and his nurse got in the habit of checking their cars for explosives, but other than the incident of the lost dog, which never reappeared, saw no evidence that the threats were meant to do any more than terrify them. They continued their PBB work but encountered new obstacles. The state's Medicaid payments for their elderly patients—the fees usually paid directly to doctors—were being delayed. Dr. Salvati's was a working-class practice, and this constituted a large part of his income.

"From March to September of 1975 we received about four hundred dollars," Kathy Horonzy related. "We should have been getting that much a week. We would call the Social Services Bureau in Lansing but receive no satisfaction. We were so short of money that we were on the brink of losing our office when eventually the payments came through. The delays were never explained, but we felt it was another way of pressure being put upon us.

"All this time Dr. Salvati was sticking to his point, but by June of 1976 we both stopped making statements. We had to, for the sake of survival. It was a very disillusioning experience."

One other Michigan doctor became deeply involved in the politics of PBB, a physician with impressive credentials for investigating the effects of the chemical upon human health. Walter D. Meester, clinical toxicologist, was director of the Western Michigan Poison Center, and head of the research and toxicology department at the prestigious Blodgett Memorial Medical Center in Grand Rapids.

David Salvati referred several of his patients to Walter Meester, and after a time others came from different parts of the state. Despite all his expertise on poisons, Meester also felt at a loss in predicting long-term effects of PBB. No previous experience existed of its effect upon humans, and no conclusive animal studies on which to base a judgment. He was disturbed by the sloppiness of the study undertaken by Michigan's Department of Public Health—the one which compared families from quarantined farms with a group of so-called "non-exposed" Michigan farm families—and did not hesitate to say so. Immediately upon publication of this study, in the spring of 1975, Walter Meester wrote a critique of it, pointing out that 70 percent of the "non-exposed" people had significant PBB blood levels and that this alone made the conclusions invalid.

He expressed himself bluntly. "I feel that the MDPH study . . . was poorly planned, does not conform to the standards of adequate scientific, medical and epidemiological evaluation, was incomplete, possibly biased," he commented. "The study has not been published in the scientific literature and it probably would not be accepted for publication in a bonafide scientific journal with a competent and unbiased editorial review board."

When he wrote this, Walter Meester had only just started seeing PBB patients. Like David Salvati, he began by measuring the chemical's level in their blood. Soon he came to the conclusion that fat tests were probably a more accurate indication of toxicity. He found no correlation between the two and discovered that if he made the same tests on the same patient after an interval of time, the blood and fat levels could rise and fall in baffling, unpredictable ways. There were great variations within families, with no relationship between the severity of symptoms and the amount of PBB in a person's tissues. Some patients with high levels of the chemical seemed to be well, some with low levels had debilitating ailments. Meester concluded that while no specific syndrome could be attributed to PBB, some persons who had ingested the chemical showed a clear pattern of symptoms. He listed these as tiredness, sleepiness, irritability, muscle and joint pain, nervousness, visual problems, headaches, skin rashes, decreased resistance to infection, decreased libido, and decreased tolerance of alcohol. He was also concerned about the long-term possibility of cancer.

One of many unknown factors, he pointed out, was whether (and if so, how) PBB interacted with other chemical agents in a person's body. Already it was widely recognized that almost every adult in the United States had measurable levels of PCB. If PBB were added to this, was the effect cumulative?

Dr. Meester posed another question already troubling some laymen. If the PBB in an animal's fat was below the FDA tolerance level but the animal was obviously sick, was its flesh safe to eat?

Neither the federal nor state authorities chose to be drawn into that discussion.

18

The First Crusader

TED JACKSON had another kind of illness. It was compounded of overwork, frustration, and a failing heart. His lone attempts to prod legislators into action were getting nowhere, one of his two veterinary assistants had unexpectedly left him to join another practice, and he was torn between the work he needed to do to make a living and the lobbying which had become his avocation. His remaining assistant, Peter Van Vranken, was a strapping young man who tried to relieve Jackson of the rough work. While he admired the older man's persistence with the Halbert problem, he made no secret of his feeling that Jackson should now put it behind him and get on with the business of earning money. It was a busy practice. With a materialistic man in charge it could have been really prosperous, and Van Vranken grumbled that "the public health department should have been doing the job of worrying about others, instead of Ted."

In the spring of 1975 there was not much glory in being Ted Jackson. He had given more time to investigating PBB-contaminated animals than would ever be repaid, he had provided the lead to the USDA research station in Iowa which put Rick Halbert on the road to discovering the contaminant in his feed—yet wherever the story of the PBB cattle disaster was reported, it was presented as the triumph of one man's persistence: Rick Halbert's. Farm Bureau's veterinarian, Jim McKean, complained that anyone who had put in the hours and effort which he had done on the Halbert problem "at least deserved a footnote." If so, Ted Jackson merited no less than a chapter.

"He felt like Don Quixote tilting at windmills, with the whole legislative structure and bureaucratic structure wanting to push the PBB problem under the rug," said his wife, Lois. The only recognition was within his own profession, and even that was limited. After his article appeared in his professional journal, Jackson was invited to present a paper on PBB toxicosis at the annual meeting of the American Veterinary Medical Association, scheduled for July 1975 in California. He was proud of the honor and had just sent the program committee a summary of his proposed talk, over which he had worried excessively, when he had to return to the hospital that April. Again the doctors patched him up and repeated their warnings that his heart could not stand over-exertion. On the drive home he complained of an upset stomach and asked Lois to take the wheel. She took him back to the hospital—recognizing another heart attack—and, after a period in the intensive care unit, he recovered sufficiently to return home on May 10.

His doctors' instructions were specific. Complete rest. No work. No worries. Pete Van Vranken and the office secretary, Ruth Guernsey, were to keep the practice going and then he must ease himself back into it gradually, working limited hours.

The office adjoined the Jacksons' house in Battle Creek and he could not keep away from it. Early on the morning of Tuesday, May 13, he appeared there in his night clothes, apologized to Mrs. Guernsey for his attire, and announced that he would be dressed and back shortly. He was—by 9:30 A.M.—looking dreadfully ill. Immediately he began shuffling through the office papers, bringing himself up to date on correspondence which had come in during his absence. There was a bill from an osteopath who had treated Pete Van Vranken for a dog bite on the neck, and Jackson started objecting that the man had charged too much. Ruth Guernsey wished he would not get so agitated. Before she realized it, he had walked down the driveway to the mail box—against doctor's orders—to pick up the day's delivery. Soon he was fulminating again. Here was a letter which really upset him. So far as the others could tell it was from a state official, in response to some of Ted Jackson's PBB correspondence. A salesman from a drug company had come into the office and, since he was the most likely listener, Jackson began to sound off at him about the letter.

"He was getting all bent out of shape sputtering about it," Van Vranken noticed. "He only swore when he was riled, and he was saying something about how those S.O.B.s should be put away. I guess the drug salesman left at that point. Then Ted turned to me and asked if I had given the horses in the neighborhood their second vaccination for encephalitis. I was working eighty-five to ninety hours a week to keep him from worrying about things, and I got a little carried away and told him that out of the hours that were left maybe I could figure the time to attend to the horses.

"He was cussing me out and cussing the state out, both; me for different reasons. It made me think he was feeling better."

Van Vranken strode out of the office, and finally Ruth Guernsey was able to persuade her employer to go back to the house and rest. It was mid-afternoon by then and Lois Jackson had left for her job at the Kellogg Community College library. Lois and Ted had a telephone conversation about 3:30 P.M. He asked her to pick up his watch from the repair shop on her way home. She asked if he were resting and he assured her that he was.

Ruth Guernsey got on with her office work, thankful to be undisturbed. After a while it struck her that the silence was too intense: usually she could hear Jackson moving around in the house. She thought about checking to see if he were all right but decided against it, concerned that she might wake him.

Lois Jackson was also struck by the quiet when she returned home at 5:30 P.M. She called Ted's name but there was no answer. Apprehensively, she hurried to their bedroom where she found him lying peacefully on the bed, eyes closed, a book near his hand. In words which she used when she told the story, "he had just gone to sleep for ever."

He was fifty-three, and she took comfort from her affectionate knowledge that he would not have been a man to age gracefully.

His brother read Ted Jackson's speech at the convention. Almost three years later Jackson was given, posthumously, the Distinguished Alumni Award of the Veterinary School of Michigan State University. But the tribute which his secretary remembered with greatest pride happened at his funeral.

The church was packed, not on account of his work on the PBB problem—that had not yet been generally recognized and was still thought by many of his colleagues to be Ted Jackson's peculiar obsession—but simply because he was locally known and widely respected. At the end of the service, as the mourners began drifting away, some standing awkwardly at the church door, the organist suddenly started to play with immense gusto "When the Saints Go Marching In."

Ruth Guernsey smiled through her tears.

"Dr. Jackson would have liked that," she said.

19

When No News Is Bad News

"A PITY Ted Jackson died when he did," Doc Clark remarked, months later. "He could have helped us a lot."

The two veterinarians never met but they had a consuming cause in common and formed a necessary succession. When Ted Jackson was removed from the fray, Alpha Clark was just becoming deeply involved. Except for their fearlessness, scientific curiosity, dedication to the farming community and mistrust of bureaucrats, they were entirely different—yet between them they carried a banner for the farmers when most men in their position were afraid to speak out. Ted Jackson was studious and mannerly. Alpha Clark was self-contained and stubbornly independent. His entire ambition was to be a good cow doctor, but the PBB crisis impelled him to assume an informal leadership for which he had no ambition but a natural talent. He was a loner, and he couldn't abide injustice.

During the winter before Ted Jackson died, Doc Clark realized that most of his farmer clients were in deep trouble. The cumulative effects of low-level contamination were showing up in many herds. Calves were dying, cows suffered an increase in gastro-intestinal upsets and uterine infections, their bodies seemed unable to handle stress or to respond to normal treatment, their joints were stiff. Some had enormous appetites, but took little nourishment from their feed.

"You could tell by looking at the fecal material," Doc Clark observed. "It looked like you could take it and run it right back through the cow."

He also noted overgrown hooves, stained and chipped teeth, rough hair, stunted growth, teary eyes, symptoms of premature aging and of neurological damage.

While Halbert and Jackson had observed the PBB syndrome in its acute stages, what Doc Clark was seeing—he was convinced—were the long-term chronic effects which most of the highly contaminated animals had not lived to develop. When he performed post-mortems he found swollen and discolored kidneys, lungs which looked emphysematous, enlarged thymus glands and lymph nodes, enlarged livers with a yellowish, spongy appearance "like pumpkin pie."

He was puzzled, as Dr. Salvati was with human patients, that there appeared to be no correlation between the amount of PBB found in an animal's tissues and the progression of disease in its body. He agonized over whether he should advise farmers to have their herds tested for PBB. "It wasn't right," he said, "that I should have to make the decision and be responsible for putting a farmer in quarantine and destroying his living." He felt badly if he did this, and worse if he didn't. One farmer, who resisted Doc Clark's suggestions for tests, complained to him months later, "I wish I had sent the cattle to the pit and got it all over, instead of trying to nurse them through it. I wish you had kicked my ass and made me have them tested."

Whenever it was practical, Doc urged testing. Although the settlements were less generous than they had been a year earlier, it was still thought that a quarantined farmer would "be taken care of by Farm Bureau"—especially if he was a loyal member. Doc took hundreds of fat samples from the rear ends of cows but after many negative results were reported by the MDA laboratory he became suspicious. Considering the condition of the animals, he was getting too many reports of PBB levels marked "non-detect." Most of his farmers were like Garry Zuiderveen—stuck with unproductive herds, destined to go downhill, for which the owners could not claim indemnity because they passed the quarantine test. It was more than tempting for them to unload these animals on the meat market.

In the spring of 1975 Doc Clark decided to check MDA's test reports. He took a fat sample from a two-year-old heifer on Garry Zuiderveen's farm and split it three ways. He sent a third to MDA for PBB analysis and a third to each of two independent laboratories. The result from one private laboratory was .03 ppm, which was below quarantine level, and from the other .7 ppm, which was above. MDA's finding was between the two. Doc Clark was even more convinced that the tests to determine quarantine levels were unscientific and arbitrary, that they denoted only the presence of PBB and not the extent of an animal's sickness. So he took new samples from forty head of cattle which he had biopsied earlier and, at his own expense, sent these to WARF, the independent laboratory in Madison, Wisconsin, which Rick Halbert had used. Again, the findings were different from MDA's—and in most cases showed higher levels of PBB.

After this it became Doc Clark's habit, copied by several other Michigan veterinarians, to split their fat samples and send one half to WARF and the other to MDA. These test results rarely coincided, and it was Doc Clark's experience that WARF reports usually showed a higher PBB level than the state's. Sometimes the discrepancy would make the difference between whether or not a farm should be quarantined. In these borderline cases Doc would get the WARF test done first and enclose this report with the fat sample for MDA. Then, he noticed, MDA was more likely to agree with WARF.

Only on the basis of MDA's findings were quarantines issued.

Discovery of the widening contamination was making no more than a ripple in the legislature or in the newspapers. One of the first pieces of in-depth reporting was a ten-part series, "The Poison Puzzle," published in the *Grand Rapids Press* in March 1975. It covered the plight of the low-level farmers and raised questions about public health. Dr. Salvati's early findings of liver damage were reported. So were a number of statements from health department officials, some conflicting with one another. Dr. Norman Hayner, epidemiologist at Michigan Department of Public Health, told the newspaper that "nearly a thousand persons—mostly dairymen, their immediate families, and their milk and meat customers—have been exposed directly to the chemical." Yet another health department physician, Dr. John Isbister, was quoted as saying that "not a shred of evidence has been found" to indicate that PBB in human blood could harm anyone. It was confusing, especially to urban readers. What was happening on the farms was still a faraway event which city folk barely thought about or, if they did, dismissed.

The role of the press in the Michigan disaster should bring little pride to journalists. It was inadequate to the task for the same reasons that the federal and state agencies were lacking: none of them was experienced in this kind of crisis. Faced with it, they tried to follow established procedure and where this didn't fit, they did nothing. The press was not deliberately derelict in its duty. It was accustomed to accepting the word of experts, and the experts—who knew little more about PBB than the reporters—managed to sound like men who understood what they were talking about.

Many aspects of the PBB story went unreported for this reason. A reporter would interview the owner of a contaminated farm, and after the farmer had told his story—interspersed with angry comments about the delinquency of the feed company and the inadequacy of the state agencies—the reporter would check with his contacts at Michigan

Farm Bureau and at the Michigan Department of Agriculture. They would tell him that the PBB contamination was under control and not as widespread as this farmer was alleging. They might ask the name of the farmer. Then there might be hints that, oh well, many of us in the agricultural community knew he was having problems long before PBB was heard of . . . not a very good farm manager, we understand . . . bills owed all over the place . . . needs a good insurance claim to get himself out of the mess. . . .

Legislators were likely to have the same experience. They would hear from an aggrieved constituent, a farmer, and check with the appropriate state agency to see if his complaint was valid. Few Michigan politicians knew the idiom of agriculture, whether it was good or bad if a farmer top-dressed his grain rations, or if his cows stayed down after they freshened. Long after Ted Jackson died, an administrative assistant in the state legislature came upon a long letter which Jackson had sent to one of Michigan's elected officials. It contained the technical expressions which an experienced veterinarian would use, and the assistant commented sadly, "That man wouldn't have known what on earth Dr. Jackson was talking about. Chances are he'd have sent the letter straight to MDA for comment."

Even more than politicians, newspaper writers are an urban breed. Large newspapers concentrate on city stories because therein lies circulation. Most journalists are out of their element on farms, and it was a long time before any Michigan editor was persuaded that there was a story worth chasing down dirt roads, not even marked on the state's highway map.

Even when city reporters learned the language of farming, they had difficulty making sense of some afflicted farmers' stories. Richard Lehnert, editor of *Michigan Farmer* (a trade magazine which did the best reporting of any on the PBB crisis), had an interesting theory about this. With his agricultural background, he did not have to overcome the first obstacle, but the second baffled him until he concluded, "We didn't realize it at first, nor did they, but many of those farmers had been physically affected by PBB. There was a change in their personalities. You couldn't get a coherent story out of them. They would keep going over events, and they couldn't get the chronology straight. They would forget things and repeat others. It was like listening to a tape recorder. They would repeat and repeat and repeat. When they appeared before state officials they were not very believable. They were confused and nervous and angry and apprehensive."

An experienced reporter on a Detroit newspaper commented, "Any newspaper in Michigan could have won the Pulitzer prize on the PBB story, but we all blew it. There's an attitude about the American press: everything has to be official. If you get a press release from the governor's office you accept it, but if a farmer comes into a newspaper office and says his cows are dying and that no one will listen to him, we all think he is crazy and try to push him off so we can get on with our work."

As the story developed, some urban reporters went out to the farms and became deeply involved. But their desk-bound editors, most of them, soon tired of releasing a reporter from all other work so that he or she could pursue, for months—years, it turned out—a depressing story which was unlikely to produce good dramatic copy day after day. Some very committed reporters were pulled off the PBB assignment for this reason. Some editors were probably afraid of upsetting their newspaper's contacts with state government sources. It was a fear based on fact. The *Grand Rapids Press* series consisted of thirteen thousand words of reporting, drew no conclusions, and made no charges. Yet it elicited an eight-page criticism from the governor's office, charging that the newspaper had "terrorized" its readers, "thanks to conclusions that have little or no basis in fact."

In November 1975, *Michigan Farmer* broke the story of what took place inside the Battle Creek feed plant after the PBB shipment arrived. Lehnert, then a senior reporter on his magazine, explained, "After we had been reporting the PBB incident for about ten months we felt we should write a story of how it happened. We approached Farm Bureau, and it was conveyed to us that this was a question journalists shouldn't be asking. Farm Bureau was suing Michigan Chemical at the time, and I heard that depositions were being taken. I barely knew what a deposition was. I had watched too much 'Perry Mason' and I thought lawyers kept information to themselves. Instead, I found that both sides in a case get together with potential witnesses and take down the story in the presence of each other. There are no surprise witnesses in real life.

"Depositions must be filed in the circuit court where the case is pending. I made it a policy from May 1975 to make a weekly telephone call to the Missaukee County Clerk's office in Lake City to check whether these documents were yet there. In June I was told they were, and I drove one hundred and twenty miles north only to find they weren't. I kept calling until October, when the clerk told me the depositions had just been filed. They were not sealed, and therefore available for public inspection.

"I took my colleague Paul Courter, and we found this massive stack of three thousand pages of material. It included testimony from workers at the Battle Creek plant, with references to the claims of Hulet and Szeluga that they had noticed the magnesium oxide was different but were told it was all right to go into the feed. We read the material into a tape recorder from eight A.M. to five P.M., and had our secretary do a transcript from which we put the article together."

It was a detailed, four-page account which farmers found sensational. Right after it was published, one of Farm Bureau's attorneys called at the magazine's office in Lansing. He was very upset, and a few weeks later Lehnert understood why. Farm Bureau was on the brink of an out-of-court settlement with Michigan Chemical and its officers feared that publication of the magazine story at this sensitive moment might weaken their position. There was an inconclusive four-hour discussion about the freedom of the press, the public's right to know, and the moral duty of the magazine to protect the interests of farmers—and, by extension, Farm Bureau.

"The following month Farm Bureau canceled its advertising with us," Lehnert reported. "It had been spending up to $45,000 a year. Later Courter and I entered this story in a competition run by the Detroit Press Club Foundation and we won the year's award for the best reporting in a trade publication. The prize was $300. I have often referred to it as our $44,700 story, but it put us on the map. We realized that we could no longer get material from our so-called friends, and from then on we were really investigative with the PBB episode."

Months passed before the state's two biggest newspapers, both in Detroit, tackled the PBB crisis in depth. The *Detroit Free Press* did three detailed articles, but not until March 1977—almost three years after the first quarantines. One of the articles included a rewrite of the Lehnert-Courter story of sixteen months earlier. Neither the *Detroit Free Press* nor the *Detroit News* maintained news offices in the western part of the state, which had most of the early contamination, and for months relied upon bare reports from wire services. After accepting the state's assurances that PBB posed no threat to public health, "it took us a long time to believe the farmers," a Detroit newspaperman admitted.

Out-of-state newspapers paid scant attention to the story, seldom printing sufficiently detailed accounts for their readers to fully understand the dimensions of the disaster. Americans who care about environmental causes were better informed about the dioxin contamination of Seveso, Italy, and the mercury poisoning in Minamata, Japan, than about the PBB crisis in Michigan—although this was the biggest chemical disaster and the worst man-made agricultural catastrophe in United States history. *The Washington Post* and *Audubon* (the magazine of the National Audubon Society) each published accounts of the PBB incident which were long, factual, and accurate, but not, unfortunately, destined for large nationwide audiences.

Some good coverage was given by local Michigan television stations, but only in a few areas. None of the national networks tackled the subject as a documentary, although the idea was presented to them. Bonnie Pollard, senior associate editor of *Michigan Farmer,* made the suggestion in letters to all three networks late in 1974 but the correspondence was not even acknowledged. The only television documentaries on the poisoning of Michigan came more than two years later from British and Canadian teams, primarily for audiences in their own countries. One argument used by television professionals was that sick cows are not good visual subjects, and since few viewers live on farms the topic would not attract enough interest. Also, it was left unsaid, a program which showed the deficiencies of two corporate giants might bring repercussions from commercial sponsors.

The ultimate irony happened late in 1978 when the Columbia Broadcasting System finally showed a program which was vaguely about PBB. Script writers took the Michigan story, fictionalized it, changed the name of the chemical, moved the setting to California, and presented it as an episode for "The Lou Grant Show." The television series portrayed Lou Grant as an investigative newspaperman who, in this instance, defied the bureaucracies and uncovered the cattle contamination. The episode was due to be televised in Michigan on the eve of the gubernatorial election. By then PBB was a campaign issue which could have unseated Governor Milliken, and the timing of this popular drama made his Republican organization uneasy.

Party officials complained to CBS, which agreed to delay the episode until after the November election. This decision was "in line with the network's interest in fairness," one of its executives explained.

The state legislature was of a mind with the press. As late as spring 1975 it was still a long way from acknowledging that PBB contamination was serious. Also, it had more pressing concerns. The national energy crisis had devastated Michigan's leading industry, the production of automobiles, and in the year's first quarter America's four major automobile manufacturers—all based in Michigan—had lost $200 million between them. This was the industry's worst economic disaster in forty years. Tens of thousands of employees were laid off, the state's unemployment rate rose to 15 percent, and legislators agonized over how Michigan could balance its budget—which it was constitutionally committed to do—while supporting its jobless. Any public admission that PBB was a statewide hazard would cost a lot of money, and there was none to spare.

Nevertheless, questions about possible health effects were being raised in enough quarters so that something had to be done. Or rather, shown to be done. One of the few dissenting voices in the legislature was that of Donald Albosta, Democrat, a newly elected representative who had grown up on a dairy farm and now ran a cash-crop farm near Saginaw, in the eastern part of the state. In December 1974, Albosta attended a one-day hearing of the House Agricultural Committee at which farmers testified about their PBB experiences. He had barely heard about the contamination, but after listening to these witnesses he realized "there was something terribly wrong," and kept a note of their names. The hearing itself was desultory. "I think the committee held it more to pacify the farmers than to take action," Albosta remarked. "Right after these farmers testified, members of the legislature listened to some Farm Bureau people who insisted there was no problem with PBB, if it were below the FDA tolerance level. So nothing was done. It was almost the end of the legislative session, this committee was about to die anyway, and then it would be somebody's else's responsibility."

Donald Albosta made it his. When House committees were reconstituted at the beginning of 1975, he emphasized his interest in farming and was appointed co-chairman of the Agricultural Committee with Representative Paul Porter. He looked up his notes of the farmers who had testified and spent many weekends driving to the north and west of Michigan, calling at their farms, the better to understand their complaints. This was Doc Clark's country, and Doc's name had been mentioned to Albosta as "the one person who could really tell me what was going on."

Doc Clark did—succinctly.

"The farmers are being screwed," he said.

Doc described the symptoms he was seeing in cattle and how, when he tested for PBB, he found low-level contamination in almost every herd. Even though many were below quarantine level Doc had no doubt about the sickness of the animals. Virtually all his clients traded at one of three local Farm Bureau mills, ones which, initially, were supposed to have escaped contamination.

"When I see how much PBB come out of those three mills, and when I figure there are eighty-seven Farm Bureau mills in the state, and almost another two hundred which may have sold Farm Bureau feed . . . well, it makes me wonder what's been done to Michigan agriculture," he told Albosta. "And Farm Bureau keeps telling us we don't have a PBB problem."

Albosta arranged to come back with a bus load of legislators to look at farms with low-level contamination. Their tour, on April 10, 1975, was a mitigated success. They had all heard stories around the Capitol that the complaining farmers were more likely to be suffering the consequences of their own poor husbandry than from PBB. Doc and Don Albosta wanted to show them some clean, well managed farms where cattle were sick and dying through no fault of the farmers.

They stopped at the Zuiderveen farm and Doc Clark was puzzled. Garry's animals looked better than he expected. Doc's astonishment changed to exasperation, but he kept his thoughts to himself. The legislators made polite expressions of concern.

The following week Doc went back to the farm, taking his farmer brother, Clyde, with him. There was a strong physical resemblance between them, both short and thick and lumbering, with the same extraordinarily gentle brown eyes. Clyde was quieter than Doc and less canny, giving an impression of a simpler man. In fact, he carried in his head an enormous fund of knowledge about cattle breeding. He had memorized every good blood line in the state for generations, he was an expert on Holsteins—the most popular dairy breed in this part of Michigan—and his professional advice was widely sought by farmers eager to improve the genetic strains in their herds.

"Garry," Doc announced. "You have to show us both everything. Not just the animals you showed Albosta."

Shamefaced, Garry Zuiderveen took the brothers to his last barn, the blue one beyond the feed lot, where he kept his young cattle. They were worse looking by far—emaciated, mangy, barely surviving. Doc gave them one eloquent look and burst out laughing. He laughed so loud he collapsed on the manger, almost hysterical. Clyde and Garry exchanged anxious looks. They thought he was about to have a fit.

"You dink, you dummy," Doc sputtered as he began to recover himself. "What did you do that for? Why did you only show them your best cattle and keep these out of sight?"

"I was too ashamed," Garry replied quietly. "I feel so badly about the young cattle. They look lousy and malnourished. Everything about them looks like a complete lack of animal husbandry. You know how they've been saying at Farm Bureau and MDA that some farmers haven't been looking after their cattle. I thought the Albosta crowd would look at mine and believe it."

Doc understood and yet he didn't.

"Those Dutch farmers are too proud for their own good," he complained afterwards. "If they have a fault, that's it. It's bred into the bunch of them. Can you imagine how difficult it was to help them? On one hand they needed help; on the other, they didn't want to admit they had 'syphilis.' Here I was, trying to get some of the legislators to see for themselves what was happening on the farms, and when they drive here all the way from Lansing some farmers act like there isn't much of a problem."

Doc felt that he had taught Garry Zuiderveen a necessary lesson in the art of lobbying. It was so well taken that a few weeks later he approached Doc Clark with a suggestion. Albosta had made a proposal in the legislature for a reduction in the PBB tolerance level. This, of course, would affect only Michigan, whereas the existing FDA guideline applied nationwide. As an initial step, state law required Michigan Department of Agriculture—in its role as a consumer protection agency—to hold public hearings. These were scheduled for May 29,

1975, at the National Guard Building in Lansing.

Garry Zuiderveen suggested exhibiting some contaminated cattle on the street outside the building, so that people attending the hearing could see what a PBB cow looked like. Both he and Doc felt a natural reluctance to do this, public agitation being against their instincts, but it seemed an opportune way to make a point.

They gathered six head of sickly cattle from local farms, including Garry's, and again made the tedious drive south to Lansing. Garry's nephew Ken and two other affected farmers joined them. Outside the National Guard Building they tied the cows to their trailer and pinned up Marlene Clark's homemade poster, which warned: "This is what you will eat tomorrow."

Their modest demonstration had little effect. Most passers-by simply thought it was odd to see cows in town. The hearing went badly for the farmers, with all the leading federal and state expert witnesses testifying against lowering the tolerance level. The .3 ppm guideline for meat and milk was adequate to safeguard public health, according to Dr. Maurice Reizen (director of the Michigan Department of Public Health), Dr. Albert C. Kolbye (associate director of the federal Food and Drug Administration), Dr. Robert M. Cook (associate professor of dairy science at Michigan State University), Dr. Robert Barker (chairman of the university's Biochemistry Department), and Dr. George Fries (the USDA animal research scientist who had identified PBB in Rick Halbert's feed).

Dr. Cook described experiments which his university had done on some of Rick Halbert's highly contaminated cattle—experiments which proved, in his opinion, that with good management many contaminated cows appeared to be healthy. From this he concluded that humans eating fat containing as much as 20 ppm of PBB were unlikely to come to harm: therefore he saw no rationale for lowering the tolerance level. There was also an FDA health survey of PBB-exposed animals, which concluded that low levels of PBB did not appear to damage cattle. The validity of this study was later questioned for the same reasons as the health department's study of farm families (an uncertainty that the control group was free of PBB), but at the time of the hearing both were generally accepted as persuasive evidence.

Dr. Walter Meester expressed his fears about long-term effects of the toxin, but among the experts his was such a lone voice that it sounded capricious. Two of the state's most successful agribusiness farmers, Larry Crandall and Blaine Johnson, testified against lowering the guidelines. Their farms had been heavily contaminated at the same time as Rick Halbert's; they too had been generously compensated and were back in business. All three farmers may have suspected that they still had residual levels of PBB on their farms (certainly Rick Halbert did) and more stringent regulations could be disastrous for them.

Larry Crandall told how a biopsy of his own body fat showed a 3 ppm level, ten times higher than the guideline, yet he felt no ill effects. Blaine Johnson also testified to his family's fitness, despite heavy exposure to PBB, and suggested that complaints of PBB sickness might be psychosomatic "especially if being sick would help to settle a claim."

Rick Halbert was at the hearing, rationally pointing out that the cost of testing all the state's farm animals to conform to a lower tolerance level would be prohibitive. It was not what Doc Clark and his companions had hoped to hear from him: after all the detective work Rick had done, they expected him to be the farmers' champion—not aligned with those who didn't want to face another round of PBB testing. Doc, of course, knew nothing of Halbert's months of frustration, and Rick understood none of Doc's.

Garry Zuiderveen had not meant to speak at the hearing, except to draw attention to the emaciated animals outside the hall. But he grabbed the microphone in anger after Blaine Johnson suggested that some of Missaukee County's dairy problems might be due to malnutrition from low quality feed.

"Don't nobody ever accuse me of underfeeding my cattle!" he exploded.

In a corridor at the back of the hall Doc had some angry words with George Fries, whom he overheard telling a newspaperman that he anticipated no health hazard from low-level exposure to PBB. Doc asked him tartly if he had difficulty sleeping at night.

At the end of the day the farmers loaded their sick cattle and drove sadly home. They had made an impact of sorts. "From then on," Doc observed, "everybody at the bureaucratic level ceased to like us."

After considering evidence at the hearing, Michigan's Commission of Agriculture came to the predictable conclusion that the FDA's PBB guideline was safe.

20

Conference in a Kitchen

WALTER MEESTER'S testimony impressed Alpha Clark, and he felt he could learn much about the toxicology of PBB from this man. "I wanted to know," Doc said, "exactly what it was got into them cows."

Back in McBain, he asked Marlene to telephone Meester for an appointment. Marlene did most of Doc's secretarial work. She also ran the household, managed the accounts, unobtrusively mothered him, and—on those rare occasions when he took time off—uncomplainingly accompanied him on long drives to neighboring states so that he could indulge in his favorite relaxation, looking at other breeds of cows. She knew Doc did what he had to do, so, whenever she could, she did it with him.

They drove to Grand Rapids together to call on Dr. Meester, a good two hours on a monotonous road. Meester showed Doc what a PBB printout from a gas liquid chromatograph looked like. It was a graph with a series of low peaks and one dramatically sharp, tall peak about a third of the way along from the left. Meester explained that this was the peak against which other tests for PBB were measured. If a fat sample from a cow were injected into the gas chromatograph and it produced this identical peak, the presence of hexabrominated biphenyl —the chemical name for Firemaster BP-6—would be confirmed. Doc was sharp enough to spot that while these tall peaks in different BP-6 graphs always matched, there were differences in the shapes of the little ones. He asked why.

It was obvious to Meester—an experienced toxicologist (of whom there were few)—but it was a revelation to Doc Clark. To recapitulate, a biphenyl molecule consists of two benzene "rings" hooked together like two beads on a chain. In the chemical diagram of the molecule, each "ring" is drawn as a hexagon joined at one of its six points, and with hydrogen atoms attached to the other five. When biphenyl is brominated, some of these hydrogen atoms are replaced by bromine atoms. A hexabrominated biphenyl molecule has six bromine atoms distributed between the two rings, leaving four hydrogen atoms in place.

This is basic chemistry. Now came the complication which Walter Meester explained. In the bromination process the hydrogen atoms are not all replaced at once. On each molecule in the mixture it happens one or two at a time, not always in the same order. Under the intense heat of the reactors, the first bromine atom to "go on the ring," in the jargon of scientists, usually does so at the opposite side of the hexagon from the point where the two benzene rings are joined. This is the number four position, given the fact that the hexagons are joined at number one. The next bromine atom is likely to land next to it, at either the third or fifth position; the one after that has a predisposition to replace a hydrogen atom closest to the point where the two benzene rings join—at numbers six or two. In the manufacture of hexabrominated biphenyl the process is stopped here. But the bromination could continue until all ten hydrogen atoms have been replaced by bromine atoms, or it could stop sooner, producing different versions of PBB.

Sixty-odd percent of the molecules in Firemaster BP-6 had the same characteristics. The bromine atoms were fused at points two, four, and five on each of the two rings which formed the biphenyl molecule. Chemically, this is described as 2, 4, 5, 2', 4', 5' hexabromobiphenyl. But the bromination process is not always precise, and in some molecules the permutation is different. In determining the toxicity of Firemaster, this was one important factor, Meester explained.

Then there was another. Variations in the heat of the reactor or in the "cooking time" produce some molecules which have more or less than six bromine atoms, again in shifting arrangements. In making an industrial chemical this is of small importance, provided the overall mixture comes out right. Hence the subtle changes in shape of the low peaks on the gas chromatogram of PBB.

Doc Clark listened without comment. For all that Walter Meester might have judged, this scientific background was irrelevant to a veterinarian. In fact Doc was tremendously excited but, suspecting that Meester didn't want to get involved in the politics of PBB, he showed no reaction. But on the drive home (Marlene at the wheel) he could not keep still in his car seat.

"Holy jeepers!" he exclaimed. "Did you hear what I heard? We ain't dealing with one compound in PBB, but with a whole mixture of compounds, and every one of them has its own physical, chemical, and toxic properties. Nobody's looking at them. When the labs test for PBB they are testing for the 2,4,5 isomer, comparing the hexapeak of the sample with the hexapeak of the control. Even then they're not measuring the compound that went into the animal, but a fat residue sample a year, two years after contamination. D'you understand?" Marlene nodded.

Doc saw the implications immediately. The laboratory tests for PBB toxicity produced only random results. All they were telling was how much of a certain isomer of the PBB molecule was in a certain part of the animal on a given day, with no indication of what tissue damage had been done or how the rest of the compound reacted. He realized that there was no way of determining the toxicity of PBB until the compound had been taken apart and every one of its many components carefully analyzed. No one, he felt sure, had even begun to do this—certainly not MDA, which was responsible for animal testing and consumer protection.

They had taken the last turn-off to McBain—a straight stretch of rural road, almost eerie in its evening isolation—before he was able to calm down. Then he turned to Marlene abruptly. "Babe, we have to find out what this chemical is really made of, and I ain't at all sure who's going to help us."

In conversation with a Farm Bureau official at the Lansing hearing, Garry Zuiderveen was promised that "if you can get quarantined we will take care of your animals"—meaning, he assumed, they would be sent to Kalkaska and he would be compensated.

"After that we tested and tested," Lois Zuiderveen recalled. "We finally realized that it was the better looking animals which tested over the limit because they were still storing PBB in their fat."

This discovery put Garry Zuiderveen's farm in quarantine—as well as that of his nephew, Ken—in the first week of June 1975. But it did not help them. "In the early days if you had one cow over the level your whole herd would be taken to the pit," Lois said. "It did not work that way for us. We were told, 'It's a whole different ball game now,' So we were stuck. We could not sell the cows, we couldn't kill them, we could do nothing but feed them."

The "different ball game" came about because the insurance money, contributed by Farm Bureau Services and Michigan Chemical, had run out. Each conglomerate had several insurers, and payments to farmers had been delayed as, one by one (most of them taking their time about it), the insurers had tried to minimize their share of the liability. Finally the $15 million insurance pool was drained. More than three hundred claims had been settled and more than three hundred waited for relief. There seemed no hope of all these being settled unless Farm Bureau Services could win its lawsuit against Michigan Chemical, and this might take years. Watching events from the Michigan attorney general's office, Harry Iwasko felt some sympathy with the feed company and its parent organization.

"Farm Bureau had taken lines of credit from all over the country, and it had farmers pounding on its doors for settlements," he observed. "Whereas Michigan Chemical could ignore these people, Farm Bureau could not, because they were its members, its livelihood."

A few claims were still being handled by Farm Bureau Services, but sparingly and—it was charged by less fortunate farmers—with discrimination. Instead of removing quarantined herds unconditionally, the feed company sent experts to contaminated farms to make their own determinations about the merit of claims, sometimes noting in their reports the farmer's degree of involvement in Michigan Farm Bureau's statewide organization. Dairy specialists from Michigan State University's veterinary school were consulted, and sometimes they found fault with the feeding practices and herd management of a litigating farmer, with the inference that PBB was not his problem at all.

This caused dissension and bitterness among those whose claims were being held up. Almost all had been loyal Farm Bureau members but now they felt that this organization, which had been a proud part of their lives, regarded them with suspicion and mistrust. The human error which caused PBB to be tipped into the feed mixer could be forgiven. It was even understandable that Farm Bureau might have genuinely believed some of its branch co-ops to be free of contamination when they were not. But two years later, when innocent farmers were suffering the consequences of these ineptitudes—bankruptcy, illness, social ostracism—they at least expected support from the organization which they themselves had supported all these years. Instead, some of them felt, it was adding insult to injury.

Alvin Green, a gentle and mild-mannered man who knew every one of his cows by name and loved them all, was typically outraged when an expert from Michigan State University suggested that his problem was not PBB but insufficient protein in his cattle feed. Al Green stubbornly insisted that his herd was well cared for.

"It was the most beautiful dairy herd you could ever wish to look at. My cattle were the kind of cattle I could walk up to in the middle of the night and talk to. I never used to eat breakfast before I would feed my cows. At night the same way. My cattle was always taken care of before I would eat my supper.

"There could be nothing lacking in their nutrition," he affirmed. An altercation followed in which, according to Al, the expert accused him, "I believe you *want* to have PBB!"

Al Green took a threatening step toward him. "You say that once again and I'll put you in the shit pit!"

As Al Green saw it, the implication of the expert's comment was that he was deliberately neglecting his cows so that he could make some quick money on an insurance claim. The hurt and indignity of suggestions like this turned some farmers from staunch Farm Bureau supporters into enemies of their own organization and radical militants. This did not help to expedite their claims—and it was suspected by many of them that while they were stuck with debts and diseased herds, neighbors active in Farm Bureau politics received prompt settlements. Soon bitterness and mistrust developed between one set of farmers and the other.

In the Zuiderveens' neighborhood these resentments caused a rift in the membership of the Aetna Christian Reformed Church, which drew its congregation from Dutch-Americans in the area. Most had been members since childhood. Many were interrelated. On a fine Sunday evening there could be no sight more comforting in its sense of community, more reassuring in its shared relationship with the land and its Creator than these farm families of Missaukee County gathered in groups to chat as they came out of church. After the evening milking they had dressed in their best to hear a long, old-fashioned sermon about the nature of faith. They had also prayed together. Yet an awkwardness had developed about their exchanged pleasantries which was alien to these good people and to this peaceful place. They were all believing Christians, united in worship yet privately divided by envy and hurt, and none of them knew how to pick up the pieces and mend the pain.

"There were times when we felt that everyone was against us," said Lois Zuiderveen, who had been a member of the congregation all her life. "The co-op would not give us credit any more. We could not get feed unless we paid cash and we didn't have the money to settle our bills. Yet other farmers in the community, ones who still supported Farm Bureau, didn't get their credit cut off.

"There are only about thirty families in our church, and we have always been very close. Because of PBB, some of us had real financial problems and some of us had health problems, yet it got to the point where some people were barely talking to others. We were all disturbed by it and asked the minister to pray for guidance. For some time I think he was fearful of making matters worse by publicly mentioning the rift, but eventually he did—and then he was criticized for doing so.

"Almost the whole congregation was born and raised around here, but we no longer visit back and forth as freely as we used to. When there is a church supper, we all come, but we try not to talk about cattle or PBB."

Against this background of hurt and suspicion, the rest of the PBB story took place. At the centre of the conflict was the feed company, guided by the same interests and served by many of the same personnel as its parent organization—one which had become defendant, judge, and jury of the litigating farmers, while nominally remaining their protector.

By midsummer of 1975 the Zuiderveens were beginning to feel the anguish of their situation. All that Farm Bureau was then offering them was compensation for the six or seven animals which tested higher than .3 ppm, coupled with the suggestion that they send any other unprofitable cattle to market.

"In my mind they were all unprofitable," Garry insisted. Yet it offended his conscience to sell them for human consumption. Feeling they were in an untenable situation, he and Lois came to an uncharacteristic decision. They decided to hire an attorney.

Although he was an experienced business man operating a large modern farm, Garry Zuiderveen had never consulted a lawyer in his life. His kind of people had a tradition, almost part of their religion, of settling differences reasonably among themselves. It was almost an act of apostasy to go to law, and attorneys were suspect. Nevertheless, Garry was sensible enough, and desperate enough, to realize that he had small chance of recovering his losses without one.

He had heard the name of Paul Greer and warmed to it "because I knew he was a farmer whose cattle had been quarantined and I felt I would rather deal with a farmer than anyone else." So he drove a hundred miles to Greer's office in Fremont, taking with him his nephew Ken and his neighbor Jim Van Haitsma. These three owners of contaminated farms told Greer their problems—and were afterwards amazed that it was possible to talk to an attorney for three hours and not owe him money.

Paul Greer and the young Grand Rapids attorney, Gary Schenk, were still working in an informal partnership, handling out-of-court settlements for quarantined farmers. Most of the early settlements had been easy and the two attorneys were frank about the fact that they had collected handsome fees for their part in these negotiations. Several years later, Paul Greer remarked, "The Lord gave us some quick settlements. Then He said, 'From now on you're going to earn the rest.' "

They began that earning process the day Garry Zuiderveen and his companions walked into Paul Greer's office.

Greer did not try to sell them his services. He pointed out the legal steps they could take and the possible pitfalls, and told them to go home and think about it. He suggested trying to negotiate with Farm Bureau to remove the cattle from their farms and get the quarantine lifted, but he couldn't promise it would work. Frederick Boncher, a young attorney who worked for the Grand Rapids law office in which Schenk was a junior partner, sat in on the conference. After the farmers left, Paul Greer remarked to him, "They won't be back. They were just after us for information. I know how those Hollanders think. They don't hire an attorney if they can get one for free."

Fred Boncher thought differently, and the two attorneys made a five-dollar bet on it. A month later a telephone call came from Garry Zuiderveen. His approach to Farm Bureau had failed. Would Greer and Schenk please represent him?

Five dollars better off, Boncher, who was an amiable and energetic young man, not long out of law school, was despatched to McBain for an exploratory talk with Doc Clark about the state of the Zuiderveen herd. It was essential in these cases to have detailed veterinary evidence, not only on the damage to cattle but also to assess the history and quality of management. Garry Zuiderveen was present at the meeting in the Clarks' cluttered kitchen which, like their living room, was decorated with photographs of prize cows.

Marlene was washing dishes and serving coffee. Garry was trying to make conversation. Doc, wearing his farm coveralls, glowered at Fred, who, out of awkwardness, was soon doing most of the talking. He did not feel he was making much headway. Eventually Doc brought out a gray file box from which he reluctantly produced the laboratory results of some of the fat tests which he had split between MDA and the WARF laboratory.

"They're from the same animal. What do you think?" Fred noted the discrepancies while Doc, head cocked on one side, watched through narrowed eyes. There were other papers in Doc's file box which he seemed unwilling to show. His displeasure at a junior attorney calling on him, instead of Greer or Schenk, was evident.

The next day Fred Boncher reported to Gary Schenk, "I didn't know what to make of him, except that he would be a difficult witness. He wouldn't open up. Maybe a little stroking by you guys would help."

Greer and Schenk flew north, and were met by Garry Zuiderveen. He showed them Ken's farm, then took them to his. The two attorneys were taken aback to find the Zuiderveens' large kitchen filled with farmers, all needing legal advice. Ten minutes after the farmers began to describe their cattle problems, a green Dodge pulled into the yard and a short square man with red hair and white coveralls got out. The farmers stood aside as he crossed the kitchen, with no more than a curt nod to any of them. He sat at the kitchen table, straight across from Gary Schenk, and glared at the attorneys suspiciously.

The conversation continued, Doc Clark remaining uncommunicative. It was unnerving to the visitors but it didn't faze the farmers. They knew his canny way of assessing whether these men were to be trusted, and afterwards they would ask his opinion of the two attorneys—and respect it.

"I sat there and I said nothing," Doc recalled, long afterwards. "I said nothing for an hour straight. Schenk was trying to impress them. I knew we were in a heck of a battle and we had to be in with the right people and make the right judgment. We had spent a lot of hours on this thing and I wasn't about to throw the effort away.

"I was impressed with Gary Schenk. I sensed he was sharp. Paul came across as sharp too, in a different way. I listened to them. I got up. At this point something in me said, 'You have found your boys.' "

Gary Schenk had these memories of the occasion, "I soon realized why he was there at the meeting. He had come to eyeball me. I was dressed to the nines in my three-piece blue suit, my scales-of-justice cuff links, and a maroon tie. He smelled of blood and horseshit, and he thought I was some silver-spooned lawyer. He did not know it then, but I came from the same kind of background as he did."

Once Doc started talking, conversations bounced back and forth across the room. The meeting which had begun soon after lunch lasted late into that August evening. "The farmers asked us to take their cases," Gary Schenk related. "I said no, not until we look into them. We would first want to discuss them with Dr. Clark and we would need their approval for this, which was standard procedure. I think it really shook them to see we were not out to make a fast buck. Eventually all these farmers became our clients."

After a while the talk dwindled to an animated dialogue between Doc Clark and Gary Schenk, the others listening. Doc revealed some of the contents of his gray file box. At his own expense he had been getting animal tests done outside the state. He did not trust the MDA laboratories or the veterinary school at the state university. He felt that each worked too closely with Farm Bureau. After his experience with the split fat samples he believed that the analysis done by the WARF laboratory in Wisconsin was more reliable than MDA's testing, and he had gone a step further.

A few weeks before the meeting in the Zuiderveens' kitchen he had telephoned the diagnostic laboratory at Purdue University, Indiana, and announced to Dr. Farrell Robinson, a respected animal toxicologist and pathologist there, "My name is Alpha Clark and I am a practitioner in Michigan. I have a problem. Can you help me?"

He had been impressed by the warmth of the response—and by the fact that, although Purdue was only a hundred miles south of the Michigan border, no one there had heard of PBB. Dr. Robinson agreed to take half a dozen sick cattle which, after examination, he would have killed so that he could take tissue samples from every part of their bodies. Marlene was careful to get the necessary approval from MDA for the cattle to be shipped across state lines—anxiously aware that these unwelcome activities of her husband were thus being brought to the attention of the agency which licensed him to practice in the state.

Only Marlene and Doc understood what risks he was taking. The moment he started working with those two attorneys, Doc knew he was making himself even more vulnerable.

He told the attorneys how far his private research had led him. The post-mortems at Purdue on cows with low levels of PBB, three of them below the tolerance level, showed chronic toxicosis which had caused skin and bronchial problems and damage to the lymphatic system. He expressed his belief that there was a chemical contaminant of PBB which might be responsible for some of the symptoms. Gary Schenk was intrigued; this had also been a theory of his. Then Doc described his observations—how some cows seemed able to survive with the toxin in their systems until they needed to mobilize their fat to cope with stress.

The attorneys were fascinated by the information which he had assembled. Alpha Clark was unlike anyone they had ever met, with two distinct sides to his personality. Part of him—the only part most people saw—was a simple rural fellow of homely appearance who spoke in the careless syntax of the countryside. He wore his work coveralls so consistently that acquaintances who had known him for years assumed he did not possess conventional clothes. In wintertime he added a well-worn lumber jacket; in summer he rolled up the sleeves. Ingenuous in the ways of the world, he would blush if anyone told an off-color joke in mixed company. The earthy speech of the farmyard was, of course, another matter.

The other side of Alpha Clark had a keen scientific mind. He could expound for hours on animal anatomy, genetics, toxicology, biochemistry—and then he spoke another language: his explanations were lucid, and the grammar came out right. He was, in effect, bilingual.

21

Unlikely Odyssey

In arguing their early PBB cases, Schenk and Greer had amassed thousands of pages of depositions. Back in his Grand Rapids office, Gary Schenk sorted out some documents which he thought would interest Doc Clark and mailed them to McBain. A few days later Doc telephoned him.

"What are you doing?" he asked casually.

"I'm sitting here worrying about taking these cases," Schenk replied. He was indeed. The out-of-court settlements for heavily contaminated herds had been more or less straightforward. Now it was necessary to prove that animals which could legally be sold for food were an economic liability to the farmers, and he foresaw long legal arguments requiring scientific evidence which he did not possess. Also, as he told Doc, he had very little chemistry background himself and he didn't know what a healthy cow should look like. "I haven't seen a good one since I got involved in this mess," he explained.

"Why not come up here and talk about it?" Doc suggested.

Gary Schenk spent two days driving around the countryside with Doc, visiting farms, learning about cattle, watching him treat sick cows, assessing their joint knowledge about PBB. He volunteered that his law firm was prepared to invest some of the money it had made out of the early settlements in taking cases of low-level farmers to court, on a contingency basis. Doc at once seized this opening.

"I thought we should convert that money to research," he explained afterwards. "I was convinced we were not dealing with the right compound. Gary Schenk had found out that it was not Firemaster

BP-6 which got into the animals but FF-1—which was what happened to BP-6 after it had been sent to the chemical plant at Batavia, Ohio, to be ground up. We did not know what chemical contaminants it picked up along the way. I felt sure there was one contaminant which was causing symptoms of X-disease, and I suspected others."

It was a revelation to Doc to learn about the rules of pretrial discovery; how, in advance of a court hearing, each side can shape its case by requiring the opposing attorneys to hand over information in their possession. By this device Gary had obtained a list of all the chemicals manufactured by Michigan Chemical in the past several years, plus other documents which Doc had not dared to hope he could get his hands on. Among the data which Doc had amassed, Gary was most intrigued by the detailed explanation of how a gas liquid chromatograph worked, and how, in checking for BP-6, the state laboratories were measuring only the tall hexapeak on the graph and ignoring the others. Both agreed it was essential to their case to find out what these smaller peaks represented.

The odyssey of Schenk and Clark in pursuit of the chemistry of Michigan Chemical's version of polybrominated biphenyl is an essential element of this story. It was as lonely and tenacious as Rick Halbert's search for the contaminant in his cattle feed. It took as much determination and energy, received as little help from government agencies, and led down almost as many blind alleys. It, too, was a long slow building process: a few facts gathered here, a name picked up there, a scientist in another place persuaded to do some necessary research—each piece of information added to the others until, set down where it should be, a picture of the whole began to form. Halbert found out what the basic chemical was. Schenk and Clark were determined to discover what was in it.

Although a hundred miles apart, they developed a habit of meeting once a week for dinner, to compare notes and determine strategy. Gary Schenk had been present when the depositions of the feed company workers were taken, so he had a good idea of how the mix-up happened. Doc Clark had a better understanding of the chemical. As a first step they sent some feed samples to the WARF laboratory in Madison, Wisconsin, for analysis and were excited by the report that "we have come up with something at the heptapeak and it isn't PBB."

Paul Greer, Gary Schenk, and Doc Clark visited Madison to discuss this. The laboratory had recently acquired a mass spectrometer and its chemists were eager to do more detailed analysis. Schenk showed them the list of compounds manufactured by Michigan Chemical, and, after checking their formulae, a WARF chemist suggested that the unidentified contaminant might be Tris or a hexabromocyclododecane (another flame retardant designed for certain plastics systems).

In the course of conversation, the chemist mentioned that several years earlier he had helped to investigate an epidemic which killed millions of chickens in eastern and midwestern states. It was called "the chick edema factor" because the birds developed fluid in the heart cavity, edema, and liver necrosis. The disease baffled agricultural experts until the discovery that an ingredient of the chicken feed had been fat from animals whose hides had been treated with pentachlorophenol (PCP), a chemical which was found to contain a dioxin as a contaminant.

"What is the toxity of dioxin?" Doc Clark asked.

"Parts per trillion," the chemist replied.

This piece of information was as overwhelming to Alpha Clark as the discovery he had made in Dr. Walter Meester's office. If the chickens died after contamination at such very low levels, and if the cause of their death was not the chemical itself but one of its contaminants, and if that contaminant could be fatal in such minute amounts. . . . "Well," Doc Clark recalled, "then my mind really started registering."

He was unable to sleep that night. He stayed up reading every reference he could find to dioxins in his textbooks. In one sense, it was a false trail because no subsequent researcher found dioxin in PBB. Yet he was on the right track. It set him and the attorneys searching for experts who might be able to isolate the possible contaminants of PBB.

As they read through the scientific literature, the names of Dr. Robert Baughman and Dr. Matthew Meselson kept recurring. These Harvard scientists had developed an analytical system for detecting some chemical contaminants as low as parts per trillion, and had been responsible for finding residues of a dioxin in 2,4,5-T, the chemical used by the U.S. military for defoliation during the war in Vietnam. This had been a significant discovery. Dioxins form a class of the most deadly of man-made chemicals, but no one suspected their presence in 2,4,5-T, which had been regarded as relatively harmless to humans. Before Baughman and Meselson refined the technique, it had not been possible to "wash out" chemical contaminants at such low and precise levels. Doc Clark and Gary Schenk were immensely excited when they learned this. They had understood that the .3 parts per million guideline for PBB had been set by the FDA because that was the lowest measurable level for a chemical contaminant. They had no idea that contaminants of the contaminant could be isolated in amounts that were minute beyond their grasp.

Gary made a telephone call to Harvard, without result. "They wouldn't talk to me because I was an attorney," he related. It was his first lesson in the fact that research scientists are not anxious to get involved in litigation.

Marlene fared much better. She had a quiet, authoritative telephone manner and when she would announce, "I am calling for Dr. Alpha Clark of McBain, Michigan. He would like to speak with you," it sounded to some scientist in a university a thousand miles away as though Alpha Clark was a name he ought to know. A long and useful conversation would ensue, the scientist doubtless imagining a distinguished expert at the other end of the wire, sitting behind a handsome desk in a well-run office. In fact Doc was most often settled at his kitchen table, wearing his coveralls, surrounded by a confusion of papers, veterinary journals, and unwashed dishes, some of it spilling over to the counter where, unperturbed, Marlene—a short, comfortably built woman—would be preparing food.

Doc talked to both the Harvard scientists at length. They had heard about PBB and agreed that it might contain contaminants. Dr. Meselson, a distinguished microbiologist, suggested that a colleague of his, Dr. Patrick O'Keefe, might be able to do the necessary research. Doc had Marlene call O'Keefe that same afternoon. "Get an appointment with him as quick as you can," he urged her. This time she went into some detail:

"I told him we had a hundred or so farmers in our area who were hurting physically and emotionally, that Doc was a GP, not attached to any university or government agency, and wanted to talk to him about the product. I made an appointment for ten o'clock the next morning for Doc and Gary. I didn't tell him that Gary was an attorney. We were delighted he would see them so soon. Often the people we called would say they were not free until next week or next month, but we did not have time for waiting like that."

For some of his travels on PBB cases in Michigan, Gary Schenk had had the use of a twin-engine private plane belonging to his chicken-farmer client, John Williams. Eager to get to Harvard in a hurry, Doc told Gary, "See if you can borrow that plane. If not, I'll get tickets on a commercial one." Although Harvard was almost a thousand miles away and Doc's income modest, the PBB quest was so urgent to him that he did not stop to consider expense.

John Williams gladly loaned his plane again and Doc left home at 2:30 A.M. to drive to Fremont, where he met Gary. They took off at 4:00 A.M. and almost missed their appointment because, just before landing time, the plane's radar signal reply device broke down and the pilot was directed to a different airport, which necessitated a long taxi ride to Harvard. They reached O'Keefe's laboratory, dishevelled and breathless, at 9:50 A.M.

Gary did not reveal his profession until they were well into the conversation; then found that it was immaterial to O'Keefe. Doc did much of the talking.

"I was trying to tell him how those cows acted," he said. "I went through the whole thing and explained our problems with the animals. We sat there for two hours trying to convince him we were not a couple of idiots. Then we all went to a restaurant for lunch. 'Some Cellar,' it was called. They ordered beer and sausages with sauerkraut, so I ordered something I was used to, hamburger and Seven-Up, and we continued to talk. Eventually Gary said, 'You think we are crazy, don't you?' And he said, 'Yes, but I think you are right.' We felt we had found a friend."

It was the same kind of luck that Rick Halbert had when chance led him to Allan Furr—that of finding a researcher free to do the work and fascinated by the project. Only the previous day, just before Marlene telephoned, O'Keefe had read an article about PBB in *Environment,* the journal of the Scientists' Institute for Public Information. It was the first he had heard of the contamination of Michigan, and he was fascinated. It was also a slack period for him. He was waiting for a new research project to be funded and had a month or two to spare before the grant was likely to be approved. He said he would be glad to do some tests on Firemaster but did not have time or money for lengthy research.

Gary mailed him two kilos of Firemaster FF-1, which he had obtained from Michigan Chemical, plus a check, which O'Keefe returned. O'Keefe also suggested other scientists who might be helpful.

At the top of his list was Dr. Renate Kimbrough, a medical research officer and toxicologist at the federal government's Center for Disease Control in Atlanta. She had been analyzing the effects of environmental chemicals and pesticides since 1962, long before most scientists acknowledged that these products could be a threat to animal life and human health; and she had investigated an incident in Missouri which bore some resemblance to the cattle plague in Michigan. It involved the mysterious deaths of horses at Moscow Mills, a small place near St. Louis. She discovered that salvage oil was used to control dust on the track where they were exercised, that chemical waste had been mixed with the oil, and that the waste contained dioxin.

Doc Clark and Gary Schenk did not know it, but before they approached her she had been informally consulted about the PBB contamination by experts at Michigan State University, and by the state's Departments of Agriculture and Public Health. She had advised them that once people had ingested PBB they may continue to be exposed to its effects, since most of it would remain in their systems. Her experience led her to suspect that PBB was even less likely to be excreted from the body than PCB. As a consultant to Michigan Department of Public Health, she had recommended a thorough study of the effects of the chemical on the Michigan populace—one which was broader and deeper than the limited survey of a few farm families which the department had thus far undertaken—and she had helped to formulate an application for federal funding to the FDA and National Cancer Institute.

"We were turned down on grounds that PBB was only a short-term incident confined to a relatively small part of the country," she said. "The decision did not surprise me. It is always difficult to get support for this kind of work. But if you stay with it, and get other people involved and let them think it is their idea, eventually you get somewhere."

Dr. Kimbrough did stay with it and, after more than a year, in 1976, the federal money was forthcoming. She also did some PBB research of her own and found that PBB, like PCB, could cause liver nodules in rats which were probably precancerous. In late 1975, when Doc and Gary flew to Atlanta to talk with her (using the same borrowed plane, and seizing the opportunity between snowstorms), she had as good an understanding of the human health hazards of PBB as anyone in the country—and the frustration of being able to do little about it.

"She really helped us to understand the problem," Doc said. "She explained to us how the different isomers of the chemical can have different melting points and different toxicities. Depending upon how the atoms are arranged on the biphenyl ring there is the possibility of forty-two kinds of hexabromobiphenyl alone. It was a tremendous revelation to us."

Doc had no problem comprehending this, and interpreted it to Gary as they talked excitedly on the journey home. "I've always been good at chemistry," he explained. Doc's most troubling recollection of the visit to Atlanta was a comment made to him by one of Dr. Kimbrough's co-workers, "There's a smoking volcano in Michigan and before long it's going to erupt."

This spurred the two on to more inquiries. They made a trip to the Cincinnati Chemical Processing Company in Batavia, Ohio—the plant to which Michigan Chemical had sent about 140,000 pounds of BP-6 to be ground into FF-1. It was a small company whose function was to pulverize chemicals for a number of manufacturers, and the two men were anxious to find out whether there was a potential for cross-contamination. They also wanted to know the composition of Floguard, the calcium polysilicate which was added to BP-6 to help break up its large, sticky lumps before grinding. They found few answers, and their inquiries raised new questions.

"It was just a little bitty outfit that did grinding and mixing for a lot of customers," Doc reported. "We spent half the day there. They were handling chemicals for plastics and steel companies. They added this Floguard but could not tell what was in it. They got their files out, and it bothered us to find some discrepancy between the amounts of PBB that went in and out. Some of the FF-1 was sent direct to some of Michigan Chemical's customers in Indiana, Ohio, and Canada. It would be hard to trace where it all ended up."

The more Doc and Gary found out about the chemical, the more they realized how little was known. And yet how much. "MDA had been saying that there was very little information on halogenated hydrocarbons," Doc remarked, "but we found there was quite a lot. It was a matter of pulling it all together."

They kept their mission secret, afraid "the other side" might try to divert them if it found out. In the office of Law, Weathers, Richardson and Dutcher—the Grand Rapids law firm of which Gary was a partner—only a few people were told, and then only essential facts. Code names were used to mask the identity of scientists who had agreed to do private research on PBB. Patrick O'Keefe became Peter O'Toole, Renate Kimbrough was Mary Poppins ("she reminded us of Julie Andrews, who played that part in the film").

They were not being paranoiac. Their involvement was known and they were being harassed. Gary had received a number of anonymous telephone calls at his home, threatening him with nameless reprisals if he didn't drop the cases of the low-level farmers. By this time he and Paul Greer had about a hundred farmer clients with outstanding claims, amounting to several millions of dollars: no other attorneys in Michigan were anywhere near as deeply involved.

"The calls would come in the middle of the night, and I would pick up the phone, half asleep," Sandy Schenk related. "One night there was a threat against our two kids, and Gary would not tell me about it for three days. We were so worried that we went to the children's school and asked the nuns not to let them out for recess."

None of the threats was carried out, but the calls continued, and for a long time Sandy was fearful to let the children out of her sight.

In McBain a local official who knew Doc Clark as a friend advised him, "If I were you I'd carry my rifle in the car and keep it loaded." Another day Doc was stopped on the road by a farmer of his acquaintance. "I think you should know," the farmer said, "that some people in Lansing are after your ass."

"All this scared the living daylights out of me," Doc reported. For about a week he carried his hunting rifle along with his veterinary supplies, but he felt uncomfortable doing so, and faintly ridiculous. The thought crossed his mind that if he were assaulted on a rural road he would probably grab the gun in a hurry and shoot himself by mistake. Thereafter he left it at home.

He was never attacked physically, but saw plenty of professional harassment. It came back to Doc that a senior MDA official had told a meeting of county agriculture agents that two veterinarians in the state were adulterating fat samples, presumably to make them appear more contaminated than they really were. Although the two were not named it was clearly implied, according to Doc's information, that he was one of them. The other was alleged to be a practitioner in Michigan's Upper Peninsula who had also been openly critical of state inaction on PBB.

Doc also came close to losing his licence to practice in the state. In the late summer of 1975, Dr. Farrell Robinson and his team of veterinary pathologists at Purdue University had asked him to send a few more sick cattle for analysis. Fearful of possible entrapment, Doc asked one of the farmers who was offering some of his contaminated cows for the experiment to telephone MDA for permission to ship the animals out of state. Marlene Clark was careful to make a separate call with the same request. Up to then, in such a case, Marlene's call alone had been sufficient. This time both Marlene and the farmer understood that the request had been approved by MDA, but the farmer took the precaution of taping his telephone conversation.

Several months later federal government attorneys and USDA officials in Michigan made it known that they were investigating Doc Clark's two shipments of a total of eighteen head of cattle to Purdue. It was alleged that the shipments may have been illegal, because Doc lacked "the proper documentation" to move the animals across the state line to Indiana. Dr. Charles Cole of MDA's animal health division stated that he had neither granted, nor been asked for, the necessary permission, although Marlene insisted that he was the one she had asked. Reporting the investigation, the *Grand Rapids Press* headlined its news story "PBB Testing May Lead to Vet's Arrest." It was the worst kind of publicity for Doc Clark, and his ability to continue practicing in the state hung in the balance.

The farmer's tape recording was produced, and the investigation abruptly ended. The absurdity from the beginning was that any case against Alpha Clark hinged upon a technicality: whether he had the permit verbally, on paper, or not at all. It was in no way associated with the animals' state of health. The fact that they had been sent directly to Purdue to be killed for laboratory analysis was never questioned—and yet a man identifying himself as a federal agent went to Dr. Robinson's laboratory and seized the test results. It seemed incredible to Gary, representing Doc, that federal authorities should devote so much energy to so trifling a matter. The investigation came at the request of the U.S. Department of Agriculture, and its Lansing office (in the same building as MDA) was given the test results. Suspicions of a Farm Bureau-MDA liaison became sharper.

"It was on the eve of the Zuiderveen case coming up for trial, and Doc was my best witness," Gary Schenk observed. "He and I had just got together, and we were accumulating scientific evidence. I felt that someone was trying to put him out of action. And when the full power of the federal and state governments can be used to try to trap one little McBain veterinarian, it is clear to me that these people have lines into the power structure which we could never match."

Virtually everyone who had the courage to speak out publicly on the PBB issue encountered barriers of opposition at various levels of the agricultural establishment. The structure of this establishment is powerful and political; in a heavily agricultural state like Michigan it involves Farm Bureau, the county agriculture agents, the state Department of Agriculture (and by extension USDA), and the land grant colleges (in this case the vast complex of Michigan State University).

It was a network built up over many years to protect the interests of family farmers, and historically it served them well. But as the number of small farms dwindled and as those remaining became larger and more mechanized, the agricultural bureaucracy became the servant, spokesman, and lobbyist of agribusiness. Furthermore, no section of it functioned independently of the others. Their interests interlocked. Robert E. Braden, administrative director of Michigan Farm Bureau, made the point that Farm Bureau Services was "completely separate from" the main organization, each having its own legal identity and financial structure. Nevertheless, it shared some of the same officers, and Farm Bureau held 58 percent of the feed company's voting stock. The rest was owned by farmers in the state who loyally regarded this as their cooperative. But in practice it wasn't.

Many senior officers at Michigan Department of Agriculture were, or had been, active Farm Bureau members. This was also true of the five-member State Agricultural Commission, appointed by the governor, which in turn had the power to appoint the director of MDA. Any proposed legislation affecting farmers usually had the blessing of MDA before it went before the legislature, and Farm Bureau's approval was a traditional ingredient of that blessing. In Michigan there also appeared to be close ties between MDA and the district offices of the two federal agencies involved in the PBB crisis—FDA and USDA.

The power lines stretched into, and beyond, Michigan State University—one of the nationwide chain of land grant colleges, set up under a nineteenth-century federal act which granted the use of public land for state-supported agricultural and technical institutions. Michigan State University's dependence upon the legislature for appropriations made it almost impossible to keep politics out of research, particularly agricultural research. MSU's research into the cattle contamination was limited, late, and, for the most part, unsympathetic to the farmers' plight. One of its earliest studies on the health of low-level herds was done by Dr. Donald Hillman of MSU's Department of

Dairy Science; to the amazement of the affected farmers, it concluded that it was not the PBB in their animals which produced the symptoms of toxicity, but an excess of iodine!

Iodine is used by dairy farmers to prevent hoof rot and as a teat dip. Small quantities of it are also added to feed to prevent goiter. After investigating seventeen problem herds, Hillman concluded that many animals with alleged PBB problems were in fact suffering from excessive iodine intake and intestinal parasites. The many afflicted farmers who ridiculed this study were not surprised to learn that Hillman's research was funded by Farm Bureau.

"He may have been right about the iodine," commented one independent agricultural expert. "A lot of it is used on farms. But he did not even hypothesize whether the iodine and PBB reacted upon one another. It was as if PBB never entered those cows."

Months earlier, Hillman had attributed Halbert's problem to moldy corn. Later he blamed some of the farmers' difficulties on an excess of fluoride.

Despite the considerable resources of MSU, it was a long time before scientists there could get independent funding for PBB research. By then, most of the significant work on PBB was being done outside Michigan, much of it at the instigation of Doc Clark and Gary Schenk.

Dr. Steven Aust, professor in MSU's Department of Biochemistry, explained, "We could have mobilized more help at this university, but it would have taken financial support, and we didn't have it. None of us professors is guaranteed money to do research. One goes out and seeks contracts and grants from people who need help." Consequently, throughout the PBB crisis, most MSU scientists were occupied with other projects, many of them trivial by comparison.

Even when the crisis was many months old, aside from a small amount of federal funding, there were still no independent sponsors for PBB research. The only ones prepared to pay for it were Farm Bureau and Michigan Chemical—and the studies they ordered, which showed minimal cattle damage, were immediately suspect among farmers. Theoretically, the university could have mobilized its own PBB research, but in the practical terms of Michigan politics it would have been unwise to take this initiative without the tacit approval of one of the state bureaucracies.

The association between MSU and MDA was close. MDA's Geagley Laboratory, used by the state inspection service, occupies five acres on the campus of MSU, and MDA's three veterinarians work out of the university, using some of its facilities and calling upon its expertise. The offices of MDA, the campus of MSU, and the headquarters of Michigan Farm Bureau are all within easy reach of one another, in the Lansing area. Over the years an intimate relationship developed between the staff at all three places—"the kind of chumminess," a legislative aide remarked, "that comes of working together in barns and fields, rather than behind desks." It followed that staff members moved back and forth between the three, swapping and upgrading jobs and taking their contacts with them—to the extent that it was sometimes hard to tell where one organization left off and the other began.

Their combined strength worked for the good of agribusiness: bigger and more mechanized farms, crops whose yield was increased by the development of new chemical fertilizers and pesticides, dairy herds whose feed was boosted with nutritional additives so that they could produce much more milk than earlier generations of cows. Somewhere in the shuffle the interests of small family farmers were often forgotten, although most of these farmers remained—until the PBB crisis—fiercely loyal to Farm Bureau. Even after some became disenchanted it was hard to break the ties, because Farm Bureau supplied them with essential services at an economic price: home and hospital and life insurance, fertilizers, cattle feed, tractor tires, baling twine, and batteries. It advised them of developments in farming. It was also their lobbyist in state politics and, through the American Farm Bureau Federation, at the federal level.

"It was much more than a membership organization for us," said Garry Zuiderveen. "It was a philosophy, a way of life. To question anything that came from Farm Bureau management was like questioning the Bible."

The power lines of the agricultural relationship reached out to farmers through the university, as well as through Farm Bureau, and over the years the two had become closely intertwined. The university's Extension Service, part of the land grant college system, provided a statewide network of county agents whose mission was to keep farmers abreast of agricultural developments. Although their salaries were paid out of state and federal funds, these agents often functioned as grassroots Farm Bureau organizers, which led to their being political activists at the same time as public servants. This alliance was not unique to Michigan. It happened throughout rural America. The county agent who spread the mischievous rumor about Doc Clark was not, apparently, passing down information from his superiors at MSU, but repeating a story he had heard through the agricultural network.

One example of how this mutually supportive relationship worked is contained in the memorandum of a telephone conversation between B. Dale Ball, director of MDA, and Donald R. Armstrong, executive vice president of Farm Bureau Services, on the evening of May 28, 1974—a month after Firemaster was found in the cattle feed. This memorandum, in Armstrong's awkward handwriting, was obtained by Gary

Schenk and Paul Greer during the legal process of discovery. Later it became a court exhibit whose authenticity Armstrong confirmed on oath.

One of Ball's reasons for calling Armstrong at his home was to discuss the fact that five days earlier, state inspectors had found that the feed company's equipment was still contaminated, despite the fact that the plant had already undergone a cleanup. Thus, Farm Bureau Services may have mixed even more feed before cleaning up again. According to Armstrong's note to himself, Ball "indicated customers could come back on department and on us if they knew we produced feeds after knowing we had contamination in equipment." It seems to have been just a warning between friends. Despite its knowledge, MDA made no move to force the feed company to close down until it was thoroughly decontaminated, or to insist on a recall of the feed which had just been mixed. So far as is known, its director simply told Armstrong that he had better watch out.

When it came to proving that their contaminated herds were unsafe and worthless, many low-level farmers felt that MDA and MSU had closed ranks with Farm Bureau and become their common enemy. This was the adversary which Doc Clark and Gary Schenk took on. This was the reason they instigated all their scientific research outside the state of Michigan.

"We had to get experts who were not politically involved," Doc Clark explained. "The people at MSU wouldn't help us. They kept sending back lab reports saying that the cows were suffering from malnutrition. They weren't really lying. You knock out the liver function and you upset the metabolic processes and you'll see malnutrition all right. No, they weren't lying but, boy, they sure stretched the imagination.

"That's what we were up against. I can tell you, it was a long road, and it was not a very even road to be on. I am not sure Gary and I would be around today if the other side had realized how far we would get."

The quest took them to some of the most experienced scientists in their field. Many of their names were discovered by Marlene Clark and Sandy Schenk, searching through medical and veterinary journals while their husbands traipsed all over the country. In between taking her two young children back and forth to school, ever anxious about their safety, Sandy spent hours in the medical libraries of Grand Rapids hospitals. Being young, vivacious, and uncertain of the categories in which to search, she did not fit the librarians' image of a person with legitimate use for medical literature. When they tartly asked the reason for her research she felt obliged to be vague, and that was another obstacle. "It was like being lost in an ocean," she said.

Nevertheless, she turned up some relevant material. She found a report about the *yusho* incident in Japan, and Doc was excited to discover that it contained a reference to the contamination of PCB with naphthalene, something he suspected with PBB because the cows' symptoms matched the naphthalene-induced X-disease. Several weeks later his suspicion was confirmed.

He heard the news from Gary Schenk who learned it in a particularly dramatic fashion. One day after calling at Paul Greer's office in Fremont, Gary was driving north for a strategy session with Doc. Passing through the small community of Baldwin on the quiet straight road that cuts through Manistee National Forest, he was stopped by state police.

"Anything wrong?" Gary asked innocently. He knew he had been speeding.

The reply was unexpected. "We don't think so, but there's an urgent message for you to call your secretary."

The original message had been relayed to Sandy who, realizing its importance, had telephoned the state police with a description of her husband's green Corvair and the route he was taking. Gary phoned his secretary from Baldwin and was told that he should immediately call "Peter O'Toole." Even she had not been told the man's real name, on account of Gary's suspicions that his phones were tapped.

The news from Patrick O'Keefe was astounding. He had found at least 40 ppm of naphthalene in the Firemaster sample. Gary hurried to Doc's house to pick him up for dinner at a roadside restaurant near McBain. Doc sensed his excitement. Gary had meant to keep his news until they sat down to eat, but on the way to the restaurant Doc dragged it out of him.

"Stop the car!" Doc demanded.

As he recounted the story, "I got out of the car and I started crying. It was a very emotional moment. I thought I had an answer, and I had worked so hard to get it. O'Keefe had told Gary that he thought he could isolate other contaminants. It was a tremendous breakthrough, like when the people at Purdue found chronic toxicosis in our cows."

The search was far from over, but they had gone an important step of the way. Several years later scientists were still trying to discover all the components of Firemaster FF-1, and which of them produced the different symptoms. There was also the possibility of a reaction with other toxins which the animals had previously ingested. A few Michigan dairy farms had silos (made by a company which had since gone out of business) in which PCB had been an ingredient of the sealant, and small quantities of this had been known to leach into cattle feed. The combined effect of the two similar chemicals was not known.

What mattered at this moment, however, was the proof that it wasn't "pure PBB" which went into the feed, although "pure PBB" was the standard against which animals were being tested for quarantine. Later O'Keefe's work took on added importance. His study inspired others to do PBB research—to the extent that this compound, which at the peak of its production represented less than 1 percent of the total sales of all American-made fire retardants, became the subject of close scrutiny by a number of scientists. But for the contamination of Michigan this would never have happened. The chemical would have passed, as most do, into commercial channels and possibly into the environment without any real understanding of its actual properties.

Gary Schenk speculated, "If it had not been for Paul and me, and if we had not met Doc, the whole goddamn thing would have been swept under the rug after the claims of Halbert and the few farmers around Fremont had been settled. Doc had something we did not have; he understood, long before most of the so-called experts, the effect of low-level contamination.

"The people at MDA and FDA had a simple concept of poison: you take something toxic and you immediately get sick. That is acute toxicosis. They could suggest that, on an acute basis, PBB was about as harmful as table salt because you have to jam a hell of a lot into you to kill you right away. That was true of the PBB contamination of 1973 and 1974. But what we were looking at a year or two later was a chronic toxicity problem. And they never spent the money on researching this because they did not understand the process."

The two men had different motivations. Gary wanted to accumulate scientific expertise so that he could take the case of one low-level farmer to court, win a suit against Farm Bureau Services and Michigan Chemical, and so create a legal precedent which would lead to all the other low-level farmers getting settlements. The two companies had financed animal studies which concluded that no PBB toxicosis existed in the low-level herds. It was persuasive evidence, and he knew he had to combat it.

"Without the help of dedicated scientists outside Michigan, we had little hope of getting very far," he observed. "The traditional methods of pretrial discovery would never have given us the information we managed to assemble."

Doc was concerned about the more distant future. He explained, "I wanted to get as much research done as possible, and enough in the medical literature, so that even if it did not benefit us it would help the next generation. I was very concerned about human health. I knew that once the cattle cases were over, our job had only begun."

While the quest continued, Paul Greer prepared some cases for trial, with the help of the staff at Gary's office. Fred Boncher did the leg-work, visiting contaminated farms, checking the farmers' records, gathering evidence for their claims.

Doc organized two teams of veterinary researchers, both from outside Michigan. Through a contact of Garry Zuiderveen he met Dr. David R. Helland, a veterinary pathologist in Rockford, Illinois, who assembled a small team to go into Michigan and take tissue samples from cattle in about eighty low-level herds. Dr. Harry Cook of Trinity Christian College, Illinois, and Dr. Kent Van Kampen, veterinary toxicologist in Salt Lake City, assisted. Meantime, the research at Purdue University continued. All these specialists gave their services at little or no cost, on the understanding that they would report exactly what they found, without prejudice to either side in the dispute. All their reports confirmed severe and lasting damage to low-level herds.

Doc Clark and Gary Schenk also pursued the human health aspects. "We went on and on, never satisfied, adding to our knowledge," said Doc. From Dr. Harry Cook he learned of Dr. Stephen Safe, professor or biochemistry at the University of Guelph, Ontario, who had analyzed the varying toxicities of the different isomers of PCB and related chemicals, including PBB. When he heard what had happened in Michigan, Dr. Safe quickly agreed to do more research on PBB.

"It does the same kind of things that carcinogens do: it reacts with critical body chemicals," he found. "Compared with most chemicals it is excreted at a fairly slow rate, and it leaves side effects which are associated with cancer. Anyone with a PBB residue may be affected ten or twenty years after exposure, and I don't think there's anything anyone can do to stop the process."

A quiet-spoken, diffident-looking Canadian, Safe was angered by the complacent attitude of the Michigan bureaucracies.

"It bothers me when they say there are no adverse effects, that there is only a teaspoonful of PBB left in the whole state. That's bloody irresponsible. Depending upon how you analyze the chemical, you can make it look good or bad. I can take a known toxic chemical and devise some tests that will make it look good. PCB is not very soluble in water, so I can do some tests in water and they won't show much."

He was just as critical of the theory that because symptoms were so varied, it was absurd to attribute all of them to PBB. "People have different sensibilities," he argued. "When you do a study of a disease you look at populations, not people, and you find trends. But within any population there are fantastic differences."

Stephen Safe told Doc Clark about another scientist, also outside the United States, who had done work with PBB—Dr. J.J.T.W.A. Strik, a senior staff member in the toxicology department of the Agricultural University at Wageningen in the Netherlands. Anjo Strik was the Dutch toxicologist who had sent to Michigan Chemical for a sample of PBB in 1971 to use in the research for his Ph.D. thesis about the effects of halogenated hydrocarbon residues on wildlife. Experimenting with Japanese quail (a species chosen because it is sensitive and breeds quickly), he found that PBB seriously disturbed liver function, causing porphyria, and that low doses over a long period caused more damage than single high doses. Liver porphyria can be either congenital (a condition which was believed to have plagued the House of Hanover and caused the "madness" of George III), or it can follow the ingestion of some toxic compounds. One of its symptoms is a change in the color of urine to a dark purplish red.

Doc wrote Anjo Strik and learned some of this from Strik's reply. It added another important piece of evidence. Doc remembered a puzzling symptom which one farmer, Roy Tacoma, had described to him: "Roy told me that when his cows peed on the snow it was this strange reddish color."

Strik offered to help in any way he could. Previously, his experiments with PBB must have seemed, even to him, an esoteric piece of research which was unlikely to be put to general use. Suddenly, however, Strik's findings were of value to an entire U.S. state—and he was glad to share his expertise.

Gary and Doc received a similarly warm response when they called on Dr. James R. Allen, professor of medical pathology at the University of Wisconsin Medical School. An eminent researcher who had done PCB experiments with primates, he was an obvious authority to be consulted about likely effects of PBB on humans, but none of the federal or state officials had done so. The first inquiry Dr. Allen received out of Michigan came from Doc and Gary.

"We spent half a day discussing it, and I told them that more than likely they would run into the same kind of problems that we had defined with PCB," Allen stated. "Loss of weight, hair loss, acne, reproductive abnormalities—difficulty in conceiving, and a high rate of spontaneous abortions among those who do conceive."

He had fed some monkeys as much as 300 ppm of PCB in their diet, and some as little as 2.5 ppm. He found that even the lowest doses produced the same symptoms of toxicity: they were simply less dramatic and took longer to appear. He predicted that PBB would react in a similar fashion.

James Allen was dismayed by the bureaucratic delays in responding to the PBB disaster.

"In Michigan the state and regulatory agencies waited too long," he commented, months later. "It became like a little Seveso, and it was as badly mishandled."

He commented of Alpha Clark: "A practicing veterinarian would normally have no knowledge of chemical structures, or of the interaction of various compounds. I was amazed at this man's understanding of what was going on."

Allen cared as deeply as Doc about human welfare. Although extremely busy, he put his work aside and did some primate studies with Michigan Chemical's Firemaster FF-1. He found it to be slightly less toxic than PCB (which surprised him—he expected it to be more so) but capable of producing similar symptoms. Monkeys fed as little as .3 ppm on a daily basis suffered weight loss and reproductive problems. Allen was deeply disturbed about the prospects for the future.

"We cannot predict what may be the residual effects of PBB deposits in tissues fifteen or so years hence," he said. "They could be cancerous or mutinogenic. They could promote cancers which are already latent in the body, so that instead of these cancers taking ten years to develop they may take about two. There could be many ramifications which are not apparent at present.

"We cannot close the books on the Michigan case until we see another generation."

PART THREE

THE HUMAN FACTOR

If my husband was a cow they would have killed him. They would have sent him to Kalkaska and buried him, because the PBB in his fat was over the tolerance level.

—BETTY MOTZ, farmer's wife

This is a very different child population. We see a complete range of symptoms in these farm children, and it is very disturbing.

—MASON BARR, pediatrician

My own family doctor looked me in the eye one day and said, "Are you sure you are not poisoning these children?"

—CAROL CURTIS, farmer's wife

22

Death of a Herd

THE world of state politics was beginning to give some recognition to the farmers' distress, but still showed no awareness of the risks to urban consumers. In July 1975, as Doc Clark and Gary Schenk were beginning their joint research, a special investigating committee of the Michigan Senate published a report on the PBB situation. The first of its kind, the document described the farm disaster with sympathy and expressed concern over the delays in compensation payments. The health of farm families who had ingested large quantities of PBB was "an area of great concern," the committee conceded, but it saw "no reason, from the evidence available, to believe that the general consuming public might, in any way, be endangered by PBB." This remained a rural problem, to be assuaged by financial aid to farmers.

An attempt to provide money was made by Donald Albosta and his co-chairman of the House Agricultural Committee, Paul Porter. In the summer of 1975, 580 low-level farmers had pending claims, and all were in financial difficulty. Albosta put a bill into the legislature to give them low-interest loans which would tide them over until the courts settled their cases. By this time legislators had become aware of the farmers' distress and the bill won unanimous approval in both legislative houses. But where was the money to come from? It was Albosta's idea to borrow from the Veterans Trust Fund. This contained about $15 million which had been set aside from a state budget surplus of better years to administer veterans hospitals in Michigan.

The bill languished in the governor's office for a few weeks. Then, unexpectedly, the governor vetoed it. His main objection was that, legally, the transfer of money from the trust fund was "of doubtful validity." A few weeks later Michigan faced the prospect of a $300 million budget deficit: with the Detroit motor industry in deep trouble, the state's revenue had been overestimated, and the cost of coping with economic depression was greater than had been anticipated. The constitutional requirement that the state budget be balanced was finally met by considerable juggling in the budget director's office, where the farmers' interests were forgotten in the more pressing concern that Michigan's public schools might have to close if money was not found for them. The budget people stretched the fiscal year to eighteen months and borrowed substantially from the same Veterans Trust Fund which the governor had at first refused to touch.

"I think the governor felt that Farm Bureau would take care of the farmers, and that the side effects of his veto would not hurt him politically," Albosta commented. "It seemed that every time we tried to do anything on PBB it was opposed by Farm Bureau or the governor's office." It was September 1975, and Albosta was still fretting because the hearings on his proposal to lower the tolerance level had come to nothing. Disturbed that the farmers' case had still not been heard, he decided to form a small sub-committee of the House Agricultural Committee to gather evidence. "I thought it would be the best way to get the facts to the legislature," he said. "The PBB episode had become more and more confusing to legislators. They knew that something was wrong but they didn't know whether to believe the farmers, Farm Bureau or MSU, and they didn't have the time to check it out." Most legislators combined their public service with jobs of their own, and depended upon the state agencies (in this case MDA and the Department of Public Health) to provide them with specialized information.

As a Democrat in a predominantly Democratic legislature (the governor was Republican), Donald Albosta anticipated no difficulty in getting the Speaker of the House to approve expenses for his hearings. He had ambitious plans for about fifteen sessions in different parts of the state, with committee powers to subpoena records and witnesses. He wanted legal counsel and court reporters to make verbatim notes of testimony, and because all this would make the Democrats look good and the governor look inadequate, Albosta—who had yet to learn state political realities—was confident that his requests would be granted.

But the Speaker, Bobby Crim, had to cope with other influences. The visits to his suite of offices, across the street from the Capitol building, by lobbyists from Farm Bureau and Michigan Chemical increased sharply. Michigan Chemical had hired one of the most influential lobbyists in Lansing, a man who had been a friend of Crim for ten years.

"There was some shoe leather worn out going over Crim's doorstep." Albosta observed. "The next I heard was that I was denied funding for the committee. I really didn't know what was going on, and I was completely tied without money for legal expenses." Compelled to pare down his plans, he resolved that his committee would hold some hearings anyway.

Even within his own party, Albosta's motives were suspect. He had been less than a year in the legislature, but already it was being rumored (correctly) that he planned to run for Congress. Some colleagues speculated (incorrectly) that his real aim in pursuing the PBB issue was personal publicity. He had a flamboyant manner, a carelessness for detail and was not always convincing in his assertions. This, and the fact that Bobby Crim's office showed no enthusiasm for his PBB project, got his inquiry off to a slow start and a poor press.

Even without counsel, Albosta believed that his committee had subpoena power. He thus attempted to get the Department of Public Health's documents relating to its study of families on quarantined farms. Specifically, he wanted the names of those who had been tested and those used as controls. The attorney general's office advised the department's director, Dr. Maurice Reizen, that he was not obliged to hand over these records, on grounds that their publication would violate the confidential relationship between doctors and patients.

Albosta had no difficulty persuading Doctors Salvati and Corbett to testify on the human health risks. He also invited Dr. Walter Meester, the Grand Rapids toxicologist who had seen many PBB patients. Meester declined to testify. By then it was February 1976.

"It's a year too late," Meester stated. "Every person in Michigan is now contaminated with PBB."

The onset of cold weather in late 1975 had forced some Michigan farmers to make a choice which, at any time in the past, would have been inconceivable. Burdened with unthrifty cattle, they had to decide whether to send these animals to market or to give up farming. Farmers who had been losing income for months could not afford the costs of winter feed for unproductive animals. Some farmers had gone the way of the slaughterhouse, and continued to do so. Others saw it as a moral decision. Was it right to put produce on the market which they would not serve to their families?

The dilemma divided households. The Green family argued about it often. Their farm at Chase, where Alvin and his wife Hilda were partners with their son, Douglas, and his wife, Donna, was in terrible shape. In better days it had been a productive family farm—not an ambitious operation like the Halberts' but one which comfortably supported four adults plus Doug and Donna's four children. Al Green was an old-fashioned farmer who really loved his cows. If one were sick he might spend all night, stretched out on the hay in the barn with her in case she needed attention. From the spring of 1974 when their first cow suddenly died, Al did a lot of his sleeping in the barn, becoming more distressed as mature cows died in labor and calves were born dead or deformed.

Over eighteen months the Greens lost half their herd through sickness and death. They were deeply in debt, and tissue samples sent to an independent laboratory confirmed that the remaining animals had symptoms of PBB toxicosis. Yet test after test made by MDA for PBB was marked "non-detectable." Although they did not know it until much later, PBB was in the tissues of every member of the Green family —no doubt as the result of consuming their own contaminated products. They all felt ill. This, added to anxiety over the farm, made them irritable with one another. Hilda and Al argued so much with their son that when both his legs were broken in that senseless accident in the tractor it relieved the tension for Doug to be in the hospital.

Hilda and he had been at loggerheads for weeks before the accident. She kept telling him that it wasn't right, the way the cows were humped up and staggering into the barn walls, and he would scoff, "Mother, you're just being a fuddy-duddy. They'd be fine if you'd stop fussing over them."

Hilda was annoyed by his attitude, refused to milk the cows any more, and turned her considerable energies to pestering various state officials on the telephone. She hoped that if she made enough of a nuisance of herself the farm might be quarantined and a case for compensation established.

A plump, white-haired woman with a lovely pink and white complexion, Hilda Green looked like everybody's favorite grandmother. But with her family and livelihood at stake, she was obdurate and persistent. By the summer of 1975 her husband was sick, her son was hobbling around on crutches, all her grandchildren had health problems, and her daughter-in-law (who wasn't feeling well either) had taken the same stand as Al: the kindest thing to do for the cows would be to shoot them. Doug objected, thinking of the debts which could be settled if the animals went to market. It was the only solution which made financial sense, and it was legal.

"No way," said Al stubbornly. "They're not fit." Hilda agreed.

The last impasse was reached when Donna, the quiet member of the family, told them. "I won't put in another winter like last. If you want to keep those cows you'll have to find someone else to milk them. I can't go into that milking parlor any more and see them looking at me with those sad eyes, as if they're saying 'Please help me.' I cannot bear to find any more of them lying there dead."

Doug was still tempted to sell the remaining cattle. Equally, he respected his father's strong sense of morality. All four adults had given up hope of the animals' recovery and, financially, could hold out no longer.

Hilda requested permission to dispose of the herd at Kalkaska and received a call from the governor's office which seemed to convey approval. Later the same morning she had a conversation with Harry Iwasko in the attorney general's office, who advised her of the legal position. Kalkaska was only for quarantined animals, with no exceptions. Iwasko's office had been negotiating with Farm Bureau, which was in charge of the state burial pit, for Kalkaska's facilities to be made available to farmers like the Greens. But since neither the state nor Farm Bureau was willing to take responsibility for condoning the slaughter of unquarantined animals, negotiations had broken down. Hilda had already been informed that it might be a violation of public health statutes to bury the cows on the farm. Frustrated and furious, she told Iwasko what she thought of him, and hung up. She and Al then hired a contractor to dig a huge burial pit on their farm.

The Greens were the first Michigan farmers to shoot their own cattle and they did not make a very good job of it. None of them knew, back then, how hard it is to kill more than a hundred large animals with bullets. Subsequently, other farmers who had to shoot their cattle profited from the Greens' mistakes.

In a last gesture of defiance, Hilda advised several Michigan newspapers and television stations that November 10 would be the morning of the execution, and invited them to attend. She also told Gary Schenk and Paul Greer, who had become the family's attorneys. Saddened that they had not yet been able to get a settlement for the Greens, Gary and Paul helped in the only way they could. They asked several of their farmer clients to assist at the execution. It was too dreadful an undertaking for Al and Doug alone.

November 10 was a bleak, bitter day, with wind and sleet. The Greens' barnyard was winter gray except for a rectangular pit, at least 120 feet long, where the newly turned earth showed up as yellow sand. Driven by a fierce wind, the rain was coming down almost horizontally, freezing as it fell. About twenty-five people had assembled near the barn, including a dozen farmers dressed in the bright orange jackets which they wore in the deer hunting season to avoid shooting each other. It was perhaps an essential precaution on this dismal morning. For a long time they stood around, blowing on their frozen fingers, waiting. Hilda Green stood by the barn, sheltering under a clear plastic umbrella, wanting to watch, yet afraid to do so. Al could not bring himself to leave the farmhouse. He was only fifty-eight, but he looked old and afraid, and he showed some of the visitors the sores on his legs which would not heal. Recent tests taken in Dr. Walter Meester's toxicology department revealed PBB in his system, he said, and he was worried about the consequences.

Outdoors there was a sizable gathering of people. "Just gawpers," Doug said, "and I shall never forgive them. Some of them thought we were stupid to shoot cattle which we could sell for meat. Then there were the farmers who came to help, some I had never seen before in my life. They were friends of friends. One said to me, 'I don't know how you have the guts to do this.' He was Garry Zuiderveen, I found out later. I didn't know then that he had sick animals himself."

The farmers helped to herd the cattle from the barn to the pit. The animals went peacefully. Then the men in orange lined up at the edge of the pit, like a firing squad, and aimed their hunting rifles. A few cows dropped instantly, but, with the gunsights misted by rain, some of the marksmen missed. A cow is not easy to kill. An accurate shot to the brain is necessary for instant death. If the aim is a little bit low and a bullet goes into the neck, the cow will continue standing there—as some did—with blood showering from the nostrils. One animal with half its jaw shot away tried to clamber up the sloping sides of the pit and others followed in panic. Sickened, Doug Green hobbled indoors to join his father. Hilda retreated to the milk-house.

"We've got to get this over quick," Garry Zuiderveen muttered. Etched in Fred Boncher's memory is a picture of Garry dropping to one knee in the mud aiming his automatic deer rifle with deadly accuracy: "He went bang, bang, bang and killed five, just like that."

Other farmers herded escaping cattle back to the pit. Donna Green, the only member of the family left watching, ran to help them. "I knew I had to," she said fiercely. "I had to. It was something I had to do."

Among the spectators were two stockily built men from the same mold, one with hair redder than the other. Tears were running down his cheeks.

"I got to go home," Doc Clark said to his brother Clyde. "It's my job to save cows, not to see them killed." Still weeping, he climbed into his car and drove away.

In less than an hour it was all over, carcasses piled in the pit and the barnyard silent.

"It was horrible," said Dick Lehnert who had been invited to record the event for *Michigan Farmer.* "All of us had this funny hollow feeling inside. It was so cold and my stomach ached. I went back to the farmhouse. Al Green was there and he didn't want to hear about the shooting. Some food was served. I was not hungry but I drank some coffee. I could not stop shaking, and must have looked as bad as Al Green. There was this lump inside me like my digestion had stopped, and I went on feeling that way for two days. Nobody who has spent all his life around dairy cattle can stand a sight like that."

Hilda's grief was softened by the kindness of fellow members of her women's society at the United Methodist Church in Chase. "They brought a kettle of bean soup and sandwiches and all kinds of pies. They stayed until it was all done and they fed the people who had come to help us. But we all had this empty feeling in our stomachs. We had not wanted to shoot those cattle but they had to be disposed of somehow, and it was the only way we knew to keep them off the market."

Doug's self-control had been shattered by the sight of a cow which, with blood pouring from her nose and mouth, staggered to the barn "just like she'd always done," then of her own volition returned to the pit to die. "I never heard one bellow," he said, "not like you normally get from animals in pain. Those cows were numb. Their nervous systems were gone. They had no feeling whatsoever."

Another incident struck Doug, because of all his earlier doubts, like a private revelation: "You know how it tells in the Bible that on Good Friday it clouded up from noon to three P.M. Ever since I have been a little boy, Good Fridays was murky like that. All the time the shooting was going on it blew hard with a cold driving rain. Everyone was soaking wet and frozen, and it was like a miniature cyclone. Yet almost immediately the shooting was over the sun came out and everything was calm. It was the weirdest thing."

"How did it go?" Lois Zuiderveen asked when her husband returned from the shooting.

Garry shook his head. "I could never do that to our animals," he told her.

For weeks he and the attorneys had been badgering Farm Bureau to keep the verbal promise of one of its officials—to dispose of his and Ken's quarantined cattle at Kalkaska. They sought a court order to force the issue, but were denied on grounds that Missaukee County Circuit Court could not prejudge the case without full knowledge of the facts.

"We were not surprised, but it was worth a try," said Garry. Circuit Court Judge William R. Peterson urged the Zuiderveens to get their case ready for court and scheduled a trial date six months hence—for May 3, 1976. Garry determined there was only one way to get through the winter and mitigate some of their continuing losses. It was to amalgamate operations by moving Ken's herd to his farm; thoroughly steam-cleaning Ken's barns and equipment to remove all the contamination, and jointly buying a new herd which Ken would care for until their claims were settled and Garry could clean out his place. It was a generous gesture of the uncle's to take the worst side of the bargain, and it gave him more sick animals than he could house through the winter. The move was precipitated by the fact that Garry had been told of a herd for sale in Indiana and he felt he could negotiate a good price. It meant going deeper into debt, but Garry saw it as the best chance of rebuilding the two farms. However, it also meant destroying some of the sick animals.

The cows were still providing a little income (the Zuiderveens' milk was never quarantined, only their herds), so it had to be the calves. All of them were in poor shape and the only reason for keeping them was as evidence. But it was costly evidence. Thin and pot-bellied, they ate voraciously yet would not grow. At the end of November, Garry and Lois made one last plea to Farm Bureau to take at least these young animals. Again they were refused. Reluctantly, Garry acknowledged that he would have to shoot them.

It was not as bad as the Greens' situation, he told himself. It meant shooting about 175 calves and it was a kindness to many of them to put them out of their misery. He felt he could never kill the older animals, which he almost regarded as his children. "The very idea of destroying our milk cows, which we worked with twice a day, was so repulsive to me that I would have done anything to avoid that kind of confrontation with my conscience," he explained.

He did not wish to bury the calves on his land, fearing contamination of groundwater supplies as well as legal action, so he made the hard decision to leave their carcasses lying on the pasture in the hope that soon he would be allowed to haul them to Kalkaska. They could stay there several months without becoming offensive. The Michigan winter is a long season of frequent blizzards and deep snows, and it would be like preserving the bodies in a freezer.

He now knew that there was a better way of doing the deed than another massacre—to shoot the calves one by one, in several sessions. This was harder on the farmer, but with the help of a few neighbors, Garry and Ken did what they had to do. The bodies lay on a snow-covered pasture all winter until, after a long hassle, Michigan's Department of Natural Resources took them to Kalkaska. It was a grisly task, prising those bodies out of the ice.

Other shootings went on through the winter and into the following spring. Most were of quarantined herds which had also been refused burial by the state. The pit at Kalkaska should have been closed a year earlier, and already contained thousands more carcasses than had been intended. Oliver Cassidy of Big Rapids shot every animal on his quarantined farm after pleading in vain for permission to send them to the pit. He had been a farmer for forty-two years. "Now I'm wiped out," he said. Twenty miles north in LeRoy, Eli Argersinger and his son John shot their remaining ninety-eight animals and left their bodies to freeze on the pasture. The older man said it was like killing his own children.

Although several such incidents were reported in Michigan newspapers, they made little impact on the urban populace. State officials, from the governor down, continued to make reassuring statements about the safety of retail food supplies, and most people assumed that farmers who shot their animals were crazy or, more likely, after some insurance money they didn't deserve. The agony of the farmers and the moral courage of those who went deeper into debt rather than put tainted food on the market was never understood by most of the nine million people in Michigan.

"If the consumers would have taken a stand with the farmers, this would all be over," Eli Argersinger observed, after killing his last animal. "But the little farmer is left to fight it alone." Like many others, the Argersingers had been told by county agents that their animals suffered from malnutrition.

Theirs, however, was nothing compared with the harassment and humiliation heaped upon a "little farmer" in Michigan's Upper Peninsula, Richard Edington of Pickford. Edington was a young man, struggling to make a livelihood and doing quite well at it when PBB struck his herd. Already paying off loans, he did not have the capital to withstand the loss of income. Unable to get his cattle quarantined, he was also unable to keep up his payments to the federal Farmers' Home Administration, and in November 1975 this agency threatened foreclosure on his herd.

Edington argued that the cows were worthless and offered to shoot them, as the Greens had just done. He was ordered not to do so and told that the herd would be repossessed by the FHA.

"Do what you gotta do," Edington responded. "I want no part of it. But they must be slaughtered. I don't want anyone eating them."

The herd was taken to Michigan Livestock Exchange to be auctioned for beef. Edington made sure that some of the Michigan newspapers knew about this and, after a prominent front-page story in the *Grand Rapids Press,* ("U.S. Sells PBB-Laced Herd for Beef") two buyers canceled their purchases. The Livestock Exchange was in a serious dilemma. Because of the unfavorable publicity, the animals were unsaleable, but under state law any livestock which went to public auction had to be butchered within forty-eight hours.

The agricultural establishment came up with an answer. The ninety-four remaining animals from the Edington herd were taken by MDA and put on an experimental farm in the south of the state, where dairy scientists from Michigan State University were brought in to study them for the next few months. The purpose was to determine whether below-tolerance levels of PBB really did affect animal health. Dick Edington suspected what answer they would find, even before they began. For months he had complained to experts at MDA and MSU about symptoms in his herd—weight loss, breeding problems, stillborn calves, lost milk production, elongated hooves, swollen joints, attacks of "the blind staggers"—all indicating PBB toxicosis, yet was told that this could not be the cause. At the outset of his troubles, Edington said, two MDA veterinarians assured him (incorrectly) that PBB could not be to blame "because it did not go over the bridge." (The Upper Peninsula's only road link to the main body of the state is by a toll bridge across the Straits of Mackinac, between Lakes Michigan and Huron.) Later, Dr. Donald Hillman, MSU dairy scientist, propounded his favorite theory by suggesting on the telephone that Edington had probably been giving his cows too much iodine. It seemed unlikely to Dick Edington, now facing bankruptcy, that these two organizations would undertake a costly study to prove themselves wrong.

In this atmosphere of bureaucratic denial, urban ignorance, and rural misery, Donald Albosta's little committee finally got on the road.

The Albosta hearings took place in the worst of weathers. Only three sessions were held—in Grand Rapids, Cadillac, and Sault Ste. Marie—and on every occasion, in February and March of 1976, there was either sleet or a blizzard. Yet all three meetings were packed with farmers. They struggled through snow drifts and drove up to two hundred miles, desperate to find out more about a problem which many thought was his alone. None had any idea, before the Albosta hearings, of the extent of PBB contamination. They had relied on their veterinarians, county agents, and the experts from Farm Bureau and MDA. Over and over they had heard that theirs was a unique case, the only complaint of its kind. The number of veterinarians in the entire state who had stuck out their necks like Alpha Clark could be counted on one hand.

Albosta's hearings were a catharsis for these farmers. They came with written speeches in their pockets, prepared to describe experiences which they believed to be unique. But as the day-long sessions got under way, they were at first consoled to hear others tell similar stories; then, as the stories mounted in number and anguish, they became enraged that so many had suffered so much for so long, and so little done to help them.

The Cadillac hearing, in the heart of Doc Clark's country, was the most emotional of all. It was held in a motel restaurant, and despite driving snow and a wicked blizzard, the large room was crowded. The atmosphere was either hushed as witnesses testified, or in uproar as the collective outrage of the farmers was directed at the few state government officials who had had the courage to show up. "It was almost a lynching mob," Garry Zuiderveen reported.

Lou Trombley was one of the most outspoken. The hearings became almost a crusade for him, and he drove up and down the state to speak at all three. He had been roused to action only two weeks earlier, sitting at his kitchen table reading the *Osceola County Herald* over lunch. He was struck by a headline, "Cancer Researcher Tells of PBB Danger," and read with shock a report of Tom Corbett's testimony before one of the first meetings of Albosta's committee in Lansing.

Plagued by unproductive cattle which were below quarantine level, Lou Trombley had sold 104 animals for slaughter only three days earlier. "I listened to MDA and Farm Bureau," he recalled. "They advised me to sell the cattle and replace them. They said it was low-level and they was all right for people to eat. I believed them. At the St. Louis cattle market I had one cow die on the sale scale. I have a receipt for it; they gave me ten dollars. That was on a Monday, and I was going to sell the rest of the cattle on Friday, more than a hundred head." But on Thursday the local newspaper was delivered.

"Why should a cancer researcher come out with a story like that if it were not the truth?" Lou Trombley wondered aloud to his wife, Carol.

Trombley—a dark-haired man of French-Canadian stock—was a person who felt passionately and acted immediately. He canceled the cattle sale and telephoned the *Osceola County Herald*'s editor. The next issue of the paper carried an interview with Lou Trombley in which he claimed that, to remain solvent, he was being forced to sell contaminated cattle because MDA would not impose the quarantine which would qualify him to sue Farm Bureau Services for reimbursement.

"I was scared to death," Trombley recalled. "I thought people in the county would crucify me. But after that paper hit the streets Al Green called me up and told me about his problems. I had heard rumors about Al Green—that he had starved his cattle—from my county agent. But it was Al who told me about the Albosta hearings.

"So I went to all of them. I wanted to hear what problems other farmers were having. I saw people limping there, women who had to help their husbands into their seats; men who told about the nightmare on their farms with their pigs and cows. I sat there and thought about the things happening on my farm, and to the health of my family. All this went through my mind, and I got mad. I completely exploded.

"There was the Department of Agriculture, the Department of Health, the governor, our elected officials covering up what two corporations were doing to the people of Michigan, telling farmers to sell sick cows to the public to save themselves financially."

The stories which Trombley heard were pitiful. There was plenty about the condition of the cattle, but it was the testimony about the farm families' health which shook him—as well as all the members of Albosta's committee.

"It became very clear to me after the first hearing that we had to lower the tolerance level," Albosta summed up. "These people were testifying under oath, and they told of serious health problems. We had heard a little about health problems before this, but only from doctors. Now I was hearing it from those who were affected. You didn't have to be an expert to realize how widespread it must be."

One farmer who complained of a rapidly worsening pain in his muscles and joints testified that "a lot of us act and feel like some of our contaminated cattle." The simile was readily understood by farmers and their wives in the audiences, who made similar comparisons.

Mrs. Carol Curtis, a farmer's wife and former nurse from an isolated area of eastern Michigan, told of her five-year-old daughter, who had intestinal problems, muscle weakness, blurred vision, even temporary blindness.

"It was very frightening. She would say, 'Mommie, I can't see you, I can't see you, where are you?' and things like that. We worried constantly about her. . . . She spent a week as an inpatient in hospital and they performed many tests. . . . She was seen by a pediatrician, a urologist, an orthopedic specialist, and they said they couldn't find much wrong with her. . . . And we were still drinking milk and eating meat from our farm."

By then her younger child was also ill, Mrs. Curtis related. "My own family doctor, whom I had worked for, looked me in the eye one day and said, 'Are you sure you are not poisoning these children?' And this was before we had any indication that we were contaminated."

Carol Curtis broke down in the middle of her testimony, and was applauded at the end of it.

Bessie Cassidy, an older woman whose husband had recently shot his remaining animals, told of loss of balance, visual disturbances, severe headaches, diarrhea, "aches and pains all over," extreme exhaustion and memory lapses.

"I've lost most of my strength," she said. "I can't tighten a canning jar. It's hard enough to seal the lid and I have canned all my life. My nerves are shot. . . . There are times I can't grip a cup or a glass. . . . I can't get a hold of it."

She paused, exhausted by her own litany of complaints. "Oh shoot, I'm sorry," she said, and sat down, embarrassed.

All the pent-up frustrations of the farm families were poured out at those committee hearings. Even the most personal anxieties were shared.

"We're going down under," stated Betty Motz, farmer's wife from the south of Michigan. "We had a farm that we had paid for. We had to borrow and borrow and borrow on this farm to keep ourselves going."

Referring to the state health authorities, she added, "They couldn't understand why we were having health problems, because we had a low-level herd and my husband only had .359 in his fat sample. But if he was a cow they would have killed him. They would have sent him to Kalkaska and buried him because the PBB in his fat was over the tolerance level."

Some farmers had suspected that not all the quarantined cattle were going to Kalkaska, and had alleged that some of the best-looking were diverted to meat markets. This had been denied by MDA and Farm Bureau; although, tested individually, some of these animals could have been under the tolerance level. Bonnie Hughston—Doc Clark's neighbor who helped to run her husband's livestock trucking business—told Albosta's committee an intriguing story.

She said that her husband had taken thousands of cattle to Kalkaska, and the routine never varied. He would receive a telephoned instruction from Farm Bureau to go to a farm, pick up a designated number of animals and take them to the burial pit. But on a recent occasion the instruction was different. This time her husband was advised, "Go to the Jim Fish farm, load the cattle and then we will tell you where to go with them."

James Fish was a neighbor of Rick Halbert—a farmer with a herd of prize Guernseys, one of the best in the country. He believed his low-level contamination to be due, at least in part, to the fact that the Battle Creek plant had sent his feed (not 402) on the same truck as Rick Halbert's highly contaminated load. How ever it happened, a superbly bred herd which he could never replace in his lifetime was irrevocably damaged, and Bill Hughston had been commissioned to take some of his animals to Kalkaska.

Hughston sent three trucks to the farm, his wife said. On one he was told to load seventeen head of cattle for Kalkaska. On another small truck he was told to take two animals to Michigan State University for research. After some delay his driver was instructed to take the remaining thirty-seven animals—the best looking—not to Kalkaska, but to the cattle market at La Porte, Indiana, for slaughter.

Hughston ordered his driver to wait while he queried this, and he argued for hours on the telephone between Farm Bureau and MDA. He was suspicious that he might land in the same kind of trouble as his neighbor, Alpha Clark, had done when he sent sick cattle across state lines to Purdue University. Mrs. Hughston testified that the Farm Bureau official continued to insist that this load of cattle should go to Indiana and that "if you get into trouble we'll back you up." (There was derisive laughter when she related this comment at the hearing.)

"How in the world can we haul quarantined cattle for slaughter?" Bill Hughston asked.

According to his wife, an MDA official responded, "We will lift the quarantine temporarily." By that time the argument had gone on for thirteen hours. It was 9 P.M., too late to get to La Porte for that day's market, so Hughston's truck was redirected to Coldwater, Michigan, about thirty-five miles south of the Fish farm. The driver followed this instruction "against our better judgment," Bonnie Hughston stated. There, she said, the animals were sold to a packing plant for human consumption.

Two days later, she added, her husband received written confirmation from MDA that the quarantine had been lifted on these thirty-seven animals. The final twist to this episode, Mrs. Hughston related, was that this notice on an MDA letterhead was mailed in a Farm Bureau envelope.

No one from Jim Fish's farm at Hickory Corners was at the hearing to comment on her story. But later Jim Fish's wife and business partner, Alice, confirmed it. "We had tests on all the animals that left the farm that day and they were all close to the tolerance level," she said. "If we had wanted to market those just under the level we could have asked for a release from quarantine, but it was not our wish to do so. I received a check for nine thousand dollars from a meat packing company in Coldwater. I did not know what to do with it. I felt like it was a hot check. Those animals were sold without our consent but they were long gone before I received the money. We felt very badly about it because it was not our wish that they went into the food chain."

As he listened to all the testimony, Lou Trombley decided what should be done. In a few days' time, the annual Farmers' Week would be held in Lansing. At each of the hearings he proposed to cheering audiences how they should celebrate the occasion.

"We'll take a truckload of dead cows down there. And we'll carry them into the governor's office. We'll take him these cows with the big old abcesses that's leaking all over the place . . . the cows that are going on people's tables right now. . . .

"We'll tell the governor we got a message for him. He either buckles up and listens to what we're telling him, or he can call out the National Guard because the farmers of Michigan is going to march on Lansing. And we're not going to stop. This is a centennial year. Our forefathers were a bunch of dumb farmers like we are, maybe, but by God they started this country and we're going to take it back. And we're not going to leave it to the chemical companies and Farm Bureau and the rest of those people. . . ."

The last of Lou Trombley's words were drowned by applause.

23

A Moral Judgment

THE farmers of Michigan had never attempted a demonstration such as this. They were unorganized, except through Farm Bureau, and congenitally opposed to making an exhibition of themselves. Many could not accept that it was an appropriate way to make their voices heard. It was not dignified, they were afraid of being discredited, and of violence developing.

"It was not the kind of thing we were raised to believe in," Clyde Clark's wife, Coralie, readily admitted. Their carefully bred herd, product of Clyde's long expertise in animal genetics, was badly afflicted by low-level contamination. All six members of the family had been ill—the four children with severe skin eruptions—and the scrapbook which Clyde kept so proudly, with all its snapshots of prize cows, was beginning to collect captions which would soon read like an obituary column: "Krishna died . . . Scarlet died . . . Rudy, calf aborted . . . Daisy, shot her . . . Dolly's heifer calf died . . ."

"We had to do something drastic to bring this to the attention of the people," Coralie explained, "and unless the farmers worked together as a group nobody would listen to us."

On March 20, 1976, the Saturday before the Lansing demonstration, dissident farmers made their first attempt to form a statewide organization. It was the direct result of the Albosta hearings and took the form of a mass meeting on the campus of Central Michigan University, Mount Pleasant. This was in the center of the state, and about three hundred farmers came from distant and scattered areas. During a day of speeches and debates they crystallized their aims: immediate lowering of the tolerance level from .3 ppm to .002 ppm (by this time, further advances in laboratory techniques made such a measurement possible); immediate settlement of all outstanding claims; complete testing of all livestock and removal of those, living and dead, which showed signs of contamination; a ban on shipment of meat and milk from contaminated stock within seventy-two hours; the instigation of immediate research on human health problems; and a full investigation by the legislature of the Departments of Agriculture and Public Health.

The farmers elected a committee to develop this program. It was evident that those who had done the most investigation of the human health problem were a group from the Upper Peninsula who had made an informal survey of families on quarantined farms. They had sent a questionnaire to more than five hundred farm families and had received some alarming replies, indicating far more damage than the state's Department of Public Health was reporting. Headed by a Sault Ste. Marie business man called Hank Babbitt, these volunteer investigators had persuaded their county medical society to take fat samples from people undergoing surgery. Analyses of the samples showed that PBB contamination had spread into urban areas. The volunteers had also started to test PBB levels in human breast milk. In all this they had the full cooperation of Dr. James Terrian, director of Chippewa County Health Department, who—as a direct result—incurred the displeasure of his boss in Lansing, Dr. Maurice Reizen.

Hank Babbitt's personal story was so startling that it should have alerted every doctor and consumer in Michigan. He had nothing to do with farming; his wife bought the family's food from urban distributors. "My herd was children, not cows," he said. Three of his four youngsters were outstanding athletes who consumed enormous quantities of protein. "We bought milk by the gallon, cheese by the thirty-pound block, eggs by the crate," Babbitt stated.

In late 1974, soon after the first highly contaminated meat and dairy food must have been sold, the Babbitts noticed their children's performance falling off badly. Scott, the youngest, then aged fourteen, was the hardest hit. He lost thirty pounds between November and Christmas. His hair began to fall out. His coordination became uncertain. His personality changed—"from open eagerness to surly withdrawal," said his father. Hospital tests failed to produce a diagnosis but, in March 1975, a photographer friend who had taken some pictures of contaminated cattle remarked, "My God, Scott has the same symptoms as those cows!" Hank Babbitt asked afflicted farmers about cattle symptoms, and was convinced.

He had experience in journalism and public relations, and he was an organizer. He soon found another non-farm family—the Hartley Coles—whose children were having similar health problems. Hank and

Donna Babbitt, Hartley and Sharon Cole managed to get MDA's "secret" list of quarantined farmers and sent out the questionnaire. "We soon realized that a lot of people's health had been affected, and we wanted to bring out all the facts," Babbitt said.

Correlating symptoms, they developed a clearer understanding than almost anyone else had of the human health damage already done by PBB. Babbitt made another disquieting discovery. Having realized that the worst period of public exposure to PBB was between the late summer of 1973 and the early summer of 1974—those nine months before PBB was identified in Rick Halbert's feed—Babbitt checked the records of Michigan's professional athletic teams. He knew that athletes would have been consuming more meat and dairy products than average people, and he was shocked—but not surprised—to discover that the performance of a great many dropped dramatically when food contamination must have been at its worst. He also had a conversation with a trainer for one of the Michigan sports teams. "He was concerned about the high incidence of joint problems and the lack of coordination," Babbitt said. "He suspected lead poisoning. Then he asked me about symptoms of PBB toxicosis.

"If, after that, any team managers did suspect PBB, none of them wanted to talk about it. Professional athletes are marketable commodities, and if it were suspected that they were seriously damaged, the management's business could have been ruined."

Hank Babbitt's information added another dimension to the developing farmers' movement. Babbitt was eager to join the fight. "I had three concerns," he explained. "My son; to stop the spread of this stuff; and to expose those responsible." When he attended the Mount Pleasant meeting, he had already drafted a legal complaint against the chemical and feed companies, the first by an urban resident, alleging more than a million dollar's worth of damages to Scott's health.

He persuaded many of the farmers of the direction their movement should take. It should concentrate on consumer education and political action. Nothing would be done, he pointed out, so long as PBB was thought to be a purely agricultural disaster. All the residents of Michigan must be made to understand that it was their problem too.

Lou Trombley had a different place in the movement. He was a militant populist. His demonstration outside the state's Capitol was a turning point for the farmers, the beginning of public awareness that they had something important to say. It was reported by most Michigan newspapers and produced strong visual material which local television stations could hardly ignore. The demonstration began around 11 A.M. on Monday, March 22 and lasted most of the day.

Michigan's Capitol is a handsome, white building with an unusually tall dome. Standing in the middle of a great green square, it dominates Lansing. Modern office buildings, set back from the lawns, surround it, and no one doing business in the city can avoid this central area. All roads converge near the Capitol.

Here, by the legislators' limousines, farmers parked their trucks laden with carcasses of diseased cattle. Some bodies were already decomposing. Ignoring police who attempted to move them, farmers paraded on the Capitol steps with picket signs.

Lou Trombley shot one of his sickest cows for the occasion. He chained her body to the front of his truck, and on the back loaded a coffin draped with the American flag to which he had pinned the slogan: "Is this our future—PBB?"

"Sure she stunk," Trombley said of the cow. "She stunk when she was alive. She was worthless. The police tried to run us out of town, but we stayed. We did not get unruly. I thought the governor would talk to us when he saw that we cared about not poisoning consumers, but he didn't have the guts to show up."

Not many people who were in Lansing that day realized why the farmers were demonstrating. "I kind of laughed," said a young attorney who worked for the legislature. "I thought they were a bunch of crazies. I did not understand what they were saying, not until long afterwards. But what they were doing piqued my interest a bit. They got a substantial amount of press play, and before that very few urban newspapers had written about PBB except for the *Grand Rapids Press.* Nothing in the western part of the state. This was the start of some really good inquiring reporting."

Immediately after his inflammatory speeches at the Albosta hearings, Lou Trombley received some anonymous telephone threats. The demonstration outside the Capitol had further consequences. As Trombley related, "A few farmers in my area and the operator of my local feed supply company called a 'Stop Lou Trombley' meeting. They wanted to shut my mouth and shut it permanent. I was hurting their market. They had a fair turnout, probably twenty farmers opposing me. I was not invited, but I walked right in for a few minutes, said my piece, then excused myself like a gentleman and left. They were more interested in the dollars they might lose than the damage they might do. But it fizzled out on them, because they never did stop me."

It was not in Trombley's nature to be temperate, not when justifiably aroused. He made a number of public declarations to the effect that the governor and officials of Michigan Department of Agriculture should be put behind bars. This went down well with some farmers, but did not help his credibility with the press. Carol Trombley lived in fear for her husband.

"I have been scared many times, and so has he," she said months afterwards. "But if you believe in something, you do what you have to do." Quietly supportive, Carol was embittered by the slander that was spread about her family. "We had neighbors who said we just wanted to get on the band wagon; that we wanted to make money. Well, if there is money in this I have yet to see it, and if this is a band wagon, I want off. We were banging our heads on a wall trying to help everyone in this state. And when you lose friends you have had for seventeen years, it hurts."

More than two hundred miles north, Hank Babbitt was campaigning in a different way. Seeing the need for political action, he and Mark Hale—a beef farmer who had been working with Babbitt's group—decided to lobby Morris Udall, Democratic candidate for the presidential nomination. On May 17, 1976, Udall was campaigning at Marquette, 180 miles west of Sault Ste. Marie. Babbitt and Hale drove there, extracted a promise from Udall that he would support the farmers' cause, and began the lonely drive back.

At 12:45 A.M. on a deserted stretch of road, crossing swamp land, they were struck by a car traveling in the opposite direction. Babbitt was convinced it veered to their side of the road as it approached them. "Mark was driving his father's pickup truck," he stated. "We were both thrown out, and it rolled over us. I dragged Mark away from the truck, because there was gasoline running all over the place. One of my early thoughts as I staggered up and down the road, looking for help, was the recollection of Albosta's warning to me, 'Watch out for hit-and-run drivers.' Yet it seemed inconceivable that anyone could do this deliberately."

At 3:15 A.M. the first car came by. It contained a couple who had been camping. They had blankets and lanterns, and by 4:00 A.M. they were able to get an ambulance. Hale and Babbitt spent weeks in the hospital and were effectively removed from the anti-PBB campaign just as it was getting organized. Mark Hale had dislocated shoulders, three broken vertebrae, and head injuries. Hank Babbitt suffered a dislocated shoulder, two broken ribs, and ruptured kidneys. The driver of the other car was found some distance down the road with a broken leg.

"He was not badly injured," Babbitt stated. "A policeman who came to the scene made a strange remark. He said, 'He almost missed you.' Perhaps he wondered whether this driver had tried to run us off the road but collided with us instead. The driver was fined, but no charges were brought. I signed a complaint, but by that time the state police could not find him to serve a warrant. It was all very strange.

"The car was registered in Florida, but the address on the registration did not exist. The driver was a man of twenty-eight who said he had moved to Marquette for a job in the graphic arts department of Northern Michigan University. A lot of camera equipment was in his car. Later, a check with the university showed that they had not even received a job application from him, and he could not be traced."

Seven months passed before Babbitt was fit to return to work, and by then events had overtaken him. Recuperating together, he and Hale did collect some interesting evidence in October 1976. "We took a flight heading south from Sault Ste. Marie and I took aerial photographs. It was after the leaves were off the trees and before the first snows, and I photographed quantities of dead cattle which farmers had hauled into the woodlots at night. This was illegal, of course. You could see the bones strewn and the carcasses stacked. Only three farms in Chippewa county were ever quarantined, but many more were contaminated. This was how some of those farmers got rid of their problems while they stayed in business.

"You hear about the thirty thousand animals buried at Kalkaska, but that was the number destroyed by the state. There's no way of telling how many thousands more were buried on farms, or left to rot out of sight."

Shortly after this Babbitt was directly affected by another serious accident. His million-dollar lawsuit, on behalf of his son Scott, was in the hands of an Upper Peninsula law firm, Wisti and Jaaskelainen, which was handling several similar cases. Andrew Wisti, senior partner, had assigned a twenty-seven-year-old attorney, Kermit Bryant, to do preliminary research. Bryant worked on this intensively for several months. He often stayed on the job until very late at night and, driving home at two o'clock one morning, he swerved into a tree and was killed.

Inevitably there were rumors that the steering mechanism of his automobile—which was a new one—might have been tampered with, but Wisti had the vehicle examined and nothing suspicious was found. Bryant's widow, Bonnie, revealed that her husband had been blind in one eye and sometimes misjudged distances. He had worked an eighteen-hour day and was exhausted. Apparently he swerved to pass a stationary vehicle, pulled out too far, and slithered in mud at the side of the road. That, she believed, was how the accident happened.

Nevertheless, rumors persisted. Babbitt became even more uneasy when Wisti dropped his case. "Bryant had devoted a lot of time to it and I couldn't replace him. It takes a special breed," Wisti explained.

Babbitt never ceased to wonder whether there was some connection between his accident and Kermit Bryant's, or between Bryant's accident and Wisti's withdrawal. "It could all have been coincidence," he said. "But you could drive yourself crazy thinking about it."

While Hank Babbitt recovered from his own accident, the PBB Action Committee was formed without him. It was only a shadow of what it might have been and its effectiveness was diminished by the well-intentioned, but misdirected, tactics of its leading officers. They were chosen from the farm community which had been brought together by the shooting of Al Green's herd—moral and courageous people, but politically inexperienced. Hilda Green and Louis Trombley headed the organization. Patricia Miller, a young woman who had recently started a group called Consumers for Clean Food, was elected secretary. Pat—wife of a beef farmer whose animals had been contaminated—was the quiet, organizing type who followed legislative action, promoted lobbying, and produced a newsletter which kept farmers apprised of developments on the PBB front. Hilda and Lou were effective, although at times intemperate, public speakers. For months the PBB Action Committee bound the farmers into an active, cohesive community whose voice was heard whenever an opportunity came to testify about the effects of PBB. Some of its officers went to Washington to talk with White House officials and to appear before Congressional committees. However, the committee did not make the impact it deserved, and it failed to impress most newspaper writers and politicians.

At one point Pat was responsible for the Action Committee's publication of a recommended diet for the people of Michigan, based on the premise that everybody in the state already had a certain amount of PBB stored in the tissues and should avoid adding more. It recommended substituting vegetable fats for butter; cutting out cream, egg yolks, liver, hamburgers, and all processed meats; and using low-fat milk. Only cheeses made outside Michigan were recommended. It was a damning indictment of the Michigan dairy industry, but many people followed the advice.

MDA, on the other hand, was trying to restore confidence in Michigan agriculture, and at this crucial time produced evidence which convinced several legislators. After five months of studying the remnants of Dick Edington's herd, MDA and MSU concluded that the animals had not suffered from PBB contamination, but from respiratory diseases, parasites, and malnutrition: in other words, mismanagement. They found only trace levels of PBB in three of the animals, but made much of the fact that they were all in very poor condition when the study began.

Dick Edington's side of the story was not represented. His veterinarian insisted that almost a year previously, ten animals had detectable PBB levels (according to an independent laboratory), although MDA then found none. Edington had replaced the sickest cattle and said that he doubted whether these thirty-odd newcomers to the herd were contaminated. However, all the animals suffered from more than four hundred miles of truck travel around Michigan before MSU began its experiment. Edington was not surprised by their poor condition.

Dale Ball took a bus load of legislators to the experimental farm to show them how good Dick Edington's cows looked after proper care. Their city shoes protected by plastic bags, the legislators picked their way through barnyard mud and were suitably impressed. Ball's office supplied them with photographs showing the poor state of some of the animals before the experiment began. A group of more appropriately clad farmers, led by Donald Albosta, showed up uninvited and staged a counter-demonstration. They alleged that the herd was still "loaded with PBB" (Albosta's words) but had been doctored up to look good. Louis Trombley, ever to the fore, handed Dale Ball a glass of milk and urged him to drink. Ball refused. He said he was fearful there "might be things in it which weren't in it when it came from the cow." Recently, he added, he had received telephone threats on his life.

Nothing was really proved by this experiment. Dick Edington was generally acknowledged by his peers to be a decent, hard-working dairyman who knew his business. He had been dreadfully humiliated, and scores of farmers were outraged. All this time they had been told that MDA had no research funds to help them, yet for five months the agency had supported an experiment which seemed designed to prove them inept.

Not long after this, Farm Bureau Services offered to take a few of the sickest cows of Roy Tacoma—a neighbor of Garry Zuiderveen's, who had a pending claim—to determine what was the matter with them. Tacoma refused. "I wasn't going to let them do an Edington on me," he explained. Every dairy farmer in Michigan knew what he meant.

Prodded by Governor Milliken (who was being prodded by farmers and their attorneys), Farm Bureau Services settled its $270 million damage suit against Michigan Chemical out of court. It was widely reported, and never denied, that the feed company accepted between $18 and $20 million—"a mere pittance," in the opinion of Assistant Attorney General Harry Iwasko. But with many farmers almost destitute, Iwasko felt that Farm Bureau was forced to accept what the chemical company offered.

Some farmers disagreed. They thought that Farm Bureau continued to misjudge how many of them were contaminated, and that it had mistakenly estimated that $18 million was more than enough to pay all outstanding claims. Thus far, $25.5 million had been paid to about 335 claimants. Under the terms of the settlement, a new "pool" of cash was created with Michigan Chemical paying a matching share. But that was all the compensation money there ever seemed likely to be. The feed company had signed away its rights to sue Michigan Chemical again. And the corporate structure of Michigan Farm Bureau and its subsidiaries was so designed that even if the feed company were made bankrupt by farmers' claims, the assets of the parent organization were protected.

However, some money was now freed to pay quarantined farmers who had waited for months without relief. One of the first to be offered a settlement in this new round of payments was Garry Zuiderveen.

The timing was maddening. For almost a year the Zuiderveens had been unable to get a cent of compensation. Now, after months of unusual scientific research, his attorneys had collected enough evidence about PBB for a court trial. They were eager to go through with this, believing that a favorable judgment would pave the way for quick and fair settlements of their other outstanding farm claims. Garry Zuiderveen's would have been a good test case. His farming methods were superior. His records were impeccable. His personal integrity was widely acknowledged. His manifestly honest face and natural candor would have impressed any judge. "He was my ideal farmer," said Gary Schenk, who, by this time, had learned a lot about agriculture.

The attorneys had one serious concern. Garry's health had been affected by the strain of his situation and—they felt sure—by PBB contamination. He had had trouble with his vision, with his perception of distances. At times he looked pale and shaky, despite his muscular build and natural tan. Lois's health and that of their three children had also been affected: they had suffered joint problems, lack of energy, depression. All of them felt better than they had a year earlier, and none of them liked to talk about their health, but Gary Schenk and Paul Greer worried whether Garry Zuiderveen was well enough to go through the strain of an important trial. The attorneys were therefore disappointed for themselves, but relieved for their client, when a last-minute settlement was offered.

The offer was adequate for Garry Zuiderveen and his nephew to start new herds, but it did not cover all their losses. By now their quarantine had been lifted, and Garry was left with 178 unprofitable dairy cows which he could legally have sold for slaughter. At the time, they would have fetched between $50,000 and $75,000.

Garry needed that money. Soon he planned to take his son into partnership, and to give Garry Junior and his new wife the big farmhouse which he and Lois had taken over from Garry's parents. A generation earlier, the old couple had moved out and built themselves a smaller home next door. Before long it would be Garry's and Lois's turn to move, and then a third house would be put up on their land. Lois was looking forward to this. She had plans for an efficient ranch house which would serve into their old age, one with a view across the fields to the homes of the last and next generations. It was part of the predetermined pattern of their lives, hers and Garry's, and there was security in knowing that the farm would eventually pass to their son, and on to a son of his, yet unborn. Farm people marry young and must establish themselves early. It was therefore natural for Garry and Lois, in their forties, to be thinking about a partnership with their son, twenty years before their own retirement.

The money raised from sending the cattle to market would cover the cost of building the new house. "In order to repay him for all he had lost, it became almost necessary for Garry to sell those animals," said Gary Schenk. "We hoped the defence would take them as part of the settlement, and have them humanely destroyed. But they said we could keep these cattle and they would pay us so much less. The clear implication was that Garry could raise the rest of the money by selling them for beef.

"Most of Garry's remaining cows had signs of toxicosis, even though they were no longer quarantined, and he was convinced that they carried a body burden of PBB and would be a danger to public health if slaughtered for human consumption. Those were the issues on which we were prepared to go to trial.

"All these months that Doc and I had been collecting scientific information, we had tried to share it with the farmers as it came to us. We had met with our clients constantly, so that among other things they could understand what was happening to their animals. They also developed an understanding of the nature of the chemical and the effects it could have. Garry Zuiderveen had a particularly good grasp of this.

"He asked me what were his options. I told him he had three. He could keep the animals and try to work with them. He could sell them. Or he could shoot them. I did not make any recommendation.

"He came to my house on the night the checks were signed and talked about those cows as though they were old friends. He said he could not bring himself to shoot them. I came to the conclusion that this wily old Dutchman was not going to give up those cows if he could make a few bucks. He is a very good business man. I thought it a foregone conclusion that he would sell them and I was very, very disappointed. Yet Garry Zuiderveen was so important to me as a friend that if he had said to me that the only way he could manage financially was to sell those cattle, I would have said yes. But he did not ask me that. He only asked me about his options.

"Later that evening Doc arrived for dinner. He felt the same way as I did. Just crushed. We had no right to tell Garry what to do but we knew it would be a very damaging blow to the morale of the community if he sold those animals. And it just shows you how we of little faith can be so easily lost because later that week Garry called me and said, 'It's done.'

" 'What's done?' I asked.

" 'The cattle,' he said. 'I shot them all. We buried them on the farm this morning.'

"I felt like I had been reborn. The interesting thing is that the defense had hoped he would sell them so they could say, 'That shows you how these farmers are when they get a little money in their pockets; that's how serious they really think the problem is.'

"The same concerns motivated Al Green."

"It was the most difficult decision I ever made," Garry Zuiderveen reflected. "I could have taken that money and had a brand new house, debt free. And MDA was encouraging me to sell all those animals that were below the level. There was a time when I might have done that, but not after I learned, from the scientific evidence that Doc and Gary had gathered, how the poison could persist and affect people.

"It was April 27, 1976, when we destroyed the last animals left on the farm, and it was a dark day in my life. I shall never forget it."

Garry knew that he would have to bury them on his own land, and because of the warm weather he would have to do it immediately. He got approval from his county health department to dig a pit a mile north of his house. "It was twelve feet deep and half a mile from the nearest building," he said. "The surface water would be no problem because it was solid clay below the sand, so we figured it was as good a burial site as we could get.

"On April 26 we destroyed about seventy of the young calves, right there in the pasture. They all died within fifty yards of one another, and they did not run or get scared. Our neighbors came to help us. On the morning of the twenty-seventh we hauled the milk cows from the barn to the burial site on three stock trailers, and put them in six or eight at a time. Within twenty seconds after they were unloaded we shot them with high-powered rifles. This finished them instantly and they did not suffer.

"As they died we pushed them in a corner with a bulldozer. Then we slit the rumen cavity of every animal so the gas could escape. We hauled one hundred and eight cows out that day and were done by two o'clock in the afternoon. I remember shooting thirty-five and putting them the other side of the silo. My Dad would not look at them. Tears running down his face, a man of seventy-eight.

"It bothers me to talk about this. I could just sit here and cry. Those fellows don't know what they put us through. We should never have had to kill our own cows; we were too emotionally involved. I remember opening the barn door and letting the last six animals out. My son Garry shot the last one, and as he did so a tear ran down his face and he said, 'Dad, we're done.' He was shaking. He had not shaken all day. He was nineteen years old and he had taken it like a man.

"If we had not been religious people we could not have made the decision we did and held ourselves together. I take no credit. It was the bringing up we had. In essence we are all our brother's keepers. I knew that, from the information we had at the time, it was the only decision we could make and still face ourselves in the mirror."

24

Politicians Versus Scientists

GOVERNOR MILLIKEN was growing uneasy. Guided by his Departments of Public Health and Agriculture, he had insisted for almost two years that there was no risk to human health, but evidence of low-level damage to herds could no longer be ignored. Were those animals really safe to eat? Were the people of Michigan running the risk of future cancer? Was the PBB guideline low enough to protect those who already had a build-up of the chemical in their bodies? Or were the low-level farmers trying to find an excuse for their own ineptitude? Who was right: they or the agricultural establishment?

By-passing the two state agencies on whom he had relied, the governor appointed a panel of six independent scientists to consider these questions. They were to examine the evidence of damage to low-level herds and to recommend whether this justified a change in the .3 ppm tolerance level. The current FDA guideline applied to the entire country, even though it seemed irrelevant outside Michigan. The governor wanted to know whether the contamination warranted more stringent regulations within the state. It was not a simple issue of public health. There were economic considerations. Another reduction in the tolerance level would mean yet another round of farmers' claims. Also, the whole image of Michigan agriculture—not just the dairy industry—could be harmed irrevocably, only months after the state's automobile industry had been seriously crippled. It was an impossible dilemma.

The scientists were to examine only the health aspects, but even the choice of panel members became a political issue. The select group to be named included Dr. Walter Meester and Dr. Lynn Willett.

Meester was an obvious choice because he was the only toxicologist in the state who had direct experience with PBB-contaminated patients. Willett was the dairy scientist from Ohio State University who had just completed a controversial study—one which found nothing seriously wrong with cattle who had ingested low levels of PBB. Meester objected to Willett's presence on the panel, revealing that his herd study had been heavily financed by Michigan Chemical and Farm Bureau Services—a piece of news which was gleefully seized upon by the affected farmers. Willett reacted by objecting to Meester's appointment. Meester's knowledge would have been useful to the panel but he was obliged to withdraw, along with Willett.

None of the members of the governor's commission, as eventually constituted, had had any experience with PBB. Perhaps this was an advantage. They could not be accused of bias, and all began the inquiry with open minds. Dr. Isadore Bernstein became chairman. A scientist with an international reputation, he was professor of biological chemistry at the University of Michigan's Medical School in Ann Arbor, and professor of environmental and industrial health in the university's School of Public Health. All other members of the governor's panel came from outside Michigan and were internationally recognized experts in their fields: pathology, physiology, toxicology, pharmacology, veterinary toxicology, and cancer research.

They met several times over a period of six weeks. One specialist the panelists consulted was Dr. Renate Kimbrough from the federal government's Center for Disease Control. She reiterated what she had told Doc Clark and Gary Schenk: her animal experiments with the closely related chemical PCB had produced liver nodules which were generally acknowledged to be precancerous. There was some scientific hair-splitting as to whether this particular type of nodule could positively be identified as a precursor to cancer, but the overwhelming opinion on Dr. Bernstein's panel was that it could be. (Subsequent research by Dr. Kimbrough reinforced this conclusion.)

Prior to this, Dr. Thomas Corbett was the only medical expert who had talked publicly about PBB and cancer in the same breath, and nobody in authority had listened to him. But the Bernstein panel was unanimously persuaded. "We rationalized it like this," its chairman explained. "On the basis that PCB was a carcinogen in animals, and that the PBB mixture put out by Michigan Chemical was understood to be five to ten times as toxic as PCB, it was a reasonable presumption that PBB would also be a carcinogen. It was also hard to believe that you could contaminate milk, and not contaminate people with a substance which is sequestered in the body fat."

Isadore Bernstein presented his panel's report to the governor on May 24, 1976. Although Bernstein made the laconic remark that the six weeks spent on the study "was an indication, not of the speed of our work but of the absence of information about PBB," the panel carefully considered all the known facts about the chemical's potential effect upon humans: It was highly persistent in the environment. It accumulated in body fat. It was very slowly excreted, and most effectively in mother's milk. It crossed the placenta to the fetus. Exposure to low levels of PBB over a long period could be as serious as short-term exposure to high levels, and a consequence of both might be cancer.

Cautiously the panel noted that at present there was no evidence of a specific disease caused by low levels of PBB. However, the thrust of the report was on the long-term dangers of adding more PBB to the tissues of people who already carried a body burden of the chemical. It therefore recommended to Governor Milliken that Michigan should impose its own tolerance level, as close to zero as possible.

Since a zero guideline was unrealistic, the panel suggested new levels which, with sophisticated equipment, could just about be measured: .005 ppm for meat and .001 ppm for milk (five and one parts per billion). This meant a tolerance level which was sixty times lower than the current .3 ppm for meat, and three hundred times lower for milk. The panel urged that these new guidelines be constantly reviewed, with the aim of making them lower still as measuring techniques became more refined.

Governor Milliken's sense of shock is not on record, but it must have been profound. Whatever he had expected the panel to recommend, it could not have been as stringent as the reality. Only three weeks earlier he indicated that what he had expected from the scientists was a finding which would "help restore a sense of credibility and public confidence"—meaning, presumably, that they would find the existing tolerance level to be adequate.

"There have been many rumors and half-truths . . . but it is our very strong opinion that the guidelines which we are currently using are guidelines which are designed to protect human health and safety," the governor had affirmed. He was answering citizens' questions in an interview program broadcast in five Michigan cities and expressed confidence that the state had the PBB situation under control.

In the same interview he praised the experience and dedication of the scientists on his advisory panel, and emphasized the "great deal of homework" they had done on PBB. He could hardly ignore their recommendations after that, although his office was prompt to point out that in addition to the 30,000 cattle and 1,500,000 chickens which the state had already destroyed, another 736 farms would have to be quarantined. This would probably mean killing an additional 19,000 cattle, 55,000 chickens, and 400 swine. Although troubled about the economic consequences, the governor felt bound to endorse the scientists' report.

This meant a reversal of his position, but, fortunately for him, others would have the responsibility of deciding how far the Bernstein panel's recommendations should be carried out.

Milliken's choice was simple. He could ask his Department of Agriculture, as the state's consumer protection agency, to call a public hearing at which this and other relevant testimony would be heard—the same kind of hearing which had been held a year earlier. Afterwards, on the basis of all the evidence, Michigan's policy-making Commission of Agriculture would recommend a figure for the tolerance level. The alternative was to let the legislature vote on it. There were risks to the latter course. This was a highly controversial subject, and a decision would inevitably be delayed by political debate, with small chance of a vote before the summer recess. So the governor decided on a public hearing.

Plans for a hearing were discussed around the governor's conference table immediately before release of the Bernstein report at a press conference. Isadore Bernstein and his fellow scientists were all there. So were Dale Ball and some of his senior staff members from MDA.

Bernstein stated that he had a professional committment to go to Japan for a month in the near future. He gave the dates and said he would be available to testify at any other time. If the hearing had to be scheduled during his absence, any of the other five panel members could be called upon, he added.

The public hearing, with Dale Ball again presiding, took place in Lansing on June 10. Isadore Bernstein was in Japan. The first Bernstein heard of it was when he returned to Ann Arbor and found the outdated invitation on his desk. He was appalled to discover that no one from his panel had been present to explain its report. This report was merely entered into the record of the hearing without comment.

"Subsequently I wrote to each member of the committee, and learned that one was called by phone from the governor's office and one was contacted by letter before the hearing. Neither of them realized that if they did not attend, the panel would be unrepresented. The other members said they were not contacted," Bernstein related.

Legislators sympathetic to the farmers felt strongly that the session was deliberately scheduled at a time when Bernstein would be out of the country. Other scientific witnesses were called, almost all opposed to lowering the tolerance level. Most of them lacked direct experience with PBB, and most used the same arguments: although it was theoretically possible to test down to the very low levels which the Bernstein panel recommended, it was impractical to do so on a mass scale. The refined laboratory equipment was lacking; the margins for error increased as the level of PBB diminished; the cost would be prohibitive. The presumption that PBB was a carcinogen was questioned. It did not necessarily follow, some witnesses argued, that PBB and PCB would have similar effects upon human health. Some speakers raised doubts as to whether PCB really did cause cancer in humans.

Dr. M. L. Keplinger, a toxicologist from Industrial Bio-Test Laboratories—a group from Northbrook, Illinois, retained by Michigan Chemical to do toxicology studies of PBB—carried this argument to an extreme. In her PCB experiments, Dr. Kimbrough and her co-workers had found that the chemical induced cancer when she used a PCB compound on a strain of female rats. "They don't even know if they get the same effect in male rats," Keplinger complained.

He and other scientific witnesses felt that PCB and PBB were human health hazards only in high doses, as in the the *yusho* incident. Most people in Michigan had been exposed to very low doses, it was argued, and were therefore not in danger. The fact that both compounds were known to accumulate in the tissues was barely mentioned. The unfair comparison was made between PCBs which had been widespread in the environment for forty years, apparently without giving anyone cancer (although who knew?), and the three-year exposure of Michigan people to a much more intensive contamination of PBB.

"PCBs are in everybody's fat in this room and I don't know of a case of harm it has done," said Dr. Frederick Coulston, professor of pharmacology, pathology and toxicology at Albany Medical College, New York. His presence at the hearing was unexpected. His credentials were impressive. He had served since 1961 on the Expert Committee on Food Additives and Pesticides for two United Nations Agencies—the World Health Organization and the Food and Agriculture Organization. His dogmatism was therefore surprising.

Describing the Bernstein panel's conclusions as "scare tactics," Coulston urged Dale Ball (who needed no urging), "Do not listen to those who tell you that doomsday is going to occur and we will all die of cancer from this compound. This is nonsense, pure speculation of the highest kind."

After he read the hearing transcript, Isadore Bernstein came to the phlegmatic conclusion that "it would not have made one iota of difference if any members of our panel had been there. There was such a tremendous amount of evidence from the other side, and apparently much importance was given to this scientist from the World Health Organization who said that our analogy with PCB wasn't valid, and that cancer in animals does not necessarily transpose to man.

"If we had had the data on PBB carcinoma which Dr. Kimbrough produced more than a year later, it would have been a lot harder for all of them to argue."

"My impression was that the hearing was pretty well stacked," commented Representative Francis Spaniola, a Democrat who had served on Albosta's committee. "In my opinion, Farm Bureau and MDA are pretty much the same outfit, the ties between them are so strong. And the governor, I think, did not really want the tolerance level to change. It presented too many difficulties. Early on, the governor told me that he did not think PBB was much of a problem, and that the press was attempting to blow it out of proportion. In fact the press was not doing the job it should have been doing.

"I don't think the governor was an uncaring man. He believed what his advisers at MDA told him. He was hoodwinked by his own people. Then he was in such a deep trap, how could he pull himself out?

"As for the people of Michigan, they got the information too late. If they had known they were being poisoned they would have risen up in indignation, but when they eventually found out they had already ingested the stuff."

Aside from the Bernstein report, which Coulston and others did their best to discredit, little information was at hand about PBB's effect on public health. It would have been reasonable to expect impartial guidance from the federal Food and Drug Administration, which had set the existing tolerance level, but FDA witnesses at the hearing had their prejudices too. The FDA's Washington headquarters relied upon its district office in Detroit for information on Michigan's PBB contamination. This office worked with MDA, performing similar functions of monitoring public food supplies. They consulted frequently to avoid duplicating food inspection. "It is a very cooperative relationship," an FDA official in Detroit said. So when MDA staff argued that a PBB guideline of .3 ppm was adequate and that farmers who complained of sick cattle at lower levels might be finding excuses for their own incompetence, their friends at FDA echoed the theory. Also they were not anxious to have their judgment on the safety level proved wrong.

Dale Ball's view, expressed some time after the hearing, was uncompromising. "I felt that .3 provided adequate protection, and I still do," he said. "The state law says that we follow the federal guidelines unless, after a public hearing, the necessity for lowering them is clearly established. It wasn't. The Bernstein report was based on a belief that the level should be as low as you could possibly test, the lower the better. But we have never heard of a poison where such a low level would hurt you. Many herds that appeared healthy were killed and buried, and I think a great many were killed needlessly. Aside from the animals that had very high levels of PBB, I doubt there were any more herd problems than you would have had with that number of cattle in normal times."

Alan Hoeting, director of FDA's Detroit division, shared this view and passed it along to his superiors in Washington. One of them, Dr. Albert C. Kolbye, the FDA's associate director of science, made the trip to Lansing to testify at the MDA hearing, as he had done a year earlier. He acknowledged that very low levels of PBB were probably still in some food, but only to the extent that "you may have a one in five hundred chance of eating a meal that might have some detectable residues."

Dr. Kolbye stated, "FDA believes that there is no scientifically sound reason for us to modify our current guidelines for polybrominated biphenyls in food. The position is, we believe, totally consistent with FDA's mandate to protect public health."

If the tolerance level were lowered, he argued, entire herds would have to be destroyed. "This could only be avoided by sampling each animal involved, which would impose a tremendous burden on the already strained laboratory resources of the state of Michigan." It could have been an MDA official talking.

Leaders of the state's agricultural establishment took up this theme. Dr. Donald Isleib, chief deputy director of MDA, stated that if all cattle and milk in Michigan were tested at the levels recommended by the Bernstein panel, it would cost about $733 million for the testing and $325 million to cover the farmers' ensuing losses. Even if this kind of money were available, the testing equipment wasn't. Speaking for Farm Bureau Services, Donald Armstrong announced that the feed company could not pay compensation to farmers for animals below FDA guidelines. If the state of Michigan adopted a more stringent tolerance level, it would have to take up the additional liability.

Elton Smith, president of Michigan Farm Bureau, spoke at length. The Bernstein recommendations were neither necessary nor defendable, he maintained, and could ruin Michigan agriculture. "Farmers from all over the state have reported to me that they were appalled at the conclusions and recommendations given by that panel," he said. Farm Bureau had ensured that a number of appalled farmers would be at the hearing, and they applauded his statement.

Dr. J.A. Hoefer, acting dean of the College of Agriculture at Michigan State University, reinforced the arguments from Farm Bureau and MDA. "There are approximately ten thousand animals slaughtered daily in Michigan. The sampling and analytical problems [of a lower tolerance level] in the meat and animal area would be horrendous," he rationalized. "Acceptance of the zero detectable tolerance concept, in this very imperfect world in which we live, would be a disaster, the proportions of which I cannot imagine."

Rick Halbert spoke at the hearing, once more emphasizing that a lower tolerance level would put an impossible financial burden on the state. Although he never openly criticized them—as other agribusiness operators did—Rick was not popular with farmers suffering low-level contamination. Once they had admired his tenacity, but now they hoped he would crusade for them as he had crusaded for himself, and Rick wasn't willing. He had become a respected figure in state agricultural circles, and his word might have meant a lot. But he seemed unable to accept that contamination below tolerance level could ruin a herd. He had worked too long with chemicals, and his own farm interests were at stake. Not understanding this, they felt that if he couldn't speak publicly to help them, he might at least keep quiet. Rick did not function like that. After all, he had more knowledge about PBB than most.

A few dissident farmers were there, Lou Trombley among them, outspokenly telling MDA and the state Department of Public Health that "you've done a terrific job of covering this up, and that's all you've done." Dr. Thomas Corbett repeated his concerns about public health. There was a brief and moving statement from Dr. F.L. Graubner, a country physician who owned an interest in a Black Angus herd, recently released from quarantine. "The cows kept on dying just the same, in fact they had the same symptoms as the cows that were above tolerance," he said. It might take twenty years before anyone knew whether PBB was carcinogenic to humans, he added, "and because I made a pledge to preserve life and not destroy it, ladies and gentlemen, as a doctor, as a farmer, as a concerned citizen, I can not and will not foist any contaminated cows of mine on the unsuspecting public's dinner table."

Dr. Louis Blesch, a veterinarian with a practice which straddled the Michigan-Indiana border, made a dramatic point. Only his Michigan clients had PBB problems, he reported. "I have not had a single case of this type of toxicity or poisoning in the state of Indiana, not one. They've been solely confined to my Michigan clients who use Farm Bureau Services feed. For this reason, I seem to discount other influencing factors, such as iodine and so forth, that have been mentioned."

Of the four veterinarians who testified, one other—Dr. Susan Jacoby of Constantine—took the same line as Blesch. Two sided with the agricultural establishment (one of these, surprisingly, was Ben Heckhuis, who had tended Peter Crum's highly contaminated herd). Doc Clark did not participate. It was not his practice to make public statements. He knew he was more effective working in the background and, as he put it succinctly, "I ain't about to get into a peeing contest with Kolbye."

Dr. Kolbye spelled out how FDA saw its role in a crisis of this kind, and to many in the audience it was a revelation. It appeared to be governed by the statute book rather than by common sense. Kolbye cited the Delaney Amendment to the federal Food, Drug and Cosmetic Act, which stated that no substance known to induce cancer in animals or humans could be *intentionally* added to food. There was no intent to add PBB to Michigan's food supply, he explained: it happened by accident. Therefore the Delaney Amendment did not apply, and FDA could not ban PBB-contaminated food below a guideline which it (the FDA) had already set.

This left no doubt that if Michigan lowered the tolerance level, Michigan would have to pay the price.

The result of the hearing was a foregone conclusion. Dale Ball recommended to the Commission of Agriculture, and the commission recommended to the governor that the tolerance level could safely remain where it was. Scientific testimony at the hearing was predominantly in favor of this, Ball said—quoting out of context one sentence in the Bernstein report, "No specific disease or symptomatology in animals or man can presently be associated with exposure to low levels of PBB." The intent of the report was thus turned around and, for all practical purposes, Isadore Bernstein's group need never have met.

"It was incredible to me that these top flight scientists could look at the entire volume of work done on PBB and related chemicals, make their recommendations, and have them totally ignored," commented Fred Fry, a research analyst in Speaker Bobby Crim's office. He attended the hearings with Edith Clark, also on Crim's staff.

"Edie and I were so depressed at the end of that day that we went to have a pizza together," Fry recalled. "Several scientific witnesses had argued that there was no accurate way of measuring a PBB level as low as the Bernstein panel recommended. Yet Dr. Douglas Rickert gave some fascinating testimony which nobody followed up. He was a pharmacologist at Michigan State University who said that his laboratory had developed a very specific method for measuring PBB as low as one part per *billion.* Edie and I went to his lab to have a look, and we took Spaniola with us. We were all convinced that it could be done but that MDA didn't want to know about it.

"After that there was no question in our minds that we had to do something to get the tolerance level lowered. Bernstein was so cautious and thoughtful, and his report gave such good reasons for doing so. But the only way to do it now was through the legislature, so Edie and I began working on a bill to bring the tolerance level down. We put it together quickly, because there was just a chance of getting it passed in that summer of 1976, before the legislature recessed for the elections. The Speaker never does introduce a bill, by tradition, and Albosta was about to run for Congress, so our office asked Spaniola to sponsor it. He was a competent legislator and we all agreed it was worth the fight to try and get it passed."

Having changed his position about the tolerance level, Governor Milliken made it known that he disagreed with his Commission of Agriculture's recommendation. However, he made no move to change the structure of the Commission. Its five members were directly appointed by him and he could have instituted proceedings for their dismissal, but he did not do so. Instead he attempted a compromise, recommending to the legislature that the guideline should be lowered from .3 to .1 ppm—nowhere near as much as the Bernstein panel recommended but enough, perhaps, to placate the critics. The costs to the state of bringing the guideline down to various different levels had been estimated, and the governor's proposal was probably as far as his budget department felt it could go.

Francis Spaniola's proposed bill went further. It called for a .02 ppm guideline and random testing of animals on the farms. The two were not really rival proposals but the governor's bill got no serious attention from the Democrat-controlled legislature, which left the Democrats to push Spaniola's bill. By this time it was July, and in about two weeks the legislature would recess for the election campaign; the Democrats would have to work fast.

Shortly before the elections a copy of Spaniola's measure went to the state Democratic party's advisory committee on agriculture, on which Representative Paul Porter was an influential member. Porter, with Donald Albosta, was co-chairman of the House Agricultural Committee and had been outspokenly concerned about low-level contamination. Now he faced a conflict. Up for re-election in a predominantly Republican district, he felt that the rural voters in his area were sympathetic to Farm Bureau. If the Democrats made an issue of lowering the tolerance level at this time, Porter risked defeat. It was important for his party to retain Porter's seat so, after some high-level discussions, the Spaniola bill was dropped. Some Democrats who knew about this were livid, but kept silent for the sake of party unity.

Both Porter and Spaniola were re-elected by large majorities. Porter's political actions thereafter seemed to be more reflective of Farm Bureau's interests than of his earlier concerns for low-level farmers. Albosta ran for Congress and was defeated, which put him out of political office until he won a congressional seat in the fall of 1978. Porter was defeated in 1978 and returned to being a full-time dairy farmer. In the same election Spaniola won a third term—all of which may say something for the ultimate wisdom of the electorate. But in 1976 the politics of PBB were such that there was no more chance of getting action out of the legislature than out of MDA. Michigan Farm Bureau, mobilizing its loyal members into an intensive lobbying campaign against a lower tolerance level, had the satisfaction of success on both fronts.

While all this was happening, Edith Clark was making headway in another direction. She had worked for Speaker Bobby Crim since January 1975 when she had walked into his office looking for work, a victim of the automobile industry's recession. Laid off from her job in the public relations department of Oldsmobile, she had heard that Crim was adding to his communications staff. Edie was a bright young woman in her late twenties, the classic image of an all-American college graduate: blonde swinging hair, a fresh face which looked "finished" without make-up, alert eyes behind fashionable glasses. Energetic and articulate, she had a quick understanding of political processes.

Crim's method of operating was to assign members of his staff to research and inform him about political issues so that, where necessary, he could develop legislation. He asked Edie to pay particular attention to the House Agriculture Committee; nobody else on his staff knew much about it. She had grown up on a farm on the shores of Lake Michigan, loved the countryside, and welcomed the assignment.

She had barely acclimatized herself to her new job when the state's Department of Public Health published its report finding no health effects on families from highly contaminated farms—the study in which, as Walter Meester pointed out, the control group was also contaminated.

"I thought, 'My God, this is bad. This is a bunch of very sick people and someone is taking a calculated risk with them,'" Edie related. "Over the summer of 1975 I kept thinking about that report. Then the governor vetoed Albosta's low-interest loan bill, and all the other political maneuvers happened. I told Bobby Crim that I thought the PBB issue was so important that I should work on it full time, and he agreed." Bobby Crim was not a dedicated environmentalist but he was an astute politician. He excelled at predicting which issues would become important to the legislature and where the Democrats could pick up votes. By summer, 1976, he had no better understanding of the PBB situation than most Michigan politicians, but Albosta's committee hearings, which he had done so little to encourage, had made him more aware of the extent of the problem. He therefore agreed that Edie Clark should be encouraged to dig into this controversy. She was relieved of her other work so that she could go where she wanted, see whom she needed in order to build up a dossier on the effects of PBB.

Edie logged thousands of miles, driving the length and breadth of Michigan. She visited farmers, heard their stories, got to know her namesake, Alpha Clark of McBain, and assimilated all he had to tell her. She inspected ravaged cattle, and listened to the chronicles of strange ailments told by farmers' wives.

"When you get this deeply involved in an issue, it takes over your life," she said. "It was all I ever did or thought about. PBB Edie, they called me around the office. But I began to suspect that if I worked long enough and hard enough I would get to the bottom of it all."

As she began to investigate, she wondered why so little had been done to halt the contamination. Then she remembered, and kept on remembering, something a history professor had told her class years earlier: The reason there were so many fatal casualties in the American Civil War was because medical research had not caught up with weapons technology. She applied the analogy to chemicals. "We know how to grind these things out every day but we do not know how to deal with the effects."

She felt deeply for the farmers. "They are the salt of the earth," she said, "and they were suffering abominations at the hands of the state agencies. The governor had no communication with the farm community and relied on these agencies. But they had underestimated the contamination, then felt bound to cover up their error."

All this she told to Bobby Crim. She also talked to newspaper reporters. There had been no one at the legislative level from whom they could get a straight story before this, and Edie Clark's frank disclosures inspired several Michigan writers to go out on the farms and see for themselves.

Around the same time as the Bernstein panel met, a Republican legislator asked Edie if she knew of a source for information about the toxicity of PBB.

"There's a doctor, Tom Corbett, in Ann Arbor," she told him. "He could probably help you."

The legislator said that he would like a personal talk with Corbett, and Edie—who wanted to meet Corbett herself—asked if she might go with him. They traveled to Ann Arbor together.

Tom Corbett had just come out of an operating room when they arrived at the university hospital. The three went to the cafeteria to talk. Corbett told of his disappointment almost two years earlier, after he made the contact with Dr. Irving Selikoff in New York. Selikoff had been willing to bring his medical experts to Michigan at no cost to the state, Corbett related, and all he needed was an official invitation from the governor, but the governor's office had failed to act. Corbett was still upset about this.

Edie Clark had never heard of Irving Selikoff but, after a few questions, she grasped the importance of his offer. Her mind raced ahead. Suppose Selikoff could still be persuaded to come to Michigan to analyze the effects of PBB. Suppose Bobby Crim issued the invitation instead of Governor Milliken. It would bring help to the farm communities, show up the ineptitude of the state agencies, and be a political coup for Crim all in one blow.

Long afterwards Tom Corbett remembered that moment. "Edie lit up like a Christmas tree," he said.

The legislator seemed interested, but only mildly. He asked Tom Corbett to send him a memo about it.

"I got back to the office and I thought I was in a corner," said Edie. "I had been invited to see Corbett by a Republican, and maybe it was unethical of me—working for the Democrats—to get on the phone to Selikoff. Bobby was in Florida, so I couldn't ask him for a decision. I waited ten days, then felt I couldn't wait any longer. So I called Selikoff and asked if he would still come to Michigan, and he said yes. I drafted a letter under Bobby's name and waited for him to return from his vacation. We sent off the letter the day he got back. Then Bobby made his announcement that Selikoff would be coming to make the first comprehensive study of the effects of PBB on human health."

Later she asked the legislator why he had not contacted Selikoff first, in the ten days after their visit to Ann Arbor. "I was waiting to receive my copy of the memo from Dr. Corbett," he told her.

Edie summed up briskly, "And that's how Republicans operate."

25

Unconventional Medicine

ALTHOUGH she had become acquainted with Dr. Irving Selikoff on the telephone, Edie Clark had no idea what to expect when she flew to New York to help him plan his visit to Michigan. She visualized a one- or two-hour talk with him, or with one of his subordinates, finishing in time for her to fly back to Lansing that evening. Her appointment was for Friday, July 2, 1976, the day before Bicentennial weekend—when the nation would stop to celebrate its 200th birthday—and Edie assumed that everyone on the Selikoff team would be going home early.

At Mount Sinai Hospital, she was directed to the Environmental Sciences Laboratory on the mezzanine floor and shown to Selikoff's office—a big, square room, so modern and clinical that it told nothing about the character of its occupant. Selikoff was seated at the head of a conference table, surrounded by some of the medical experts on his team. Affably, he motioned to her to join them. She expected to sit and listen, but soon realized that they expected her to talk.

"None of them had any firsthand experience of the incident and they wanted to know the symptoms the animals had displayed, and what kinds of complaints the farmers were talking about," Edie recounted. "I had been doing my own one-person investigation, and I was able to tell them about the medical problems. They kept asking me whether I felt the farmers would mind discussing their symptoms. They were really worried that rural people might not cooperate."

Edie told them, "That should be the least of your concerns. You will have no problems with the farmers."

"Where are we likely to have trouble?" they asked her.

"With the federal and state agencies," she responded promptly.

She was not sure they believed her. Most of their experience had been in industrial situations, investigating the health of factory workers. The unions were always helpful. This was their first intervention into a situation as politically tense and complex as the one in Michigan. Edie suspected that even Selikoff, old enough to be her father, did not yet appreciate the isolation of the afflicted, and the entrenched attitudes of those who should have been helping them and weren't. The farmers would probably be the only ones to welcome his investigation. The FDA, MDA, and the state's Department of Public Health were likely to regard Selikoff's research as a threat; it might uncover something important and obvious which they had missed. Edie explained this as best she could, fearing that she sounded overly emotional. She doubted whether she had succeeded in making her point.

The meeting went on through lunch (sandwiches around the table) and into the late afternoon. Edie was not able to leave for Lansing until the following day, missed her plane connection at Cleveland, and spent the Saturday night of Bicentennial weekend sleeping in the Cleveland terminal. In those wakeful, uncomfortable hours she worried about the assignment which Selikoff had given her. As she described it, "He wanted me to find a place where he could set up a clinic and bring in a crew of thirty-five to test a thousand people. It had to be conveniently and centrally located for farmers from all parts of the state, and because his federal funding was limited he could not pay rent. He wanted me to arrange a preliminary tour of contaminated farms for a group of doctors he was bringing in, and he wanted to meet the top people at the Michigan Departments of Agriculture and Public Health. He wanted me to appoint a farmers' advisory council, and to produce a group of volunteers who would work in his clinic. He wanted information on the PCB contamination of fish in the Michigan lakes; information on the potato crop. And these were just the assignments of one day."

She was overwhelmed by Selikoff's dynamic personality. A short rotund man with white hair, he bounced around his office on crepe-soled shoes, looking like Santa Claus in mufti. He loved to talk about his work, but the words which tumbled out of him—in the slightly nasal speech of a native New Yorker, toned to an unusual softness—were weighed and balanced, the sentences carefully constructed as though he had first polished them on paper. Although his repartee was immediate, everything he expressed had an essential place in his thought process. Edie had never met anyone quite like him. She left New York bewildered and admiring.

From then on she worked twelve hours a day organizing the farmers' clinic. Even for the Selikoff team, the size and scope of it was enormous. This was to be their first examination of whole family units which had been exposed to a chemical contaminant. It involved a painstaking new approach to epidemiology, far more complex than the standard study which had been done by the state Department of Public Health. Patients would be required to give detailed medical histories, running into dozens of pages, and to undergo physical examinations which included neurological, visual, and pulmonary function tests, as well as biopsies of their body fat to determine PBB levels.

Edie Clark soon became accustomed to Selikoff's 7:30 A.M. telephone calls to her home, with questions calling for immediate answers. She loved the challenge. "You cannot imagine what it was like working for him," she related. "He says, 'Build me a castle,' and he expects it to be ready to move in tomorrow." She was able to find him a perfect site for his clinic: an entire floor of the east wing of Kent Community Hospital in Grand Rapids. In more prosperous times the hospital administration had over-built and this section had stood idle for years. The Selikoff study of Michigan farm people depended upon volunteer patients and, despite the team's forebodings, 1,029 of them came forward, delighted that someone was doing something to help them at last.

It took four months to organize the clinic and schedule all the appointments, every one of them lasting four hours. It was put to Selikoff that it might be more politic of him to delay his mass testing until after Michigan's elections, or—as tactfully stated for the record—until after the harvest, when farm families would be more free. In the meantime Dr. Reizen's Department of Public Health, which must have felt it was being upstaged by the Selikoff survey, carried out a project of its own. It took breast milk samples from 108 women who gave birth in Michigan hospitals during August and September. These tests were taken at random among patients who, for the most part, bought their dairy supplies in urban supermarkets.

The results were a profound shock to the health department. In Michigan's lower peninsula, where farm contamination had been widespread, 96 percent of the women had detectable levels of PBB in their milk, ranging up to 1.22 ppm—more than four times the official tolerance level for cow's milk. In the upper peninsula, where virtually all the contamination was confined to one county, the level went up to .32 ppm. It was the first official indication that PBB-contaminated products must have been ingested by almost everyone in the state.

Reizen said that he and a panel of scientific advisers agonized for a whole day whether to advise Michigan mothers to avoid breast feeding. Finally they decided against doing so. They left it to women to decide, on the basis of what little knowledge they could glean from the press, if the natural advantages of mother's milk outweighed the risk of contaminating their babies with low levels of PBB. "I want to emphasize again," Reizen stated publicly, "that available evidence makes it clear that no toxic effect has been demonstrated from the amounts of PBB found in milk of Michigan mothers." It was a misleading assumption at best, based on the premise that PBB levels below the FDA guideline must be safe for babies because the FDA had ruled them safe for adults. It failed to take into account that babies were not protected by the dilution factor which was believed to decrease the general public's exposure to contaminated cows' milk; that milk constitutes most of a baby's diet; and that it comes from only one supplier. Further, mothers were not told that if they were excreting PBB in their milk, they must already have transmitted it to their babies in the uterus. Some farm women knew this from their experience with cattle. Most urban mothers did not.

Reizen's department made testing kits available to new mothers so that they could discover the PBB level in their milk. But unless this level was extraordinarily high—which was the case with some women from heavily contaminated farms back in 1974—it did not advise them against breast feeding.

A few weeks later Irving Selikoff arrived to open his seven-day clinic. Reporters asked him whether he thought Michigan mothers should nurse their babies. Promptly he replied that he was not in favor of this because of PBB's potential to build up in an infant's tissues. It was a straightforward response to a simple question, but overnight Selikoff made headlines and political enemies in Michigan.

"Just before the clinic started he gave me more assignments," Edie Clark related. "I had to get room dividers, delivered and set up. Someone to donate a TV for the nursery, folding chairs for the hallways. I had to get a slit lamp with a Goldmann applanation tonometer. It's a ten-thousand-dollar machine used by opthalmologists, too fragile for the team to risk shipping from New York. I managed to find an optical company which would loan us one, free. A television station in Kalamazoo gave us the TV, and volunteers brought toys for children to play with while they waited. I had the hardest time with room dividers, but finally an office furniture company loaned us twice as many as we asked for, and we ended up using them all.

"Selikoff told me to hire a secretary who could work accurately and conscientiously without supervision, and who did not need a salary. I called Gary Schenk and he suggested the wife of another lawyer in his firm. Chris Muldoon was her name, and she did a marvelous job scheduling all the appointments."

Edie appealed to the PBB Action Committee for volunteers to help in the clinic and got an immediate response. The clinic was a community effort where farm people with health problems had their examinations, then stayed behind to do necessary odd jobs. It became a status symbol in Michigan's rural neighborhoods to have been at the busy, makeshift Grand Rapids clinic that November week. "I am part of Selikoff's study," people boasted with the same pride that they might tell of having a walk-on part in an important film.

The Selikoff team included physicians, neurologists, biochemists, dermatologists, opthalmologists, pulmonary physiologists, a pediatrician, and a behavioral toxicologist (a specialist who does psychological testing to evaluate the effects of toxic agents on the brain). Every patient was asked to describe the kind of diet which he or she had consumed over the past five years. The answers proved highly relevant. The team found one of the highest PBB levels in a man who did not live on a quarantined farm but who regularly ate twenty-one eggs a week. The team took tests of cells, hormone levels, and fat biopsies. At the end of the examinations there was not a clinic patient who had not had every part of the body subject to thorough scrutiny.

It was Irving Selikoff's first close experience with midwestern rural folk, and he was impressed by their openness and integrity. "I had misgivings before we started," he admitted. "I did not know whether they would be reluctant to involve other people. On the contrary, they were concerned that they may have done someone harm and were anxious to tell us who had purchased their eggs, milk, and meat." His survey also included workers from Michigan Chemical who hired a bus to take them to the clinic.

One uninvited person mingled with the crowd in the hospital lobby: an FDA investigator named Raymond Hedblad. Ostensibly he went to Kent Community Hospital to talk with Selikoff and ask him, among other things, about his statement on breast feeding. Obviously FDA had not liked that any more than did the state's Department of Public Health. Hedblad was told that Selikoff was out for the day and was referred to his deputy, Dr. Henry Anderson.

While he was waiting to see Anderson, Hedblad spotted Tom Corbett talking on a telephone in the lobby. He listened in, and afterwards wrote a memorandum describing Corbett's private conversation and sent it to Alan Hoeting, Hedblad's superior at FDA's Detroit office. It related how Corbett was overheard asking a woman newspaper reporter (so Hedblad assumed) to use caution in writing about some of his statements, because he did not want to prejudice Selikoff's work in Michigan. It was the kind of request that a trusted source would make to a journalist, but Hedblad's interest in it raises questions about the role of FDA's Detroit office in the Michigan crisis.

His memorandum came to light during a U.S. Senate committee investigation the following March, under the chairmanship of Michigan Senator Donald W. Riegle. Hoeting defended it, stating that "one of the problems that has come out of the Michigan problem has to do with press sensationalism . . . of the most alarming type." He singled out the *Grand Rapids Press* as being "very much involved in this kind of activity." (Hedblad assumed, incorrectly, that Corbett's telephone conversation was with a member of its staff.) Hedblad, incidentally, left the hospital before Anderson was free to see him.

This was just one of several unexplained political activities which coincided with the arrival in Michigan of the Selikoff team. At the same time, a smear campaign spread among members of Michigan's agricultural establishment and its legislature. It aimed at discrediting the Selikoff study. As evidence of this study's alleged bias it was cited that a wife of a colleague of Gary Schenk helped to organize the clinic, and that members of the PBB Action Committee were prominent among its volunteers. Edie Clark was furious. "We got the voluntary help where we could," she snapped. "This was a clinic for sick people, not a political campaign."

The results of his investigation were so eagerly awaited that Selikoff did a very unscientific thing. He promised to publish them in episodes, as they came along. Normally it is two or three years before the conclusions of an important scientific study are published, and then only after being presented for review by independent scientists. Dispensing with this tradition, Selikoff announced that he would hand his preliminary findings to the governor within a few weeks. In the meantime, he talked freely about them to the press.

"We know that something like two thousand pounds of PBB got into the food chain in Michigan," he remarked, shortly after returning to New York from Grand Rapids. "But the company made twelve million pounds of the chemical and we have very little knowledge of where that went, except that it was sold in commerce. There may have been twenty different names for it. All I know is that it was sold in the United States and Europe and that twelve million pounds are somewhere out there. And our experience with other chemicals of this kind is that once they are loose they somehow leach into the environment."

At this stage he was still trying to sort out the diversity of symptoms which had been collected at the clinic. "Many people seemed perfectly well, had no complaints, and we found nothing abnormal on physical examination," he said. "Others had infections, arthritic conditions, neurological problems. Disturbances of sleep patterns and memory defects were striking complaints, but these symptoms are difficult for doctors. A man who loses his farm and his livelihood does not sleep well. On the other hand, insomnia can be a symptom of toxic reaction. If a man tells you that his wife has to drive him into the city because he can no longer remember the way, you do not know whether the memory loss is due to brain tissue damage or to some of the psychological pressures he is undergoing. Some farmers have had claims settled and some are still in litigation. That may have some bearing. How do you sort these things out? We should be able to, and we shall have to be very cautious in coming to conclusions."

Selikoff could not yet tell whether PBB was likely to be a human carcinogen. It might be twenty years before this was known for sure, he said. In the meantime he felt there was a possibility that science could develop "medicines that will take it out of the body." A means had been found for flushing lead poisoning from the human system "and it is by no means impossible that we will learn how to do this with PBB," Selikoff added. First it was necessary to understand the nature of the chemical and its effect upon every aspect of the human system, and that is what his clinic at Grand Rapids had been all about.

He wished he had been called to Michigan sooner. "We are going to have more environmental emergencies like this," he remarked. "My responsibility is to develop the intellectual capacity to act as a fire brigade, but we must come to the fire faster and faster. We can only work effectively if people sound the alarm rapidly, and do not wait until the flames go through the roof before they pull the lever."

Selikoff described his preliminary findings at a press conference in the Lansing Capitol on the afternoon of January 4, 1977. Outside, the fahrenheit temperature hovered around zero, but Governor Milliken's conference room was jammed to suffocation; reporters herded together, wilting under television lights. The meeting started late and Selikoff had to catch the day's last plane connection to New York, so it was a rushed, disorganized affair with the governor and Bobby Crim competing for the credit of bringing Selikoff to Michigan. The governor upstaged the Speaker, and the Speaker's staff bristled.

It had been barely two months since the Grand Rapids clinic which, Selikoff said, was "one of the most unusual I have seen in my medical experience." He explained that all the patients were either farm dwellers or people who had bought food directly from contaminated farms, therefore they might not be typical of Michigan's population as a whole. Two-thirds of them felt perfectly well, although this did not necessarily mean that, long-term, they might not suffer from PBB exposure. As for the rest, Selikoff said, "we found health problems that we did not anticipate and would not anticipate among people in general." He could not explain why some were more affected than others, except that there are vast differences in the way people's bodies react to toxins; some heavy cigarette smokers live to a healthy old age while others die young of lung cancer.

Selikoff felt there was strong evidence that PBB had an effect upon the immuno-suppresive system, damaging blood cells which normally fight disease, but his team had not yet had time to check into this. He had also discovered that the tissues of his Michigan patients contained a variety of PBB isomers, not just the hexabromobiphenyl (BP-6) which was being used as the yardstick for FDA's tolerance level, and he warned that different isomers might have different toxicities. Urgent research was needed in both these areas, he told the press conference. These points were not taken up by reporters, although Doc Clark would have latched onto them instantly.

Another disquieting discovery—which the press did seize upon—was that about 25 percent of the clinic patients complained of neurological disorders, which seemed to indicate that PBB affected the brain cells. This, too, had been observed in cows—the "stupid" behavior, the attacks of "blind staggers". It might turn out that as many as 40 percent of the clinic patients had some neurological damage, warned Dr. Sidney Diamond, the senior neurologist on Selikoff's team.

"In people the symptoms first appeared after stress," Diamond told the press conference. "Stress tends to mobilize fat, and PBBs are contained in the fat-soluble tissues of the body." Again the similarity with animals—the cows which seemed to tolerate initial contamination, then fell apart under the stress of pregnancy. The most dramatic symptom of neurological damage in humans was the memory lapses: Selikoff and Diamond had been impressed by stories of people who lost their way in neighborhoods they had known all their lives, or who had to give up playing bridge because they could no longer remember the cards from one player's turn to the next.

Selikoff told reporters of skin changes, breathing problems, swollen and painful joints. He spoke of the possible effect of PBB upon PCB, which was probably in everyone's tissues. "There is reason to believe that these are additive," he warned.

This time he was prepared for awkward questions: what about breast-feeding, is Michigan food currently safe, and is the tolerance level as low as it ought to be? He fielded the first two and faced the third head on. Having gained some insight into the politics of PBB, and with Dr. Maurice Reizen at his side, Selikoff temporized about breast feeding. "It's a very difficult question—whether the risks are outweighed by the benefits of breast feeding, and I don't think every case can be equated with the next one," he said. On the other hand, he warned, a nursing mother may have had a longer time to store a chemical in her tissues than a cow.

"Doctor, would you eat bacon and eggs from Michigan each morning? Honestly?" one reporter asked.

There was a thoughtful pause before he replied, "Well the answer is yes, I would, because I'm only going to be here for a short time."

"If you lived here, would you eat them every day?" the reporter pursued.

"Then," he answered, "I would be in the same tough position that you are. "I would have to ask not just about bacon and eggs, but the milk that's put into the bread, the butter that's put on the bread, the cream that is used to make sauces. . . . I don't know the answer any more than Governor Milliken does."

Reizen intervened by citing MDA's soothing reassurance that PBB had been virtually eliminated from Michigan food "and I'm sure some of us are doing very well on bacon and eggs."

Selikoff, however, remained adamant about the tolerance level. "Based on what I have found, I would urge the people of the state to prod Governor Milliken to get the levels of PBB in food to as low as possible," he insisted. "We should get them as close to zero as we can."

The time was ripe for Francis Spaniola to revive his bill. He introduced it into the reconstituted legislature in February 1977, in the wake of Reizen's breast-milk study and Selikoff's preliminary report. All other means having failed, it was probably the last chance to get the tolerance level lowered. Almost three years after the contamination was discovered, public opinion had finally been aroused and, for the first time, the political climate seemed right for legislative action. The choice between protecting the state's agricultural markets or its public health had ceased to be clear-cut. Michigan agriculture was losing its good name, and rapidly.

Now that Michigan consumers were aware that they probably had PBB in their bodies, they could no longer be persuaded that food containing below-tolerance amounts of the chemical could do no harm. They had learned that PBB was additive and they were frightened. In normal times most Michigan beef is consumed in Michigan. Overnight, it became the last item which people in the state wanted to eat. Within a few weeks of Selikoff's press conference, supermarkets throughout Michigan were advertising that all their beef now came from the western states—Iowa, Colorado, Kansas, and Oklahoma. "Come in for proof —we sell western beef" a Detroit packing house proclaimed in a full-page newspaper advertisement. Supermarkets pasted large notices, "No PBB—all western beef," over their meat counters.

The boycott spread to Canada, Michigan's biggest out-of-state customer—this time officially. Every year Canadians had been buying about 57,000 head of Michigan cattle, not realizing that some animals might be contaminated. The PBB incident had not impinged upon Canadian consumers, although tens of thousands of them lived a river's width from Michigan's eastern border. They could tune in to Michigan television and radio stations as easily as to their own. Yet not until Selikoff's press conference inspired the Canadian Broadcasting Company to do a documentary program about PBB did the Canadian government take action, blocking further imports of Michigan meat. This unexpected action was devastating to the Michigan livestock industry.

Michigan residents were curious about where all the state's meat was going. It had disappeared from butcher's counters. It was no longer being served in Michigan restaurants. Canada wasn't buying it . . . yet farmers were still able to sell their animals. A lot of Michigan livestock must have gone to other states where the standard slaughterhouse inspections [for tuberculosis and brucellosis] did not include a check for PBB. Farmers joked that the Wisconsin and Iowa beef suddenly being sold in Michigan as "safe" might well come from animals born and bred in Michigan, sold on the other side of a state line, and shipped back—butchered.

Suspecting this from her knowledge of the cattle trucking business, Bonnie Hughston remarked, "We kind of laughed about all this 'western' beef the supermarkets were suddenly selling. Michigan meat had to be going somewhere, and we were probably still getting it." People as knowledgeable as she did their best to avoid it. Her neighbors, Doc and Marlene Clark, raised a heifer on their own pasture, refrained from giving the animal commercial feed, and slaughtered it for their home freezer. "We tried to be as cautious as we could," Doc explained. "I knew too many farmers who wouldn't eat their own beef."

The loss of confidence in Michigan meat and dairy supplies spilled over to unrelated agricultural products. Michigan's fruit and vegetable growers began complaining that their markets were suffering. Dale Ball contended that the British were becoming suspicious of dried beans from Michigan, after eating them for years. While a lowering of the PBB tolerance level might help restore this trade, the sheer cost of it—in a new round of compensation to farmers, lost meat and dairy sales—did not bear thinking about. The agribusiness dairy farmers felt more threatened than at any time in the PBB crisis.

Farm Bureau's political machinery went into action, and Francis Spaniola felt the full force of it. His background was urban (a former high-school teacher, he now owned two fast-food businesses) and the assault was so unexpected that he scarcely knew what had hit him. Although still an active member of the House Agriculture Committee, he announced that he would put his bill into the House Public Health Committee, where he believed it would get better support. Representative Paul Porter, who had taken over the full chairmanship of the House Agriculture Committee, did not like this. The next meeting of Porter's committee took place on a day when Spaniola was having trouble with his car. Since there were only routine matters on the agenda, Spaniola telephoned his excuses. In his absence Porter unexpectedly introduced a rival PBB bill, which was unanimously approved. Spaniola was furious. "It was voted out of that committee like cod liver oil through a snake," he said.

The Porter bill proposed leaving the tolerance level at .3 ppm, but provided for animals above that level to be destroyed and their owners indemnified by the state rather than depend upon settlements from the chemical and feed companies. Owners of animals which had PBB below tolerance level would have an option. They could have them destroyed and collect compensation or keep them. This proposed legislation, which did nothing for public health, would placate the aggrieved owners of low-level herds, and permit farmers with valuable cattle which were only lightly contaminated to keep them if they wished. All this would heal the rift between the farmers and restore lost loyalties to Farm Bureau.

Michigan Farm Bureau conducted a statewide campaign to defeat Spaniola's measure and drum up support for Porter's. Through its county chairmen it had built up a telephone network which reached into the homes of its 61,000 members. As a result, state legislators received a flood of telephone calls from farmers urging them to vote against lowering the tolerance level. These farmers believed that stricter PBB controls would damage them economically. Farm Bureau also organized a "fact-finding meeting on PBB" which brought several hundred loyal members to Lansing. The "facts" as presented by the organization's president, Elton Smith, were that news coverage of the PBB issue had been one-sided, even distorted; that Michigan Farm Bureau had been subject to "emotional and political bombardment" which, if not countered, could spread and threaten other farm commodities "in which chemicals are used in production."

This was to be expected from Elton Smith. But, surprisingly, two leading speakers at this highly charged political meeting were senior officials from bureaucracies who could be criticized for taking sides: Dr. George Whitehead of MDA, and Alan Hoeting, district director of FDA. Both insisted that the .3 ppm. guideline was adequate.

Aside from his apparent approval of Raymond Hedblad's eavesdropping on Tom Corbett, Hoeting was a party to some extraordinary political activity during the complex journey of the Spaniola bill through the legislature. The bill was no sooner filed than a group of agribusiness farmers met in private to discuss strategy to defeat it. Calling themselves the Concerned Dairy Farmers of Michigan, they included Rick Halbert and others who had been heavily contaminated in 1973, well compensated, and were now back in business. Larry Crandall, one of the most vocal critics of the low-level farmers, came into this category; he presided at the meeting which took place at Emil's Restaurant on the outskirts of Lansing—an establishment which, ironically, was owned by a cousin of Spaniola. Hoeting attended, along with representatives from MDA, Farm Bureau Services, Michigan State University, the USDA veterinary service, and Michigan Milk Producers' Association.

The only record of the session (at which it was requested that no notes be taken) is an FDA office memorandum which Hoeting wrote afterwards. He recorded the opinion of Dr. Harold Hafs, chairman of the Dairy Science Department at MSU, that "the Selikoff report is deficient"; the allegation from Rick Halbert that Selikoff's investigation "was opened up to the Trombley group but not to farmers who feel they have no health problems" and that about half the farm people in the study were law clients of Gary Schenk; and the comment of another farmer that "the entire study is a farce." How the Concerned Dairy Farmers of Michigan operated after that is obscure, but Larry Crandall spoke out against the low-level farmers at every possible forum.

Meanwhile, Governor Milliken announced his support of the Spaniola bill and told his Agriculture Commissioners that they had better support it too. They did so reluctantly. Milliken wrenched the same commitment from Dale Ball, having pointed out to him that if MDA had not opposed lowering the tolerance level a year earlier, the Canadian boycott would not have happened and the people of Michigan would have no qualms about eating their own meat. Apparently, Milliken was still judging the issue in economic terms.

He also tried another approach. He sent a telegram to the top of the FDA hierarchy—Joseph A. Califano, secretary of Health, Education and Welfare in the Cabinet of newly inaugurated President Jimmy Carter. "I urge you to review the existing PBB tolerance levels," Milliken wrote. "Specifically I ask that FDA lower the tolerance levels from the present .3 ppm to the lowest reliable measurable level." It got him nowhere, and some people suspected a conflict of interest. Before his Cabinet appointment, which was only a few weeks old, Califano had been a highly paid Washington lawyer, and in this capacity had represented Michigan Chemical in its dealings with FDA during 1975 and 1976. It had been Califano's mission to try and protect his clients from another round of lawsuits which would inevitably have followed a lowering of the tolerance level in Michigan. Upon being confirmed in the post at Health, Education and Welfare, however, Califano stated that he would not participate in any matter in which his former law firm had represented a client.

All the leading Michigan newspapers supported the Spaniola bill, some of them apologetic that it had taken them so long to see the light. Joe H. Stroud, editor of the *Detroit Free Press,* told his readers, "I must say I feel a deep sense of regret that the *Free Press* did not draw this issue more clearly sooner after it began. A lot of us were slow to recognize it."

With this and other help, the Spaniola plan cleared the House Appropriations Committee (where the Porter bill died) and won passage on the House floor, 78 to 25, in early April. That hurdle overcome, it continued on another obstacle course through the state senate. There it went first to the Agriculture and Consumer Affairs Committee, which conducted seven detailed hearings in different parts of the state, much of the testimony repetitive of what had been said in House committees and at the MDA hearings. Visibly impatient with all this duplication, Dr. Albert Kolbye of FDA made his fifth trip from Washington to Michigan to defend publicly, yet again, his agency's decision to keep the tolerance level at .3 ppm.

The Committee's chairman, Senator John Hertel, also wanted Irving Selikoff to testify, but Selikoff declined to be drawn into politics. He had already learned the futility of this, the hard way. Years earlier Selikoff's brilliant research among asbestos workers, directing public attention to the high risks of lung cancer to which they were exposed, had led to important reforms in the asbestos industry—but not before industry itself had attempted to discredit his findings. So he politely told Hertel that if his Senate Agriculture Committee members would like to talk to him, they were welcome at his laboratory in New York.

Hertel took Selikoff up on this. A Democrat from an urban district, he was not particularly sympathetic to farmers. He was more worried about cost. He suspected that recontamination of the farms was likely to be discovered if the tolerance level were lowered; hence he feared that Spaniola's bill could be very expensive. Also, in common with many legislators, he felt that since the worst PBB exposure had passed there was no longer any point in changing the guideline. The committee's interview with Selikoff, Spaniola therefore suspected, might be aimed at casting doubts on the validity of the Grand Rapids study.

On Saturday, May 14, a delegation of five spent a day with Selikoff. Hertel was accompanied by Senator Richard Allen, a Republican member of his committee who was a veterinarian and a beef farmer. Each of them took a party aide, and Spaniola went along to see fair play. For a professional opinion of the Grand Rapids study, the committee invited Dr. Peter Greenwald, director of the Cancer Control Bureau in New York State's Department of Health. It was Spaniola's impression that many committee members hoped Greenwald would find fault with Selikoff's work in Michigan—Selikoff having left himself open to criticism by announcing his preliminary findings before he had compared Michigan farm families with a similar populace in another state. Greenwald, however, praised the Grand Rapids study as "an exemplary effort to determine whether or not there is a problem," urged that Selikoff be asked to undertake a long-range human study in Michigan, and recommended prompt legislation along the lines proposed by Spaniola. Spaniola was delighted by this turn of events.

Nevertheless, members of the Senate Agriculture Committee "continued to bad-mouth Selikoff," according to Spaniola. "They said they had learned nothing new from him, although he told us about his preliminary evidence that PBB damaged the immuno-suppressive system," he reported. "They did not want Selikoff. They wanted Reizen and the state university people to do their studies."

With intensive lobbying from Bobby Crim and his friends, the bill survived the onslaughts of the Senate. With a few tolerable amendments, it was enacted in August and became state law on October 3, 1977, more than four years after Rick Halbert's herd was poisoned. Spaniola's main proposal, a reduction of the PBB guideline from .3 ppm to .02 ppm, remained intact. His proviso that all animals on Michigan farms should be tested was dropped. Instead, PBB tests were to be limited to animals being sold for slaughter, which was less costly to the state.

Spaniola emerged from the fray as resilient as ever, wiser in the ways of the agricultural establishment. He went on to sponsor two more pieces of PBB legislation, one which appropriated $2.25 million of state funds for the public health study which Maurice Reizen had been urging; one which provided loans of little or no interest (depending upon the sum) to distressed farmers. For his trouble, Francis Spaniola was harassed for months. A year after his first bill became law, he related, "My phone still rings in the middle of the night, every night. It's always about the time that farmers get up to milk, about four-thirty or five A.M. As soon as I lift the receiver the caller hangs up. It's a great annoyance. I can't take my phone off the hook at night because my mother lives alone and I worry about her. But I don't regret the fight in the legislature. I only wish I had tackled it sooner."

His was a good law for consumers, but it may have hurt as many farmers as it helped. It would have been ideal for everyone back in 1974, but had been delayed so long that no legislation, no matter how tough or well intended, could be fair to all whose lives had been damaged by PBB. It did not help impoverished farmers like the Greens, who had not been able to afford to keep sick cattle alive as evidence—and it caused new suffering for farmers who had been heavily contaminated at the beginning, and who had thought their troubles were over.

At least fifty such farmers, it was estimated, were unable to pass the new tolerance tests because of residual contamination in barns and pastures which showed up in their animals. For three years they had been getting along nicely, with new herds which seemed to be healthy and productive. Suddenly their farms lost value overnight. This time they had small hope of compensation because, back in 1974, they had taken settlements and signed away their rights to claim again.

Rick Halbert was among the newly afflicted. He had no herd problems, but tests showed low levels of PBB in his soil and in the wildlife on his land. About 35 percent of the "clean" animals with which his family had restocked their farm had ingested just enough PBB residue to be unfit for sale by the new standards.

"The farm is contaminated and probably always will be," he said miserably. "PBB residue is everywhere. We took samples of it from the air filters in my house, five miles from the farm, and we don't even have animals there. It must have come from the dust on our clothing. We found it in my dad's house too. I think the entire farm property may be unsaleable, and a farmer depends upon the sale of his land for retirement.

"We cleaned out everything before we were released from quarantine. After we discovered the recontamination we decided to abandon some of our pasture land, and we may have to abandon more. We made this decision on our own. The state did not give us any guidance."

26

Trial of Strength

In February 1977, the month when Francis Spaniola introduced his PBB legislation, the battle of the low-level farmers opened on a second front. Immediately after the Zuiderveen claim was settled out of court, Gary Schenk and Paul Greer prepared to bring another farm case to trial. They wanted to avoid another last-minute settlement, and to get a favorable legal judgment which would speed payment of the other outstanding claims. For almost three years they had been gathering veterinary and scientific evidence about the effects of PBB, and the background which Gary and Doc had gained from all the experts they had consulted amounted, in Gary's view, to an intensive Ph.D. course in toxicology.

Gary was an avid learner, and these three years had done more than expand his intellectual experience. They had added a new dimension to his character. In 1974, he was an upward-striving young corporate lawyer in a conservative practice, dreaming of "a nice judgeship" in about fifteen years. PBB had caused him to become a committed crusader: a course which rarely leads to financial reward.

Eighty-four farm families were pinning their hopes on his efforts. But he felt an even broader responsibility. "As we developed an understanding of the chemical," he said, "we realized that much more was at stake than the farmers themselves." He wanted to put on record all that was known about a highly toxic substance, now loose in the environment. He hoped to expose the indifference of industry and the ineffectiveness of government. He was concerned about the health of nine million people; and he believed that if he could bring all these things into the open, future environmental disasters—since others would surely happen—might be tackled with more justice and speed.

Gary Schenk thought a lot about this. He had come of age during the civil rights campaigns, the Vietnam war, the draft resistance movement—all clearly defined issues in which right, as he saw it, eventually triumphed. His legal experience had reinforced his belief that even when the system is inadequate, the individual can effectively seek redress. Now he was troubled to find that this is not always so: that when government and industry are threatened, these two may close ranks and, in self protection, militate against the good of the public they are supposed to serve. What happened in Michigan, he believed, involved subtle shifts in the power structure so that bureaucrats ended up protecting corporate giants, not for monetary gain (that might have been easy to prove), but to preserve their own authority and status. Pitted against all this was a little band of farmers who had been abused by industry and failed by their elected officials. Gary welcomed the chance to fight for one of them in court. There, he believed, justice would finally prevail.

The next farm case which came up on a Michigan court calendar was that of Roy and Marilyn Tacoma of Falmouth, owners of a quarantined herd with low levels of PBB. Like their neighbors, Garry and Lois Zuiderveen, they were offered a settlement a few days before the trial date. The Tacomas refused.

Roy Tacoma explained: "Ours was one of eighty-four similar cases, on which Schenk and Greer had put in months of preparation. I felt I was a bit privileged to have been picked from all these for a court trial. If I had settled, where would the other eighty-three cases have gone? The defendants could have done the same thing to them—waited until each one was ready for trial, and then offered just enough to persuade them to drop the suit. That way they could have taken twenty years to settle with the farmers. I figured that it was worth spending six weeks on a trial, so that our settlement would set a pattern for the others."

Roy Tacoma was a rangily built man of few words and dogged opinions. His Dutch grandparents had settled in Michigan during the lumbering boom, and Roy grew up on his family's dairy farm. For a time he and his father worked in partnership, but their disagreement about farming methods caused Roy to take half the herd and set up independently in November, 1973—only a few months after the disastrous misshipment of PBB from Michigan Chemical and five months before the mistake was discovered. He gave these animals some of the highest quality feed put out by Farm Bureau Services, and for a short time succeeded in increasing milk production. Then his cows became sick, while his father's animals, fed conservatively on homegrown grain, continued in good health. "I began to think my dad was right," Roy said gloomily.

The difference in health between these two sets of animals from the same original herd, now living on neighboring farms, made a good case for Schenk and Greer to argue in court. Roy Tacoma had been introduced to the attorneys in Garry Zuiderveen's kitchen, that day when Garry had assembled a group of his neighbors to discuss the wisdom of hiring lawyers—and Doc Clark had ambled in to pass judgment.

The Tacomas had been Doc Clark's clients for years, but it was a long time before Roy could bring himself to talk to any other farmer about his ailing cattle. "Whether you like it or not," Roy explained, "your neighbor is your competition in the milk market. If the market goes down and you can produce milk cheaper than him, you may be able to survive when he cannot. So you don't tell him about your problems."

The Tacomas, the Zuiderveens, and all the other afflicted farmers in their community had come a long way since then. When the trial opened in the nearby town of Cadillac, two or three dozen farmers and their wives sat in the body of the court in a gesture of solidarity. Most of them continued to attend the trial day after day, leaving quietly before the afternoon session ended so they could be home by milking time. The Tacomas felt grateful and proud. But, said Marilyn during the trial, "for every one who turns out to support us there are two who will ignore us for trying to rip off their beloved Farm Bureau."

The plaintiffs' attorneys elected a trial without jury. Their decision was unexpected, because Michigan Chemical's attorney Roland Roegge had made that very request, on grounds that an impartial jury would be hard to find in the area. Gary Schenk's reasoning was different. Circuit Court Judge William R. Peterson, who would be hearing the case, was a highly regarded Michigan jurist with an impeccable record; a quiet, erudite man of fifty-three who had the intellect and patience to grapple with the complex scientific evidence which Gary was anxious to put on the record. This would be a trial of statewide importance, and Judge Peterson could be depended upon to write a thoughtful decision. Their attorneys were reasonably confident that it would be sympathetic to the Tacomas. They did not expect to get all the damages they were asking (in excess of a million dollars for cattle and income losses, and for the anxiety and grief) but any favorable verdict would be a victory. The judge, a native of Cadillac, had written a history of the area: a finely researched volume which had a modest, continuing sale at the town's only bookstore. The attorneys visualized the care which Peterson would lavish on a written judgment likely to find an important place in the annals of Michigan jurisprudence.

Evidence before a jury, on the other hand, would have to be simplified. To a certain extent its impact would have to depend upon emotional appeal. Much of the documentation about PBB's effects, which Gary and Doc had taken such pains to assemble, would have to be omitted. Nevertheless, a trial without jury would probably be shorter. The judge and attorneys estimated about six weeks. Their guesses were wide of the mark. It lasted fourteen months.

Before it began, Gary and Paul were approached by an independent attorney from the country town of Hastings, Richard Shuster. For many years, Dick Shuster had been the family lawyer of James and Alice Fish, whose nationally famous herd of pure-bred Jerseys had been devastated by PBB. He had tried to negotiate a settlement for the Fishes but was offered, he said, "only a fraction of the real value" and realized that he might have to go to trial. If so, he would need the kind of information which Paul Greer and Gary Schenk had assembled. So he offered to do a deal with them. At his own expense, he would help them on the Tacoma case if they would share their research with him. They agreed. Greer and Shuster had started in law practice together more than twenty years earlier, and both welcomed the reunion. Gary Schenk was glad to have the extra help. Fred Boncher, the youngest member of the Tacomas' team of attorneys, was dubious. He wondered whether Dick Shuster, joining them at the last minute, could have anything useful to add. "In fact he added a great deal," Fred Boncher summed up, months later. "He knew a lot about trial strategy and he was so diplomatic, so good with people, that he more than anyone was responsible for the harmonious functioning of our team."

It was a disorganized trial, by any standards. Despite all the preparation, neither side was completely ready to go to court. New evidence kept coming to light which changed the shape of the case. With unusual leniency, Judge Peterson permitted pretrial discovery to continue while the trial was in process—a consideration which, in his words, "resulted in plaintiffs turning the offices of defendants inside out and figuratively emptying their waste baskets in court." He commented, "At the beginning of the trial, I had no idea where the attorneys were going. I felt it necessary to indulge every discretionary ruling in favor of letting it all hang out, a kind of catharsis for the farming community. There was so much suspicion that people had concealed things—with some justification, I guess."

Quantities of documents from the feed and chemical companies were entered into the record: inter-office memos, minutes of closed meetings, notes of telephone conversations, private letters, PBB test results. Some pieces of paper were trivia. Others revealed in detail how, for a long time, the two defendants had much more knowledge of the contamination than they had publicly admitted.

A variety of specialist witnesses were called to explain to the court how the business of dairy farming functioned, the intricacies of breeding to develop specific strains of cattle, the complexities of feeding practices, the daily pressure upon farmers to maintain the highest possible milk yield for the most economic expenditure. It was all this and advanced science too. The Tacomas' attorneys were gratified to observe that Judge Peterson showed a keen interest in details, made copious notes, and occasionally interrupted with an incisive question. It had been a long time since he had studied chemistry, the judge admitted, and he enjoyed the challenge of mastering the essentials of biochemistry and microbiology.

Gary Schenk's firm of Law, Weathers, Richardson and Dutcher (usually abbreviated to Law Weathers) put effort and money into the trial. Week after week the partners sent to Cadillac a team of four or five attorneys, two legal assistants, a scientific consultant and two secretaries. This badly depleted the Grand Rapids office, but Law Weathers calculated it as an investment which in the long run was likely to bring good returns.

The Tacoma trial became a way of life for Paul Greer, Dick Shuster and the Law Weathers team. Every Monday they drove more than a hundred miles north to Cadillac and every Friday evening they rejoined their families in the Grand Rapids area. There was almost no leisure. Evenings and weekends were spent continuing the research and preparing witnesses. Strategy was debated night after night. Their independent ideas plus the pressure of living and working together caused brief but vehement clashes. Gary Schenk was intent on doing a detailed job. Paul Greer felt it would be more effective to bring the trial to a swift and dramatic conclusion. Nevertheless, policy decisions were always made jointly, with the attorneys huddled round a coffee table at their motel. "We looked like a bunch of minks screwing," Gary Schenk remarked.

Doc Clark, who lived only a twenty-minute drive from Cadillac, "was always there as a sounding board," Gary reminisced later. "One night, early in the trial, we were all sitting around and I was feeling particularly depressed. We were having a lot of problems accomodating to living together, and Doc said, 'We can succeed if we don't care who gets the credit.' That was a point well taken. That piece of advice really gave us the will to function as a cohesive, self-supporting unit, and any success I had in molding together all those unwieldy egos was directly due to those words of wisdom from Doc."

The team's new home and office was a wing of the Sun 'n Snow Motel on the outskirts of Cadillac. It was a modest family establishment, attractively situated on a lakeside, but it was intended for people on brief vacations, not permanent residence. There was a long row of identical double bedrooms which, for economy's sake, the attorneys had to share—a factor which added to the friction—and a small cottage which they rented as an office. Between these narrow confines and a court room, four men—Gary Schenk, Paul Greer, Fred Boncher, and Dick Shuster—spent the best part of fourteen months. The others pulled out one by one as Law Weathers found the cost too crippling.

It became almost tormenting to the four to return to the same place week after week where, even though allocated different rooms, they were faced with the same orange, beige, and brown curtains; the same orange, beige, and brown industrial carpet; the same fake oil painting of stylized orange and beige chrysanthemums. Lying on his orange-covered bed one dreary evening, studying a trial transcript by the light of his orange and beige lamp, Fred Boncher idly conjectured that if he were to shoot a bullet into the largest chrysanthemum it would make a hole in the identical flower in room after room, all the way down the block. The tedium of the trial and the living conditions brought out that kind of humor.

The attorneys for Michigan Chemical and Farm Bureau Services, working in an uneasy partnership, settled in a more expensive motel. Cost did not appear to be a consideration to their employers. As time went on, the Tacoma attorneys became increasingly fearful that "the other side is hoping to starve us out."

Gary Schenk was convinced that justice was on his side from the beginning. "The morning the trial began my guts were in my boots," he said. "I picked up a Gideons Bible in my room at Sun 'n Snow, looking for inspiration." The pages fell open to the first Book of Samuel, chapter two, and he read intently. It was full of resounding Old Testament promises about the righteousness of the weak overcoming the strength of the mighty, and he took it as an omen. He drove the two miles to Wexford County Courthouse feeling almost exuberant.

Roy Tacoma, one of the first witnesses, told his story. Having dissolved the partnership with his father, he drove half the herd to his own land across the lane and began farming alone on November 29, 1973. Immediately he began an enriched feeding program ("shoving the protein" into his cows) aimed at higher milk production. He bought his supplies from Falmouth co-op, which was affiliated with Farm Bureau Services. Although it did not stock the 402 mixture, it sold other Farm Bureau feeds which contained magnesium oxide. Shortly after he began using one of these mixtures, Tacoma's animals began to develop health problems.

He enumerated them in detail. The calf losses. The cows not breeding. The deformities—crooked legs, elongated faces, overgrown hooves which looked like snowshoes. The matted hair and the skin like elephant hide. The endless infections, the dysentery, the stupid behavior, as though the cattle were blind or retarded: "They would hold their head peculiar, like they couldn't see or were scared." After almost two years of these conditions, the farm was quarantined in October, 1975. Because this was low-level contamination, Roy Tacoma was still permitted to sell his milk, but the time came when he made a decision which he found very difficult. He and his wife had young children and, for their own family consumption, they bought milk from Roy's father. For income purposes, they sold milk from their own cows. It was considered unfit to be sold bottled, but was defatted for skim milk products.

"I am not proud of this," Roy Tacoma said. "It was of necessity."

Questioning his client on the witness stand, Paul Greer managed to bring in some direct evidence about the effects upon the Tacoma family of consuming their own meat and milk. This trial, however, concerned only farm losses, and any suit involving human health damage would be well into the future. In a private conversation early in the trial, Judge Peterson expressed regret that personal injury cases might not reach the courts before the resources of Farm Bureau Services were exhausted. Overriding the objections of one of the Michigan Chemical's attorneys, he allowed Roy Tacoma to talk about his family's health on grounds that "it related to the witness's reaction." But that was as far as it could go.

Tacoma maintained that his wife Marilyn had strange fainting attacks when she would momentarily lose her eyesight, and "uncontrollable diarrhea just like the animals." Their children had joint pains and extreme fatigue. "When little kids come home from school and their legs hurt, their feet hurt, they don't want to go upstairs. . . . I'm plenty worried. . . . In the meantime, Farm Bureau is maintaining there is nothing wrong with your animals until they drop dead."

Doc Clark testified to the condition of the Tacoma herd. Roy Tacoma was a very competent farmer, he said, with "such big, upstanding cows" that in normal times it was a pleasure to treat them. But in the past three years, Doc had made at least a hundred visits to the farm, vainly trying to cure one complaint after another. One day in December, 1974, he spent more than two hours thoroughly examining the Tacoma herd.

"We laid the records on the milk cooler and went through them," he related. "When I got done I found that Roy's calf crop was shot. The yearlings were big-headed and pot-bellied and stunted. The two-year-olds looked very rough. They moved slowly as though they were hurting. Heads down and backs up and shuffling along with their hind legs, and their hooves elongated." Doc explained that since a cow in this condition would have difficulty getting up, she was vulnerable to udder injury. "You can take the best conditions in the world, and if you got a cow with no legs under her and she can't manipulate in a nice even flow of movement, she's going to damage her udder and teats," he told the judge. To make sure his point had gone home, he added, "The udder is the money maker."

Doc was an uncomfortable witness. He didn't like being on show. He thought about wearing a suit for his court appearance but dismissed the idea "because it wouldn't look like me," and put on clean coveralls every day instead. His testimony stretched over several days and, although he was outwardly calm, he kept going only by pouring frequent drinks of low-fat milk from a thermos flask to calm the pangs of his nervous stomach. Marlene sat at the back of the court, hooking a rug, ready to refill the thermos.

Doc detailed his observations of the PBB syndrome—how the level of toxicity in an animal's fat sample seemed to bear no relation to the severity in its symptoms, and how in time the symptoms changed from acute to chronic. "My opinion," he told the judge, "is that the tests I was taking a year after the initial contamination were of a residual sample, that I was only measuring for one isomer and that has no bearing on the initial amount of the compound that went into that animal's body."

He also treated the court to a primer in the anatomy and breeding of good dairy cattle. Judge Peterson smiled as Doc passed on this piece of rural wisdom: "We want a cow that is sharp enough in the shoulders to split a rain drop, wide enough so you can run a sow between her legs, and high enough in the neck so she can reach the second row of a neighbor's corn crop." He reacted to cross-examination as if it were a personal affront, a sentiment shared by his loyal followers in the body of the court. "How dare they treat Dr. Clark like that!" Hilda Green exclaimed to another farmer's wife who sat beside her on one of the public benches, in reaction to some rigorous cross-examination from the defense attorneys. One question carried the suggestion that Roy Tacoma might have been neglectful of his cattle. "Roy is a good operator and he spoon-fed them buggers," Doc declared defensively.

One day, towards the end of Doc's testimony, the judge chatted with reporters in his office behind the court room. "You can see that man loves cows," he remarked. "He loves cows as much as people, maybe a bit more. And I don't think he can quite understand a person who doesn't feel about cows the way he does."

Doc was not so relaxed. After that same court session, he went straight to the attorneys' temporary office at the Sun 'n Snow, sank back exhausted on the sofa, and worried aloud, "D'you think I did all right? I'd hate to let those farmers down. D'you think the judge understands what a really good cow means to a man? It's like finding a good wife. They're rare and you can't replace them. And that's what happened to Roy Tacoma. He lost some really good cows and he lost the years of his life it took to breed them, and there ain't no money can buy them back."

Throughout the trial, Doc developed the habit of stopping by the courthouse between farm calls, listening to the evidence of other witnesses. He shared the public benches—narrow and straight-backed as chapel pews—with the regulars: his brother Clyde and Clyde's wife, Coralie, still hoping for a settlement; Alvin and Hilda Green, no longer able to farm and dependent upon help from their son, Doug, who was working as a truck driver; Virginia Prehm, a dairy farmer's wife who lived near the Tacomas and who never missed a day of coming to court with Marilyn; Lou Trombley, less ebullient and desperately trying to sell the shambles of his farm so that he and Carol could move out of state and begin again; Garry Zuiderveen, strongly supportive of the Tacomas and regretful that it was they and not he who had to bear the brunt of the trial; Bonnie Hughston, the Clarks' neighbor, who came from a family of farmers. Every one of them saw it as a personal battle in which the plaintiffs could have been themselves.

There was also a slightly built woman with a pale face, and deep, dark eyes who sat apart from the rest. A resident of Cadillac, she dressed in stylish city-bought clothes and was cast in a different mold from the farmers' wives, with their home-set hair styles and polyester pants suits. Like Hank Babbitt, this woman had an athletic teen-age son who had consumed huge quantities of milk and meat. The boy's health had been wrecked and, unable to get much help from doctors, this mother attended the Tacoma trial day after day to learn what she could about the effects of PBB. When Doc Clark described how the acute stages of toxicosis change to the chronic and apparently incurable, she wept silently.

Doc's concern that the court might assume that cows can be replaced like automobiles was shared by all the farmers. One evening during the trial he explained what is entailed in rebuilding a herd. "You see that cow over there," he said, nodding to a color photograph on the wall of his living room. "That cow didn't just happen. That cow was built out of somebody's dreams. She has been bred with legs that are short and sturdy so she can stand well on concrete, her front legs toe straight ahead, and her withers are higher than her rump so she can comfortably walk uphill. The eyes are set well apart to give her 180 degree vision —that way she isn't nervous. The neck is long and the nostrils huge so she can take in a lot of air. She has a good barrel that allows her lungs to function like a really efficient carburetor. There's a tremendous depth of forerib, a fine strong back, and a good suspensory ligament for the udder so that the teats point straight down. All this helps to make her a good milker. But these things don't happen by themselves. You have to know what you are doing when you build a cow."

Building a herd is a multiplication of this process. An efficient herd is not just an arbitrary collection of cows, in which damaged animals can be replaced by outsiders. Rather it is made up of animals which are related to one another many times over (calves which are sired, for example, by their own brothers) plus a few unrelated cattle whose bloodlines enhance those of the family. As new calves are born, those in which the desirable genetic traits have been intensified are kept, and those with unacceptable characteristics are sold. This breeding and culling must continue uninterrupted if the quality of a herd is to be developed.

The untimely loss of valuable cattle can be disastrous to a farmer, since there are no animals with the same genetic combination to replace them. The fact that owners of PBB-contaminated farms had to hold on to cattle which would normally be culled was doubly devastating. It not only diminished the farmers' current income. It threw their breeding programs out of balance for years to come. These were some of Roy Tacoma's problems. He did not aspire to the top rank of cattle breeders, but he and his father had developed a good working herd. Now, in Doc Clark's view, the only way for him to return to efficient farming was to scrap every one of his animals and start again.

This was to be the lot of nearly every farmer who sat in that Cadillac court room. Clyde Clark, whose prize herd had been devastated, thought he was more fortunate than most because the genes of his champion cow, Flag Valley Shamrock Bonnie, had not been lost, despite the fact that she was so contaminated that Clyde had felt obliged to shoot her. Her many descendants included a superb bull, her grandson, which had sired some magnificent cattle all over Michigan. "Some day," Clyde remarked during the trial, "that bull is going to put us back in business. We have three hundred ampules of his semen in our liquid nitrogen tank which we can keep for years, and we shall use it to rebuild our herd." Long after his death, that bull would be fathering his own great-grandchildren. Roy and Marilyn Tacoma lacked even this security.

PBB also destroyed the ability of a herd to continue functioning within its environment. When cattle are replaced by others born on a different farm, the newcomers have to build up immunities to surrounding organisms, to accustom themselves to different nutrients and enzymes in the soil. Roy Tacoma's cattle had to go through this adaptation process when he started farming alone, although they were moved only a few yards down the lane from his father's farm. Highly bred dairy cattle are also more susceptible to the effects of change and disease than cows of mixed ancestry which are put out to pasture. There was, therefore, a process of natural selection governing which animals suffered the worst damage from PBB: the purest bred were the hardest hit.

At the other extreme, farmers who clung to old-fashioned methods and who did not supplement their cattle feed with nutrients were also badly affected by the chemical. In case after case, Farm Bureau Services implied that if the farmers' feeding practices had been somehow different, PBB might not have been so damaging. Gary Schenk and his colleagues argued against this. "One of the first things you learn in law school," Fred Boncher commented, "is the dictum of the glass-jawed plaintiff. If you hit him and his jaw breaks more easily than the next man's it is your tough luck. If a farmer breeds a certain kind of cattle, or if he does not give a high protein feed and PBB makes his animals more sick than his neighbor's, this should not diminish the defendants' responsibility."

If the attorneys had been concerned with Roy and Marilyn Tacoma's cattle alone, they might have been wise to rest their case after submitting this kind of evidence. But they felt a commitment to eighty-three other farmers with pending claims, all avidly watching the outcome of this trial. "This isn't just a lawsuit," Dick Shuster remarked. "It's a mission. People like Doc and Schenk and Greer got into this for the sake of others. They're the heroes, although nobody but the farmers gives them credit."

It was true. There were other attorneys in Michigan who had PBB clients, secretly wished they hadn't, and who were waiting for a judgment in the Tacoma case before filing similar suits. In private conversation, one such attorney was blunt about the reasons for his delay. "I have a hundred other clients with meritorious cases of different kinds," he said. "How much time and effort can I devote to PBB? A law firm is like any other business with rent and salaries and expenses to pay. My firm is dependent upon me to produce a certain amount of income, and a PBB case is likely to drag on for years without bringing in a penny. We can't afford to take them on."

Paul Greer put a lot of his own capital into the lawsuit. Dick Shuster paid his expenses out of bank loans while his partner kept their practice going. Law Weathers became deeply divided, with the partners in Grand Rapids resenting the drain of money and manpower to the temporary office in Cadillac. All the Tacoma attorneys became increasingly careful about the size of their restaurant bills, the use of telephones. Money was conserved for essentials, and there were no luxuries to relieve the strain of the trial. Dick Shuster and Paul Greer (who had sworn off alcohol) would unwind after a day in the court room by taking a sober stroll along the lakeside. Gary's method of relaxing was to change into jeans and tattered moccasins and drive around the countryside with Doc, watching him treat sick cows. Fred amiably fitted in with the others' plans, which often meant working late to prepare for the next day's testimony. Once a week the local farm families would give a supper party for the attorneys. The warm companionship between these two disparate groups of people was remarkable. They were a happy band of beleaguered crusaders, convinced of the morality of their cause. Once when somebody shot a deer there was a particularly nostalgic meal of venison roasted over a spit in the woods, eaten by the firelight, while a farmer strummed folk tunes on his guitar. On the midnight drive back to the motel, the countryside black and silent under the stars, Dick Shuster philosophized about the immorality of the power structure and how, when the people's rights are threatened, new leaders must emerge from the masses, lest democracy collapse. His words caught the spirit of the evening, of the trial, of the cause which bound all their lives together.

This was never far from their thoughts. One evening, over a restaurant dining table, Gary—who loved to talk—was telling stories of his boyhood, judges and attorneys he had known, his experiences when deer shooting with Doc. Abruptly his mind switched back to the trial and, unashamedly sentimental, he remarked, "Y'know, there are so many good people out there working and thinking and praying for us that we have to win. We have to."

The Tacoma trial lasted from February 22, 1977 to May 2, 1978 with testimony amounting to twenty-five thousand pages of transcript. It was the longest and most complex trial which had ever taken place in Michigan. Several experts came from other states to testify. Dr. Anjo Strik traveled from the Netherlands to tell of his early studies with PBB (the ones which the staff at Michigan Chemical couldn't remember receiving), and of his conclusion that small doses of the chemical taken regularly over a long period could be more harmful than large single doses. He visited the Tacoma farm—a curious meeting between a Dutchman and the descendants of Dutch immigrants—and testified to finding the same type of liver damage in the cows that he had seen in his European experiments with birds.

"These were all very special people who saw a problem which was destroying some innocent farmers, and they had the courage, ethics, and moral responsibility to help, often at their own expense," Gary Schenk remarked. "Aside from their testimony, they were invaluable in the preparation of our case. One of the best questions of the trial was given to me by Dr. Gerald Snider. The defense produced studies which concluded that there was no discernible health damage to cattle fed low levels of PBB. 'What were you feeding those animals?' I asked. 'Wet brewers' grain,' their expert said. Wet brewers' grain is rich in Vitamin B and it has a high motility which pushes the feed through the gut before it can be absorbed. It's what a veterinarian would give to cattle if he wanted to minimize the side-effects of a toxin. I was able to bring that out in my cross-examination, and I wouldn't have known it but for Jerry Snider."

Snider was an Indiana veterinarian who specialized in the gynecology of cows, and who was regularly consulted by farmers from a wide area. The studies to which Gary Schenk referred were conducted at the Ohio Agricultural Research and Development Center under the direction of Dr. Lynn Willett, the dairy scientist who had made some earlier feeding studies with PCB and whose expertise had been called upon by Michigan state authorities at the beginning of the contamination. The PBB experiments were largely funded by FDA, with smaller grants from USDA, Farm Bureau, and Michigan Chemical. The financial involvement of the two defendant companies made the studies suspect in the eyes of the plaintiffs; however, they were more elaborate than any the Tacoma attorneys could afford, involving a staff of thirty-two veterinary experts and months of observation.

Farm Bureau and Michigan Chemical cooperated in another large study which was taking place during the trial. Jointly they operated a 310-acre experimental farm of eighty Holstein cows (the breed preferred by most of the state's dairy farmers), half from Wisconsin, half from Michigan. The Michigan group had been fed contaminated Farm Bureau feed in late 1973 and early 1974, and still carried body residues of PBB in 1976. Reports of the study claimed that there was no difference in health and performance between the two groups of cows.

It was difficult to refute this kind of evidence, but by questioning how the test animals were chosen and treated, who funded the studies, and for what reason, Gary Schenk and his colleagues felt that they had at least cast a doubt on the validity of these costly experiments.

There was one weakness in Roy Tacoma's case which the defense exploited. It was revealed in court that he had sold six animals under his father's name, three of which had not been released from quarantine. "I was ashamed of myself, but I was under such pressure to mitigate my claims," he told the court. Shortly after this, he added, he shot more than a hundred of his animals "lest I be badgered to do it again."

With tears in her eyes, Marilyn Tacoma told how she made her children stay indoors that July day, curtains drawn "so they would not see the cattle being hauled out of our yard."

A shy, slender woman in her thirties, she was a more eloquent witness than her husband. She described how humiliating it was, in their tight Dutch community, to be mistrusted by farm families who had once been their friends; how the church which she attended—the church of which Lois Zuiderveen had spoken—had been split between farmers who had not been compensated and farmers who had and who were still loyal to the Farm Bureau.

She added quietly, "I'm sure people in this courtroom will never know; could never put themselves in our place to know how we really felt; that we felt this was intentional, the way they were treating us; that they were trying to disgrace us. And I hope and pray that some day this will all be behind us; that we shall have a fresh start in farming. I hope we can look forward to rebuilding our cattle and rebuilding our lives that have been so badly damaged."

Meantime, she said, "I don't go out to the barn any more than I have to because I'm just sick to see these cattle, especially in the morning when I'm helping with milking, when I see these cattle come in—and I don't mean four years ago, I mean yesterday, two weeks ago, a week ago when cold weather set in this fall and they took a turn for the worse. They have snot rags hanging out of their noses a foot long. Their eyes are matted together. You can hear them breathing. Their chests rattle. . . . I can't stand it."

There were two long breaks in the trial. After the 1977 summer recess—when the judge and his wife made a trip to Europe to look at old cathedrals, and Doc and Marlene drove through some of the other midwestern states so that he could refresh himself with the sight of some healthy cows—the court adjourned again for sixty days. This second recess was at the request of all the attorneys, who felt they might shorten the proceedings by negotiating a settlement. Judge Peterson was dubious, but agreed to this unusual arrangement. Gary Schenk and his colleagues insisted that the defendants must settle all eighty-four claims, not just the Tacomas', but after two months of private meetings in a Cadillac motel they were far from agreement. Doc hung around the Sun 'n Snow, reminding Gary and Paul, by his silent lumbering presence and his shrewd sidelong glances, that they had better not let their egos get in the way of a fair deal for the farmers. One evening when he and they were eating together he got up without a word, put a coin in the restaurant juke box, and selected the recording, "I Believe in Miracles."

The court resumed, with no miracle in sight, on a crisp and sunny October morning, just as Francis Spaniola's legislation was becoming effective. Logically, this should have had some impact on the Tacoma case. Cows which had been "below the level" before the recess were now above it, but the plaintiffs were still obliged to prove that, no matter how much or little PBB remained in their herd, the animals had been irrevocably damaged.

The number of farmers in the courtroom diminished for a few weeks while the men gathered corn for winter feed, but soon they were filling the public benches again. They had the air of students returning to classes after a long vacation, happy at the reunion, drawn by a sense of duty, hoping the trial would soon end yet afraid of losing the solidarity which it gave them.

Judge Peterson's energy was unflagging. Day after day he sat behind his raised courtroom desk, taking notes, his five predecessors looking down from their gilt frames, expressing a range of judicial emotions: from left to right, incredulous, suspicious, irascible, placid, and content.

When the time came for his portrait to be hung alongside them, William Peterson would add a touch of the quizzical, with a glint of humor in the eyes. The Tacoma attorneys felt fortunate to have him judging their case.

Michigan newspapers, most of which reported it only sporadically, called this a landmark trial. It was, but not quite in the sense they meant. This was a rare public inquiry into a massive chemical contamination, in many ways unprecedented. Usually only the barest facts of industrial disasters are known; inquests take place behind closed doors, and the public can only surmise the extent of corporate failure. Essential information stays in locked files. But here it was laid out in stark detail. It was brought out in evidence that both the feed and chemical companies knew—long before the PBB incident—that their plants had a potential for cross-contamination. Michigan Chemical knew that the natural brine pumped into the plant became contaminated with toxic residues, yet it continued to use this to wash magnesium oxide and salt which went to Farm Bureau Services. Farm Bureau Services knew that its mixing machines still had traces of PBB months after they had been cleaned, yet it went on using them to turn out cattle feed. Months before Rick Halbert's discovery, the feed company knew of complaints from a variety of farmers who had used the 402 feed. It had known since 1974 that a single fat test on an animal did not really tell how badly the beast was contaminated. Michigan Chemical had known since 1971 that naphthalene was a contaminant of some of the biphenyl it used in making PBB, but its most senior chemist—who also knew that naphthalene can cause hyperkeratosis, or X-disease, in cattle—did not publicly reveal this until the Tacoma trial.

"Why couldn't those people have been forthright in the beginning?" Doc Clark asked rhetorically. "They knew almost everything that it took a lot of us up to now to find out. We went down a lot of blind alleys, spent a lot of money, and by the time we got there it was almost too late."

At the end of July 1978 the attorneys filed their final briefs. Judge Peterson took almost three months off from his court duties to go over the evidence and write his opinion. He worked in the basement recreation room of his Cadillac house, from five o'clock every morning until late at night, seven days a week. He played tennis twice a week to keep fit, and worried a little that something might happen to him or his notes before his task was finished.

Waiting for Judge Peterson's decision, Gary Schenk tried to relax with his family at their lakeside cottage, north of Grand Rapids. He had become pale and overweight from too much stress and too little exercise, and was utterly exhausted. For fourteen months he had led the trial team, and this concentrated effort seemed to have drained away his youth. Sitting on his porch, trying to summon the energy to do some fishing, he talked about how his involvement in PBB litigation had changed his life.

"It has made a mess of my career, and the practice I built up has almost gone, but there's no question of quitting. Too many people are dependent on me. I don't regret the overall battle because I have come to develop feelings I never had before, strong feelings about the environment, and the right of people to know what is contaminating and polluting it. It has also given me some understanding of who really runs this country. When renegade chemical companies poison our lands and our livestock and our people, they can hide behind their corporate headquarters as though it never happened. And the bureaucrats help them by saying there is no problem—not until it has been proved. Government does not have much ability to deal with these things independently of industry.

"We may yet have to try another four or five cases to get justice for all the farmers, and the other side will attempt to wipe us out financially. So, eventually, I think our effort is doomed to failure. It is doomed because of their superior resources, and their access to the power of government agencies. I do not expect in the long run to win this thing."

He did, however, feel confident of winning the Tacoma case. The only question in his mind was whether the judge would award sufficient in damages to have made so much effort worthwhile.

27

End of the Road

JUDGE PETERSON'S regret that no human health claim had yet come to court was widely shared. Several suits had been filed, but no attorney was willing to proceed until there had been a favorable judgment in a cattle case. Personal injury would be harder to prove. People had ingested PBB at second hand, few could remember if or when they did so, and the only dependable scientific studies of the effects—Irving Selikoff's—were far from completion. Selikoff was anxious to produce a full report as soon as possible but emphasized that he would not be drawn into litigation.

After the Grand Rapids survey of families from contaminated farms, his team studied sixty-seven Wisconsin dairy farm families. This comparison confirmed his original findings: that Michigan farm people had an unusually high incidence of physical and neurological ailments. They also had less ability to combat infections than their Wisconsin counterparts.

This lack of resistance to disease was the first symptom Doc Clark had noticed in low-level cattle—quite different from the dramatic onset of toxicosis in herds as badly poisoned as Rick Halbert's. The Selikoff team pursued this, making sophisticated tests on the blood samples taken from Michigan residents. The conclusions, announced in August 1977, were alarming. Something was strangely wrong with the white blood cells of many who had been exposed to PBB. It was a defect which a standard blood test would miss, because the count of the white cells was normal. But their distribution was awry. Some of these cells, known as T-lymphocytes because they originate in the thymus, were either missing or did not function. Since T-lymphocytes are responsible for rejecting bacteria and other foreign organisms, this meant a drastic reduction in the body's natural defenses against disease.

The team discovered this almost by accident. In previous chemical contaminations this particular blood test had never been thought relevant. One day some of Selikoff's team members had a casual conversation with Dr. George Bekesi in a corridor of Mount Sinai Hospital, and asked him if he would look at their Michigan blood samples. Bekesi was an immunologist, and professor in the Department of Neoplastic Diseases at Mount Sinai. He wasn't particularly interested in his colleagues' suggestion until they explained that some of Michigan's stricken cattle had developed thymus gland atrophy and strange wasting diseases.

"That caught my curiosity," Bekesi related. "Whenever you see wasting disease you usually suspect some problem of an immunological nature." His initial tests showed 18 out of 45 of the blood samples taken from Michigan residents to be "extraordinarily different" from those of the general population, and the remaining 27 were not quite normal.

"It was a unique and totally unexpected finding," said Dr. Henry Anderson, the physician in charge of the Michigan project. "Our initial thought was that something must have happened to the blood specimens because this kind of finding had never been described before. We still don't know how it happens." (Subsequently, medical authorities recognized that the more toxic halogenated aromatic hydrocarbons can cause degeneration of the thymus and spleen, both organs associated with the immune system.)

The team made several trips back to Michigan, and extended its survey to urban residents. After thousands of detailed medical examinations, the group gathered enough information to produce a representative study of the entire Michigan populace. Medical specialists at the University of Michigan helped with the work, and were as shocked by some of the health problems as their New York colleagues.

Dr. Mason Barr, University of Michigan pediatrician, found the profusion of symptoms in farm children "very disturbing." He was convinced that the children were genuinely sick, but physical examinations showed no correlation between the amount of PBB in their blood and the severity of the symptoms. Sometimes it was the unexpected way around: the more PBB, the fewer the symptoms. Asked what conclusion he could draw, Barr replied, "Only that I am very confused. Perhaps there is a genetic predisposition to metabolize the drug. Perhaps some people are more sensitive to it at lower levels, like some cattle."

The health of the farm children troubled him. Headaches, muscle aches, fatigue, gastro-intestinal upsets, abdominal pains, persistent colds, and general weakness were much more common among them than was normal. "This is a very different child population," he observed. "We see a complete range of symptoms in these farm children, and it is very disturbing." One day when PBB examinations were made of people in Detroit, two farm families came into the Selikoff team's clinic. "Two of my colleagues, as independent observers, were able to pick out these PBB children immediately," Barr said. "Their health and appearance was so remarkably different from the city children." That was in May of 1978, more than four years after initial exposure to the chemical.

Every member of one farm family had multiple symptoms, with the exception of a four-year-old child. "The mother told me that in infancy this child was a very poor eater, so she supplemented its diet with powdered milk," he related. "It is interesting to speculate that because of a diet in which powdered milk—probably made out of state —was the main ingredient, this child was spared."

As Selikoff's studies advanced, they were supported by state financing, but this did not come easily. Before the funds were voted there were renewed attempts in the legislature to have the work taken over by the state medical authorities. Again Selikoff remained outside the argument; it passed, and his Michigan survey continued.

Only once did he let down his political defenses. In October 1977 Michigan State University held a two-day workshop on the scientific aspects of PBB. Most researchers who had worked with the chemical were there, but underlying their professional politeness to one another were deep antagonisms. In general, those whose research funds had come from Farm Bureau, Michigan Chemical, or the state of Michigan reported that PBB did little damage and was not very toxic. Most scientists independent of state or industry funding took the opposite view. The conference program seemed to have been made up in such a way that the scientists most likely to present reassuring papers spoke in the morning session, and those whose findings might alarm the populace were scheduled in the afternoon. Reporters with daily newspaper deadlines were thus tempted to give the early speakers their greatest attention.

Journalists who were aware of this felt aggrieved, and on the first day of the conference an argument developed in the press room. A university official who had helped to organize the conference joined in, unexpectedly partial, blaming Michigan newspapers for "sensationalizing" the PBB story. It had been exaggerated out of proportion, he maintained: very few Michigan people had become genuinely sick from PBB.

Almost unnoticed, Irving Selikoff was sitting on a desk in the far corner of the room, taking in every word. "Look, look, look," he broke in. "Fourteen out of fifteen people who smoke two packs of cigarettes a day don't die of cancer, but that doesn't mean it's all right to smoke two packs a day. In this chemical world we're going to have more accidents like PBB, and people have a right to expect that someone is looking out for them; that scientists are investigating and not seeking the cover of universities; that the press is doing its job."

Selikoff said he regretted the loss of time before he was called in, and a MSU scientist intervened, "I have got to stick up for the Department of Agriculture. . . ."

"You have to stick up for the truth," Selikoff told him. "You have no other options."

Over the next several months, further results of his studies were announced. It was found that virtually all residents of Michigan had PBB in their bodies, and would probably retain traces for the rest of their lives. Two years after the Grand Rapids clinic of November 1976, little change was found in the PBB levels of those first farm families to be examined. From these figures Selikoff extrapolated that only about 10 percent of the body burden of PBB now being carried by nine million people would be excreted in their lifetimes. He could not tell whether this would have a long-term effect upon their health, but was still hopeful that pharmacologists would find a way of flushing PBB from the human system.

There were a couple of optimistic findings. The team had done sperm studies of some young farmers and found no abnormalities. Also blood samples of non-Michigan students newly residing at Michigan State University in 1977 showed no PBB in the blood serum, indicating that no new contamination was developing in the state.

Residues of PBB in Michigan Chemical workers were generally higher than those of farmers. They also had a different chemical composition. There was no precise way of telling what had happened to Firemaster after it had been metabolized by a cow, or how much contamination from other chemicals the industrial workers had sustained. Even the exact recipe of Firemaster was elusive. Brominated naphthalenes were known to be a contaminant, and other contaminants were still suspected. "We can never be quite sure that there was not some extraordinary material which went along with the PBB and caused some of the symptoms," Selikoff said. In general, the chemical workers did not have as many symptoms as the farmers, but he suspected some of them might become more ill eventually.

"PBB has all the elements of a Greek tragedy," he remarked. "We shall not know the ultimate effects for twenty or thirty years. In many people we have contamination without disease. The chemical is in their tissues but they feel well. They may be well. But it may take a generation before we know."

One worker at Michigan Chemical, not examined by the Selikoff team, had already become seriously ill. He was Thomas Ostrander, a young maintenance man whose job had involved repairing machines which bagged PBB. He began to suffer from dizziness, mental confusion, disabling fatigue and weakness towards the end of 1976. By the following March, he was too ill to work. "I felt lousy all the time," he said. "Just couldn't get enough sleep. I tried to go to work but I was exhausted."

The company sent him to various doctors who found liver abnormality and a very high PBB blood level, but could offer no hope of a cure. The sad condition of Ostrander made his workmates fearful for their own health—to which they had not given much serious thought before this—and they insisted upon his being sent to the Mayo Clinic for thorough examination. They knew about Selikoff but mistrusted his study because most of its initial participants had been farmers. They resented the press attention which farmers were receiving while they, who had been directly exposed to PBB, were rarely mentioned. They supposed that money had something to do with this. Everyone knew, they said, that farmers were rich.

So Ostrander went to Mayo, and Michigan Chemical paid the bill. The Mayo physicians' report to his employers was such that they entered a disability claim on Ostrander's behalf. Michigan Chemical's report form to the state Bureau of Workers Disability Compensation contained these entries:

> Nature of injury: PBB Syndrome.
> How did the alleged injury occur? Exposure to PBB.
> Name the object or substance which directly injured the employee: PBB.

Instead of pressing the claim, the company paid Tom Ostrander full wages, month after month, while he remained at home, feeling terrible. His union colleagues were unwilling to push the state disability claim in case Ostrander lost it, and his wages too.

As time went on, some farmers also lost faith in Selikoff. It was all very fine, they thought, that his laboratory kept publishing its research results. But what was it doing for the victims? Detailed reports of every PBB clinic patient's state of health had been sent to his own doctor, but no doctor knew how to cure the effects of PBB. They could only continue to treat the symptoms. The farm families had expected Selikoff to produce a panacea for all their ills. They were asking the impossible and had misunderstood the purpose of his study.

"It was a research project which would help them ultimately, but not overnight," Edie Clark observed. "I wish I had had the foresight to warn them that Selikoff was unlikely to find an immediate antidote for PBB. When he didn't, farmers were very disappointed. After all they had gone through, finally they had found someone who believed them and who was willing to be outspoken. They expected him to deliver a miracle, and they couldn't quite understand why he didn't."

This was the state of events in October 1978, a few weeks before Michigan's elections. William Milliken, who had served as governor since January 1969, was campaigning for another four-year term and for once there was a possibility of defeat. One of the major campaign issues was PBB, and how his office had handled the crisis. Milliken had already admitted that the state government had been too slow in responding to it. His Democratic challenger, William B. Fitzgerald, a lawyer and state senator, was exploiting this. In fact, he was over-exploiting it. One of Fitzgerald's radio campaign commercials inferred that PBB did more damage to human health than any doctor could substantiate, and he was obliged to withdraw the commercial with an apology. Nevertheless, there was so much public concern about PBB contamination that Fitzgerald stood a good chance of winning.

Among his supporters were farmers prepared to vote Democratic for the first time in their lives. In past elections the Republican party stood for all they believed in—private enterprise, middle-class morality, everything decent and conservative. They were the kind of people who went to church every Sunday and hung out the flag on the fourth of July. These observances continued, but PBB had radicalized their politics. Garry Zuiderveen saw this change as an extension of his faith.

"This experience has made bigger people of us," he said. "We can better appreciate the frustrations and the hurt of those who are regarded as second-rate citizens. For a long time I could not understand the racial unrest in this country, the student disorders of the late sixties. Now I know what happens to a person who is stripped of his dignity, why he becomes unruly."

Enough voters were thinking this way for Milliken's staff to be worried. Even Farm Bureau, which endorsed a number of Republican candidates for the legislature as "friends of agriculture," pointedly omitted Milliken's name.

Ten days before the November 7 election, Judge Peterson's decision in the Tacoma case was announced. His verdict was an enormous shock to both sides. He dismissed all the Tacomas' claims against Michigan Chemical and Farm Bureau Services, and ordered the plaintiffs to pay these companies' court costs. The Tacomas had been unable to prove, he wrote, that the low levels of PBB in their herd had been harmful or that they were justified in shooting many of their cattle. On the contrary, Judge Peterson concluded that "in small amounts PBB is not toxic," and that the greatest tragedy of the contamination had been the "needless destruction of animals exposed to low levels of polybrominated biphenyl and even of animals that never received PBB."

The feed and chemical companies hailed the judge's ruling as a triumph of reason, a vindication of what they had been saying all along. Governor Milliken expressed his pleasure at this judicial determination that PBB was not harmful in small amounts. Peterson's judgment, coupled with Fitzgerald's misguided anti-PBB campaign, was generally felt to be a decisive factor in Milliken's re-election.

Judge Peterson said afterwards that it did not occur to him that this opinion might influence the vote. He pointed out that if he had delayed publication until after the election, that could have been seen as a political action, too. He had not been an active Republican during his twenty-two years as a judge, he added, and his only concern had been to deliver the judgment immediately it was ready, before he suddenly dropped dead or his house burned down, destroying his notes.

The opinion was 155 foolscap pages long, with footnotes and appendices. It began almost as though the judge had intended to write a book —setting out the background of the Tacoma case by recalling the early contamination of the Halbert farm. Rick was so put out by a statement that he and the feed company's veterinarian, James McKean, had worked closely together in identifying the contaminant, that he telephoned the judge to point out that this was not so. It was too late, he was told, for a word of the document to be altered.

Abruptly the style of the judge's writing changed, as he turned from the narrative to the Tacomas. He did not spare their feelings. He wrote: "The claims of plaintiffs that defendants have been guilty of criminal concealment, gross neglect, misrepresentation and of working a cover-up with governmental authorities have not only been unproved but appear to be flagrantly irresponsible in view of proofs to the contrary."

In short, he found the case for the feed and chemical companies more believable than the testimony presented by Gary Schenk and his colleagues. He was convinced that Farm Bureau Services did all it could to meet the problem, and he could not accept that Michigan Chemical had been grossly negligent. The amount of PBB which was missing in inventories was much less than the plaintiffs had claimed, Judge Peterson stated. He also thought that the appearance of cross-contamination at the chemical plant may have been illusory, due to the zeal of the company's researchers in trying to test to a lower level than their instruments could accurately measure. The judge found the various studies commissioned by the defense to be much more scientific and dependable than those of the scientists and veterinarians who testified for the Tacomas. He was severely critical of the testimony of two plaintiffs' witnesses, both crucial to the Tacomas' case. One was Dr. David Helland, the Illinois veterinarian who had conducted pathology studies for Doc Clark and Gary Schenk—research which the judge described as "incompetent and dishonest." The other was Doc himself.

The judge wrote, "Plaintiff's counsel picture 'old Doc Clark' as just a 'country vet' uninterested in litigation, concerned only with treating cows, and too busy doing that to keep records or to prepare to testify. The trial picture, however, is that of a man who, on learning about PBB, leaped to a phobic conviction that it was causing every ailment seen in his clients' animals, and devoted his complete energy to the dual cause of being against sin (PBB) and aiding his clients and other claimants in obtaining financial settlements from the defendants. . . . The Court felt very sorry for Dr. Clark during his testimony as he was obviously embarrassed—a man who had apparently voiced a lot of positive and extravagant statements which now, at the moment of truth, he could neither abandon nor substantiate."

Doc's lack of statistical analyses or studies made his conclusions about herd damage suspect, according to the judge—who seemed to be making a direct comparison between the elaborate veterinary research financed by two major corporations and the observations of a rural practitioner working alone. Doc was also criticized for assuming that an entire herd was contaminated on the basis of PBB tests upon only some of the animals.

"How did Dr. Clark's recollection of his clinical and pathological observations stand up?" Judge Peterson asked rhetorically. "Not very well. He conceded that there was no correlation between observed symptoms and his autopsy findings."

Here was the crux of the judge's opinion. He could not accept that PBB poisoning was different from the classic toxicosis, with its dose-response relationship. He noted that no other halogenated hydrocarbon was known to differ from this norm. Doc Clark, he maintained, "was just trying to talk away the fact that some animals with positive tests were healthy while some who fit his syndrome tested ND [non-detectable]."

Judge Peterson was also unable to accept Helland's claim that PBB impaired an animal's immune system. "It is appropriate and reassuring to note at this point that just the opposite has been proved," the judge wrote. "The immune system is not impaired by PBB." This categorical statement was a shock to everyone who had followed the Selikoff studies: Bekesi's work on T-lymphocyte cells and the Mount Sinai team's conclusion that the amount of PBB in tissues did not necessarily correlate with the severity of symptoms.

It was ironic. But legal decisions are made solely on the basis of evidence before the court. Results of the Selikoff studies were not presented in the Tacoma trial and, if the plaintiffs had tried to submit them, this material on human health would doubtless have been ruled irrelevant to a cattle case. Judge Peterson *did* hear about Bekesi's findings —from a judicial colleague who showed him a newspaper clipping when he was in Europe during the trial's summer recess—but, the judge explained, the only immunologist who gave evidence at the Tacoma trial (a defense witness) testified that PBB did not damage the immune system.

Like some of Farm Bureau's consultants, he concluded that most of Roy Tacoma's cattle problems stemmed from his own farming methods. After breaking the partnership with his father, Roy "pushed" milk production and, in the judge's opinion, started his cows on new pregnancies too soon after they calved. "The evidence indicates that rebreeding too quickly may affect the dam, may cause more uterine infections and reproduction problems, and reduces the total amount of milk that a cow will give." The judge concluded that none of the breeding or health problems in the herd, or the additional calf mortality, could be attributed to PBB. The plaintiffs, he said, had failed to demonstrate a connection, or that they were justified in shooting any of their animals. Poisoning is, of course, one of the most difficult crimes to prove, especially when the evidence has been irrevocably buried. Nevertheless, even the defense had conceded that the Tacomas were sold contaminated feed.

What appeared to bother the judge more than any other aspect of the case was the Tacomas' expectation that their entire herd should be replaced, even though some animals had no detectable levels of PBB. Again, there was no way that anyone could determine, four years after the event, whether those cattle had once been contaminated. Most of the poison could have been excreted through their milk, leaving lasting damage to the tissues, but this could not be proved. Judge Peterson felt that the original settlement policy of the feed and chemical companies —of sending entire herds to Kalkaska, on the basis of positive PBB tests in a few animals—was overly generous. He regretted what he felt to be needless slaughter, with the result that farmers like Roy Tacoma felt discriminated against if their herds were not taken too. It was the Tacomas' misfortune, rather than an injustice, stated the judge, that by the time their farm was quarantined "the wholesale destruction of quarantined herds had come to an end."

Roy and Marilyn Tacoma had claimed damages to their entire herd. Judge Peterson felt the claim was not justified, although it was undenied that some cows had PBB levels and the herd had been quarantined. Therefore, he gave them nothing. Or rather, less than nothing. He dismissed their case, and ordered them to pay the defendants' court costs. The final blow to Roy Tacoma was the judge's description of him —"a man whom the Court believes to be fundamentally honest and God-fearing" but who was "found to have told a number of inconsistent and untrue stories."

There was "not a shred of credible evidence," the judge summed up, in the plaintiffs' charges that the defendants had acted wilfully or wantonly. The picture, as he saw it, was that "in the face of the unknown and the clamor that developed in the press" dairy farmers and veterinarians began to "look a second time at all the animal ailments which are ordinarily endured and forgotten" and suspect their cause to be PBB. In this atmosphere, he felt that some professionals had lost their objectivity.

His opinion was so unwavering and unexpected that those who had followed the trial found it hard to believe. Low-level farmers still awaiting settlements were dismayed. The Tacoma attorneys felt crushed. The judgment was made public on a Friday and throughout the weekend the Tacomas did not answer their telephone. When they were not busy with farm chores, they drove around the countryside to avoid hearing it ring. That Sunday they went to another church. They had long since learned to deal with the disapprobation of conservative members of their Dutch community, on account of their break with tradition by going to law, and eventually they would overcome the loss of their suit—but never the fact that Roy had publicly been called a liar. Marilyn and he were stricken by the shame of it.

Judge Peterson received several letters from people in the Tacomas' community. Some were severely critical of him. Others applauded his decision, and one of these included the comment: "I think God is now punishing the Tacomas, and thank you for ruling against them."

The judge smiled as he read it. "I never thought of myself as being God's messenger," he remarked.

Like a stone cast into a lake, Judge Peterson's verdict made waves which spread. Despite Governor Milliken's re-election, Republicans lost some of Michigan's rural vote and a few legislators who had supported Farm Bureau's politics were ousted—a direct reaction from farm families sympathetic to the Tacomas. Donald Albosta won his second bid for a U.S. Congressional seat, campaigning on the PBB issue.

Dale Ball agreed to resign from the Department of Agriculture. He had been under pressure to do so for months, but had refused, fearing that it might be interpreted that he was "running for cover." Now he felt free to bow out gracefully.

Michigan Chemical and Farm Bureau Services felt even less urgency to compensate farmers with outstanding claims, and Michigan attorneys became even more reluctant to take such cases to court. Its long support of the Tacoma trial had crippled Law Weathers financially, and Gary Schenk was given the option of deserting his remaining eighty-three farm clients or pulling out of the partnership. He left Law Weathers bitterly, along with Fred Boncher and two other attorneys who had worked on PBB cases. Gary and Fred set up their own law firm in Grand Rapids, with a legacy of clients unlikely to make them rich.

Paul Greer and Dick Shuster went back to rebuild law practices which their partners had been carrying alone. Gary and Paul filed an appeal, with little hope of success. To win an appeal on a nonjury trial, it is almost necessary to prove that the judge was at fault. Peterson had handed down some tough opinions over the years, but not one which could be held legally questionable.

His decision in the Tacoma case was the talk of Michigan. This was the first quarantined farmer to be told he did not have a claim, and most people were baffled by the judge's reasoning. The verdict was criticized in several newspapers, and most strongly in *Michigan Farmer* which commented, a year later, that with the passage of time, it seemed to be even more "spiteful and politically motivated—and not the reasoned opinion of a brilliant jurist." The Tacoma attorneys acknowledged that they might have improved their case by presenting it more succinctly . . . and yet, they concluded, they could not have changed the outcome. They felt the cards had been stacked against them from the beginning. Dick Shuster, the quiet, independent member of the team, sought to understand the reasoning behind the verdict.

"I think the judge made a political decision—one which was influenced by the belief that the greater good for the people of Michigan could best be served by such a verdict," Dick said. "I do not think he was corrupted. I think he was worrying about Michigan agriculture. If our evidence were listened to, a lot of farmland would have to be abandoned. It is like a battle commander who is prepared to sacrifice a few troops to win a major campaign: the judge was only putting a hundred or so farm families out of business. I think he became convinced that it would be in the best interests of Michigan, politically and economically, to act as if PBB had never happened. But the political arena has a different standard, and it has no place in the court system.

"The judge's decision has ruined many, many lives. I think it has ruined something else—the idea that you can right wrong in the courts. I have never read an opinion so vitriolic and venomous. It is a mean, nasty opinion, a totally unsympathetic opinion, and I think the other side was as dumbfounded by it as we were."

The judge went back to his everyday court business—robberies with violence, damages claims, the occasional murder—outwardly unperturbed by the storm he had caused. He was aware of the Biblical sense of righteousness within the Dutch community, and said he felt sorry for the Tacomas. "I am sure they had a miserable time after the judgment came out," Peterson remarked. "But the business I am in was not designed to make people happy."

His opinion was a month old when he made this comment. It was late November, and Cadillac seemed unusually quiet now that the excitement of the trial had left it. The courthouse car park, once filled with farmers' automobiles and pick-up trucks, was almost empty; and the courthouse itself, a solid piece of municipal architecture, seemed to have lost its importance—its columned brickwork merging with the substantial white frame houses which surrounded it. The illusion was created by a light snow fall which softened the eclectic outlines of the town, making it seem even more secure and calm than usual.

Judge Peterson was sitting in his office on the top floor of the courthouse, watching the snow. Wearing a comfortable woollen cardigan instead of a judicial black robe, he, too, seemed smaller than before. He talked about the evolution of his judgment, how at the beginning of the trial he had "misconceptions about the plaintiffs' case and about PBB which maybe the public shared—that there was no dose-response relationship and that even the smallest amount was toxic." As weeks passed and he compared his notes of the plaintiffs' witnesses with the testimony of a Farm Bureau veterinarian who had gone through the Tacoma herd, cow by cow, he was troubled by discrepancies. And as evidence mounted he became more convinced that it was Roy Tacoma's change in breeding practices, not the contaminant in some of his feed, which caused trouble in his herd.

This feeling was strengthened by a comparison between the amount of PBB in Rick Halbert's feed—as high as 3,000 ppm—with the .09 ppm in some of the mixture which Roy Tacoma bought. "It was minuscule," the judge remarked, "and the evidence indicated that small amounts of PBB are excreted from the body."

Was he saying, then, that there was no such thing as low-level contamination? The judge side-stepped the question. "I am saying," he replied, "that low-level contamination has not been proved." What of the financial losses suffered by Roy and Marilyn Tacoma by having to keep quarantined animals which could not be sold? "That question was never presented to the court," the judge replied, "It was only addressed to me by reporters."

The defense had offered to settle on the basis of the animals above quarantine level, Peterson went on, but the Tacomas refused. They wanted the whole herd replaced. Clearly, he felt they were asking for too much. He could not accept the argument that a good dairy herd is a functioning unit which can never be profitable if the health of some of its animals is destroyed.

"A question I asked myself," the judge continued, "was did the plaintiffs' attorneys try to claim too much? If they had focussed on the animals that were proven to have had some contamination would I have done differently? I do not know. . . . But as they presented it, they did not have a case to come to court with."

Other low-level farmers, he inferred, would probably do no better by going to law—not unless they had better evidence or could prove specific damage done by PBB, instead of more of the same kind of problems which were common in dairy herds.

There was remote chance of that. Even the sophisticated analyses of the Selikoff team had not revealed any new disease, only an unusual concentration of symptoms which could be associated with toxicity. Dr. Stephen Safe, who testified for the Tacomas, had tried to put on the record his belief that PBB is a carcinogen. "But I could not allow him to say so because he had no proof," said the judge.

Peterson seemed gratified that his judgment would allay the public's fears about PBB. "I think we have been lucky," he added. "We have been scared to death, but now we find that PBB may not be as toxic as people thought it was. I am satisfied it is not, on the basis of this evidence. Nowhere near.

"You are left with the regulatory problem of removing it from the environment. Kolbye spoke of the compromises that are involved after the fact; once it is there, what do you do about it? You now have the capability of removing from the food chain everything containing PCB. But if you were to do that, you would wipe out the entire livestock in the United States, all over the world. So the losers would be the farmers. . . . And I would hate to feel, as a taxpayer, that I have the right to go to the government because I have five parts per million of DDT."

It was uncanny. When he expressed his moral outrage at the verdict, Dick Shuster could have been analyzing these remarks by Judge Peterson. Except that Shuster's comments were made before the judge was interviewed.

Judge Peterson's concern—about how far it was practical to clean up the environment—was a real one. Having under-reacted to the PBB crisis four years earlier, many people in Michigan were now reacting too strongly. Since Spaniola's proposals had become law, thousands more cattle were being condemned at slaughter houses. Yet no acceptable way to dispose of their bodies existed. The Kalkaska pit had been sealed, and no community in Michigan would tolerate condemned cattle being buried near its boundaries. The dead animals were like nuclear waste; everyone wanted them disposed of, so long as it was somewhere else.

Michigan's Department of Natural Resources settled on a site near Mio, in Oscoda county. Fifty miles due east of Kalkaska, it had the same kind of landscape—great tracts of jack pine trees with swamps and streams, and more deer living there than people. The entire county had only six thousand scattered inhabitants, but they formed a solid opposition as soon as state authorities began to dig a burial pit.

They hired an attorney and took legal action to hold up the burial, which left nothing to be done with the slaughtered cattle but to store them in commercial deep freezers, pending a resolution of the dispute. For the most economical use of freezer space, the carcasses were cut up and put in barrels. Condemned cattle which had not yet been slaughtered were lodged on farms, with the state paying the cost of their keep.

The argument dragged through the courts from the fall of 1977 until the following summer, with the Oscoda County Commission and the PBB Action Committee on one side, and the state's Department of Natural Resources on the other. Responsibilities had shifted. Under the new law, the state was obliged to compensate farmers for condemned cattle and to get rid of the carcasses, and the state had to be be more sensitive to public opinion than Farm Bureau had been. It agreed to line the Mio pit with twenty feet of clay on all sides which, in the opinion of Howard Tanner, director of the Department of Natural Resources, was "overkill." These animals were nowhere near as heavily contaminated as the Kalkaska carcasses had been. Tanner's staff calculated that between two and three hundred pounds of PBB was contained in the animals buried at Kalkaska. The 3,500 or so animals destined for Mio were estimated to contain only two ounces of the chemical among them. Essentially the Mio dispute was about burying dead cattle, and hardly at all about PBB. While the argument raged, many Michigan citizens must have gone quietly to their graves with more PBB in their bodies than some of these cattle, and no one to protest their burial.

The Oscoda residents would not be mollified. They held up excavations by scattering nails on the road to the pit, and the tires of bulldozers were ripped to pieces. They made effigies of Governor Milliken, Howard Tanner, and Dale Ball and hung them from a gallows at a nearby road junction, where they solicited the support of passing motorists. This demonstration went on for months.

"After a while we gave Milliken a new suit of clothes to make him look half way decent," related Nelson Yoder, one of the local activists. "We gave all three of them glass eyes to observe our point of view, and we lashed their legs together so that, on that windy hill, they sort of danced in unison. One day the governor came to talk to us, and it was a highly dramatic moment when he stood in front of his own gallows."

Eventually, by court order, one of the three proposed Mio pits was dug and filled. Live animals were shot at the site and the barrels brought out of the freezers. By this time, it was rumored that they contained not dead flesh but active chemical waste. A few barrels were opened for inspection but still the protesters were not satisfied. They urged incineration, although there was no incinerator in Michigan powerful enough to do the job effectively. Alternately they wanted shipment somewhere—anywhere—out of the state. In the climate of public awareness which had taken too long to develop, it was suddenly impossible to bury a few spoonsful of PBB without risking a riot.

The Department of Natural Resources searched the country for a disposal site where the last barrels would be acceptable. Eventually it found a privately operated landfill, approved by the federal Environmental Protection Agency, in a desolate area of southern Nevada. It took up to three barrels to hold a full size dairy cow and, Rick Halbert noted, "some of them had sat around in rented freezer space burning up electricity for almost two years."

In January 1980, more than a thousand barrels were loaded into refrigerated trucks and sent to their last resting place. The irony of Michigan going to all this expense to dispose of meat which would probably have passed federal inspection for human consumption in Nevada was not lost on officials in both state governments. Jack Bails of Michigan's Department of Natural Resources made the wry comment: "It was cheaper than the obvious alternatives. Politically and socially it was no longer acceptable to bury cows in Michigan, but I'm sure there was more PBB walking around than there was in those carcasses."

Meantime, farmers still awaiting compensation remained in limbo. Without Law Weathers to finance him, Gary Schenk gallantly prepared to take a second farm case to another court under another judge, and Paul Greer again worked with him. They were almost ready to go to trial when Michigan Chemical offered an out-of-court settlement. This time the attorneys felt unable to refuse. The offer covered all of their eighty-three outstanding cases and precluded any future suits on existing damages, including health. The Tacomas were forgiven payment of their court-ordered costs provided they dropped their appeal, and $3.7 million was allocated to the remaining clients of Gary Schenk, Paul Greer, and Fred Boncher. Out of this sum, $900,000 was to be used to write off debts and interest which the farmers owed to Farm Bureau Services for feed bills. The attorneys took 28 percent of the whole—not a particularly handsome return, they pointed out, for the investment of four years of their lives plus legal expenses. Paul Greer, who had financed himself throughout the trial, said he did not get enough out of it to break even. Dick Shuster, who was not an official member of the team, received nothing. Law Weathers, which had financed Gary and Fred, demanded its share of legal fees, and a dispute developed between Paul and Gary as to how their portions should be divided.

The remaining sum gave the eighty-three farmers an average compensation of a little over $32,000 each. This was a pittance for most of them. Alvin and Hilda Green got $90,000—although the farm which they had to mortgage and the livestock which they had shot had once, by their estimate, been worth $400,000. Reluctant to agree to this settlement, they at first wanted to fight on. "We were forced to accept," Al related. "If we didn't, the mortgage holders were going to foreclose on the farm and machinery. The only thing we still owned, free and clear, was the house, and in order to keep that we had to settle. We didn't get a dime out of it."

Both Al and Hilda Green still had persistent health problems, but were now obliged to sign away their right to claim compensation for these. "We all got a raw deal, all the way through," he complained. "The lawyers were pushed into it, the same as we was. They were good lawyers but they were whipped. Our health is gone. Our credit is no good. As soon as you mention PBB, people back right off from you. Even the doctors here want nothing to do with it."

Their neighbor, Pat Miller, tried to find ways of raising money to keep the legal actions going, but even if the farmers had been able to contribute, their solidarity, like that of the attorneys, was shattered. There were jealousies over the different sizes of the settlements, bitterness—from a few—about the attorneys' inglorious surrender.

"It was the only way to go," said Paul Greer. "We could not hang on much longer. I could have stayed with it, maybe, for another couple of years, but my estate would have been in ruins."

Sandy Schenk spoke for her husband. "Gary has given so much of himself that there is no more left to give. It has affected our home life and our children. I do not want him to be a crusader any more."

It was the end of a road for them all. About seventy-five farm claims remained in the hands of other attorneys, scattered around the state. Gary and Paul had always promised that they were welcome to share the scientific knowledge which had been researched for the Tacoma case. That no longer held good. It was implicit in Michigan Chemical's settlement that the Tacoma attorneys keep their PBB information to themselves. Sooner than Gary had foreseen, the superior resources of those responsible for the contamination had won out.

"Perhaps it makes you fight the harder," Doc Clark commented. "But it's tragic and it ain't right. There was no justice."

On October 19, 1979, eleven days after her forty-fifth birthday, Lois Zuiderveen died of cancer. Before the disease developed, she had talked with concern about the recent high incidence of malignancies among young and middle-aged people in her rural community. It seemed almost like an epidemic. The Selikoff researchers contended that these cancers could not be associated with PBB because they were developing too soon after exposure to the chemical. But Lois and her neighbors wondered. What of the theories that PBB could trigger latent cancers which might not otherwise appear for twenty years? Or hasten the aging process? After her own malignancy was diagnosed, Lois stopped speculating. Typically, she worried more about others—like her neighbors, Roy and Marilyn Tacoma, trying to recover from the ignominy of the trial verdict—and, in her last days, about how well her family would manage after she had gone.

Lois was unusual, even within the breed of hard-working, unassuming farm wives . . . a radiant spirit who touched with her personal warmth everyone she met. A slender woman with bright, steady eyes and a smile which dimpled, she was the cornerstone of her family, the one who kept them all close and faithful when their livelihood was shattered. Yet after it was all over, she admitted she had taken longest to get over the hurt, because she not only suffered with Garry the losses on the farm, but also, vicariously, his anguish when the cows were dying and the business was failing and he felt helpless. She had carried his pain along with her own, and said nothing.

Her funeral service was held at the little white church in the fields where she was baptized and married, the church whose congregation had been divided. On this occasion there was an overwhelming spirit of unity. Almost everyone who had been to the forefront of the PBB battle in that area was there—farmers from miles around, the attorneys from Fremont and Grand Rapids. There were also Farm Bureau supporters from the neighborhood who had strongly disagreed with the stand the Zuiderveens had taken. "You have no idea how hard it was for us to come," one of them told Garry. "But we knew we had to."

The church was filled to overflowing. Some mourners spoke to one another for the first time in several years, sharing a bewildered sadness at Lois's death. Even the pastor, in the thoughtfully chosen words of his homily, Dutch accented, seemed to be struggling to make sense of it for himself, Lois's contemporary, as well as for them. He spoke of God's love and inscrutable purpose, and seemed to be talking not just about Lois but about all the wounds this community had suffered. There was a profound sense of healing in the service. On her way to the funeral, one farmer's wife had burst out, "I feel I should have done something to prevent this. I should have spoken out more, made myself heard. I should not have accepted what they were doing to us. . . ." She was wrong to blame herself, and the pastor's words helped her to see this.

They buried Lois in the small community cemetery, a corn field away from her last home—the new house which Garry could have financed by selling the stricken cattle, which he shot instead. Beyond a turn in the lane was the big old farmhouse which had first belonged to Garry's parents, then to them, now to Garry Junior and his wife. There was also an infant granddaughter. Life would go on, changed but unchanging.

On that sunlit October afternoon the landscape looked like a watercolor, with the reds and golds of dying leaves softened by mist. When the funeral was over, the mourners—at least three hundred of them—embraced by Lois's grave. It was the end of a chapter in their lives, and some of them might never meet again.

EPILOGUE

The Ultimate Experiment

In 1973, the year when the poisoning of Michigan began, polybrominated biphenyl was one of several thousand industrial chemicals manufactured in the United States. The toxicity tests to which it was subject before being marketed seemed to show that, with proper safeguards, it posed no discernible danger to workers who would be handling it or to people using products in which it might be incorporated. Such tests, on small animals under controlled laboratory conditions, are about as far as any toxicologist can go before a new chemical is marketed. The ultimate experiment—to discover the long term effects upon human health—happens only when a chemical gets loose in the environment. Or when, twenty or so years after handling it, workers fall into a pattern of illness for which there can be no other explanation.

PBB came under intense investigation because of the cattle feed contamination. Otherwise, like hundreds of other synthetic organic compounds which are created every year, it would have escaped public attention. It was never a widely used chemical. Although it had been around for three years before the misshipment from Michigan Chemical Corporation, most chemists had never heard of it—which is why it took Rick Halbert, with all his knowledge of chemistry, so long to find anyone who could identify the foreign substance in his cattle feed.

The research which has since been done on PBB would fill a fat textbook, and the careful reader would find a moral in it: the more that chemists try to analyze this—or any other similar compound—the more they realize how little they understand. Because of the Michigan accident, at least nine million people have PBB in their systems. But because of the way in which our world has developed, everyone living in an industrialized society has accumulated a peculiarly personal combination of chemical residues which may, or may not, eventually cause harm. Nobody knows whether some of these cancel others out, or whether, reacting to different human metabolisms and genetic makeups, their combined effect is additive, antagonistic, or synergistic. Given all these unknowns, the accepted concept of a standard "tolerance level" for any contaminant becomes meaningless.

Then there are variables within the chemical itself. One of the few facts about the original Firemaster mixture upon which scientists agreed was that 60-odd percent of the compound consisted of the 2, 4, 5, 2', 4', 5' isomer—that which was used as a yardstick for measuring PBB in animal and human tissues throughout Michigan. But what of the other ingredients? One scientist calculated that there are 209 theoretically possible compounds for PBBs which have this same principal isomer, and the toxicity in every case might be different. There was a collective sigh of relief when a Michigan State University researcher, Dr. Stephen Aust, discovered that the 2, 4, 5, 2', 4', 5' isomer was barely toxic—a finding which led some people to conclude that PBB wasn't so threatening to humans after all. The relief was short-lived. Taking the Firemaster mixture apart, Dr. Stephen Safe and Larry Robertson, working at the University of Guelph, Ontario, showed that some of its minor components were relatively much more toxic—from which the bleak assumption might be drawn that all official estimates about the effects of PBB contamination could have been wrong.

Perhaps it wasn't just the PBB which caused animals and people to sicken. For several years the related chemical, PCB, was thought to be responsible for all the toxic symptoms in the *yusho* victims. More careful analysis showed that the PCB which contaminated the rice oil was in turn contaminated by polychlorinated dibenzofurans, which are between 2 and 3.5 times more toxic than PCB. Meanwhile, the "tolerance level" for PCB in the United States was based on observed symptoms in the *yusho* victims. Subsequently, FDA's tolerance levels for "pure PBB," which didn't exist either, were based on this "pure PCB" calculation. Added to these faulty assumptions was the likelihood that the complex digestive system of a cow may have changed the nature of the chemical. One piece of research led persuasively to this conclusion. A 1975 study funded by Farm Bureau Services showed that animals which had been fed meat from PBB-contaminated cows and chickens became more ill than when they were given the same quantities of PBB directly. This research, begun at Michigan State University, depended upon some laboratory work from the state's Department of Agriculture —which because of "a backlog of work" was not begun until early 1978. The study came closest of any to duplicating the contaminated food eaten by the people of Michigan, yet this important document was not published until September 1979, four years after the research was begun.

Some PBB researchers, with little or no funds, did a splendid job. Others appear to have been guided or restricted by the politics of PBB, depending upon who funded their studies, what the scientists discovered, and whose interests would be served by publication of the results. During the early days of the contamination those who spoke out encountered professional difficulties, or were ignored. There were the attempts to discredit the work of Dr. Irving Selikoff. Dr. Isadore Bernstein's urgent recommendations for a near-zero tolerance level were brushed aside. Physicians who tried single-handed to alert state authorities—like Dr. Thomas Corbett, Dr. Walter Meester, and Dr. David Salvati—became disillusioned. Corbett saw his research funds being restricted and claimed he was passed over for promotion at the university after he spoke publicly about the dangers to human health.

"I tried to take the other road by going first to the authorities, but that didn't work so I talked to the press," Corbett explained. "Talking to the press is not kosher in academic circles, and it suddenly became difficult for me to complete my cancer study. I was promised more time to do research than I was eventually given, and when I tried to get some of the university staff involved—people in pharmacology, toxicology, and biochemistry—they weren't interested. They saw the kind of problems I was having.

"At one time I was proud to be on the staff of the University of Michigan. Here was the state's most prestigious institution, with a medical center which, for years, was considered one of the finest in the world. It should have been involved in the PBB crisis from an early stage of the game, but people there didn't have the guts. I lost respect for the institution, and finally, when I had an attractive offer to work with a private practice group in Ohio, I took it. I was ambivalent about leaving the university, but I had made enough trouble for my family in Michigan. For those of us who tried to sound the alarm when the contamination was discovered, it turned out to be a very bitter experience. We made a lot of enemies."

The experience of Corbett, and of the few other outspoken researchers, gives an insight into the strictures upon scientists as well as into the functioning of bureaucracies. Both the University of Michigan at Ann Arbor and Michigan State University in East Lansing have some of the country's most competent researchers on their faculties. The fact that very few of these people were called upon until the worst of the contamination was over, and that most early PBB research was done outside the state (almost none of it coming from public funds) is a depressing commentary upon the relationship between the universities and government, and upon the political considerations which can affect research grants. Scientists who choose academic rather than industrial employers usually do so because they value professional independence over income. As the political influence of industry spreads, this independence is being eroded. Michigan's two major universities did not become deeply involved in PBB research until it was politically acceptable, by which time—as Gary Schenk remarked—most of the problem had been eaten.

In Michigan, it was almost predestined that the power of industry would also prevail over the grievances of private citizens. If environmental contamination cases must be settled in the courts—there being no other tribunal—there is no practical way that small law firms which serve individual plaintiffs can match the resources of large corporations to sustain lengthy actions and appeals. Environmental litigation requires an expertise beyond the experience of most attorneys, just as environmental medicine is outside the qualifications of most doctors. The farmers' interests might have been better served if an environmentally experienced team of attorneys had filed a class action suit on behalf of all of them, but this would have cost more than the farmers could have raised. Public interest law groups have successfully organized national fund-raising campaigns for cases with less merit than that of the Michigan farmers. But no such groups came forward. It is one thing to champion a civil rights suit, where the issues are clearly black and white; another to prove poisoning, long after contamination, by a substance which appears to have no dose-response relationship and whose toxicity has not been positively established—and to take on a multimillion dollar corporation at the same time.

In this legal vacuum, Farm Bureau Services offered to submit every farm contamination case to binding arbitration. Gary Schenk refused on grounds that this would mean relinquishing his clients' right of appeal. Ultimately he had to relinquish this anyway, because money ran out. The defendant corporations foresaw that this would happen. On the other hand, the farmers did have some legal representation (indeed, of an unusually dedicated and zealous sort) which, in cases of this kind, is rare. Few victims of environmental contamination get a hearing in the courts.

The chemical and feed companies eventually became involved in a complexity of lawsuits, likely to drag out for years. The name of Michigan Chemical was so blemished by the PBB incident that its immediate holding company, Velsicol Chemical Corporation, dissolved the corporation in 1976. Thereafter the plant at St. Louis, Michigan, was operated as one of several owned by Velsicol, an immediate subsidiary of Northwest Industries. The new title had no better image. Velsicol's plants in other states manufactured several environmentally dangerous chemicals including Tris, the fire retardant which was found to be a potential carcinogen; chlordane and heptachlor, garden pesticides which the firm produced for some time after they were widely suspected of being environmentally unsafe; and leptophos, a pesticide banned in the United States but made by Velsicol for export, which allegedly caused neurological damage to some of the men who worked with it.

The Michigan Chemical plant was found to have polluted the land and water around it. Levels of PBB were measured in the Pine River twelve miles downstream from the plant. The state of Michigan imposed a relatively modest fine of twenty thousand dollars for this damage, with the tacit understanding that Michigan Chemical (now Velsicol) would leave the state. Michigan's Department of Natural Resources found that the company had not only poisoned a stretch of the river, which had to be closed as a fishing ground, but had dumped about eighty tons of PBB along with other toxic substances in the Gratiot County landfill, a short distance from the plant. This dumping took place between 1971 and 1973, but Michigan state authorities did not learn of it until February 1977. An unverifiable amount of PBB—thought to be another eighty tons—was reportedly shipped to a waste disposal site near Lewiston, New York, operated by Chemtrol Pollution Services. There were discrepancies in such records as were kept, giving rise to the suspicion that not all of this second load reached the Lewiston dump. In the early 1970s, and throughout all the years before, chemical waste was treated much like garbage. Some of it was buried with care. A lot was casually and dangerously dumped, with no clear accounting of what went where. The corporation which operated the Lewiston facility also owned an incinerator and land disposal unit near Saginaw, Michigan, and officials at Michigan's Department of Natural Resources suspected that some of the PBB waste destined for New York State may have gone there.

"If it was put in the incinerator it would have come back as a vapor which would have distributed itself over Saginaw Bay," said Jack Bails, the department's chief of environmental enforcement. "We were very concerned about this, but we were unable to get the shipping records."

The Gratiot County landfill was improperly lined, and both surface and ground water in the area were found to be contaminated. Some sense of the enormity of the problem lies in the fact that while toxic waste containing about 160 tons of PBB may have been buried or burned without proper safeguards, it was estimated to have taken only 1,000 pounds of PBB to contaminate the entire state of Michigan.

Early attempts to shift the liability to the chemical and feed companies were not very successful. The U.S. Attorney in Grand Rapids, James Brady, brought the first criminal charges against the two companies, in connection with the PBB contamination, in November 1977, alleging gross negligence, adulterating foodstuffs, and shipping this food across state lines. Over Brady's strenuous protests the companies were permitted to plead "no contest"—a plea which could not be used as an admission of guilt at future proceedings—and fined the maximum of four thousand dollars each. Brady complained about the difficulty of bringing corporate defendants to justice. "When it comes to the protection of the law they get all the benefits," he said. "When it comes to the burdens, they seem to be sheltered under the law." Almost two years later, federal authorities filed a lesser suit against two senior employees of the former Michigan Chemical for allegedly giving false information to FDA inspectors in order to minimize their company's part in the contamination.

The biggest suit against the chemical and feed companies was filed in February 1978 by the state of Michigan, seeking $120 million in damages from the two to cover the cost of cleaning up the environment and repairing some of the damage done to people and property by the PBB contamination.

None of these actions was of direct help to people personally injured by PBB. Among those who fared the worst and received the least consideration were the workers at the former Michigan Chemical plant. Rather than pay the price demanded by the state's Department of Natural Resources in the spring of 1976—a complete cleanup of the plant and surrounding area—Velsicol decided to close down the operation. State authorities gave Velsicol permission to operate until September 1, 1978, provided it met certain pollution control requirements. The standards were met, while the company made plans to move production of its bromine-based fire retardants to its plant in El Dorado, Arkansas, and tried—without success—to find a buyer for the Michigan buildings and machinery. Eventually, with Velsicol's encouragement, a group of fifteen senior employees headed by Charles Touzeau, the St. Louis plant manager, attempted to raise the capital to make the purchase themselves. As president of the St. Louis Chamber of Commerce, Touzeau was concerned about the economic effect upon a community of four thousand of the loss of three hundred jobs, along with the tax income from the biggest employer in the area. The local unemployment rate was already 13 percent. Officials in the Department of Natural Resources showed sympathy for the plan, which included an undertaking that Velsicol would pay the cost of cleaning up the area, and the new company would not produce highly toxic chemicals.

By September 1, negotiations were almost complete, except for one major difficulty. If after the agreed cleanup, other areas of past contamination were to be found, neither the employees' group nor Velsicol wanted to bear the additional cost of decontamination. Touzeau felt he could reach an understanding with Velsicol, given a little more time, but the state attorney general's office refused to extend the deadline. Touzeau was crushed.

"We were one week away from concluding an agreement," he related. "In my opinion the attorney general's office killed the deal, and it was a politically motivated action. There was a lot of public feeling against Velsicol. If only Judge Peterson's decision had come a few weeks earlier, it would have helped the atmosphere in which we were negotiating."

On August 29, 1978, Touzeau called the workers together and told them the plant would close. He was so emotional he could barely finish the announcement. "I have been here for thirty-three years," he explained afterwards, "and it is difficult to see friendships break up. You cannot realize how close a family it was."

Many of the men were unable to find work elsewhere; Touzeau was one of the few to be offered a transfer to another Velsicol plant. Others had to settle for jobs seventy miles from home. Some employers were reluctant to take Velsicol men, judging them to be poor health risks, and the question of future compensation for PBB-related diseases remained indeterminate. Those already disabled, like Tom Ostrander, were left without pay when the plant closed. Former members of their union local filed a class action suit against Velsicol, demanding $250 million in damages for alleged criminal neglect in exposing workers to PBB and other dangerous chemicals. Meantime, at the union's request, the National Institute of Safety and Health had done a medical survey of former Michigan Chemical workers and found that 75 percent of them had some form of liver disease. There were also numerous neurological complaints, lung problems, a high incidence of skin eruptions, and evidence of damage to the immune system. The report pointed out that the men had been exposed to a variety of toxic chemicals, including DDT, benzene, ethylene dichloride, hydrochloric acid, lead, methyl alcohol, methyl bromide, Tris, phenol, and PBB—all of them posing serious health hazards.

The NIOSH report, published almost two years after the men were examined, urged that "a concerted effort must be made to improve working conditions in every area of the plant." This recommendation came too late. By then the plant had been closed for more than a year. All that remained was a complex of contaminated empty buildings on the outskirts of an impoverished small town.

Heavily contaminated dairy food, seized from supermarket shelves after the PBB accident was discovered, was buried in waste dumps throughout Michigan—so were thousands of pounds of PBB-laced cattle feed. Several severely contaminated animals from the Halbert herd and others were taken to Michigan State University for experimental work and, it was revealed long afterwards, their bodies incinerated. A temperature of at least 2,000 degrees fahrenheit is needed to destroy PBB and none of the university's incinerators could produce such heat. This incident alone must have released an unhealthy cloud over the campus.

Throughout the worst years of PBB contamination, federal and state authorities acted as though PBB was somehow confined within the boundaries of Michigan. Yet there was ample evidence that Michigan cattle were sold in neighboring states; that contaminated chickens had probably ended up in nationally distributed canned soup; that PBB-contaminated Aureomycin Crumbles—the medicated feed which Farm Bureau Services made for American Cynamid—were sold outside Michigan; that PBB waste material may have been trucked as far as New York State. None of the bureaucracies made any significant attempt to trace these sources of potential contamination. Dr. Isadore Bernstein was appalled but not surprised. "The analytical problems are stupendous," he commented. "There are also a host of social, political, and economic reasons why it is better to do nothing."

Eventually, in July 1977, the federal Environmental Protection Agency picked from Michigan Chemical's list of 130 PBB customers the name of the Borg-Warner Corporation, whose plant at Parkersburg, West Virginia, had been a heavy user of the chemical to fireproof business machines and electrical housings. Although Borg-Warner discontinued using PBB in September 1975, almost two years later EPA inspectors found traces of PBB in catfish in the Ohio River near Parkersburg—an area in which catfish restaurants are popular.

After the Michigan accident was discovered, EPA officials understood—and publicly stated—that all U.S. production of PBB had ceased. But early in 1977 it came to light that in addition to the estimated 11.8 million pounds turned out by Michigan Chemical, two New Jersey companies—White Chemical Company of Bayonne and Hexcel Corporation of Sayreville—made about a million pounds of PBB between them, for export to Europe in 1976. PBB was also being made in West Germany. Although no chemical spills had been reported from White or Hexcel, PBB was found in fish, plants, soil, and water near the two plants. Traces of PBB were also discovered in fish near the Standard T Company on Staten Island, New York, a firm which used PBB in the manufacture of wire coatings. Clippings of human hair collected from barbers' shops in the neighborhoods of all three plants contained traces of PBB.

In January 1977 the federal Toxic Substances Control Act took effect. This legislation was the direct result of public awareness that the increasing number of chemicals in commercial use posed a serious threat to human health. Designed to regulate their manufacture, the act required the EPA to be notified of all new chemicals and gave the agency the authority to stop production of any which threatened health or the environment. If the Toxic Substances Control Act had existed when Rick Halbert's cows reacted to their first dose of bad feed, there would have been a mechanism for prompt identification of PBB. Or such a toxic chemical might have been banned from the beginning.

By the end of 1979, PBB was no longer a news story in Michigan. It emerged that the state had other chemical contaminations, lesser and localized, caused by careless dumping and by environmental pollution from factories, and these were of pressing concern. In fact, the Great Lakes states were generating more toxic wastes than any other region in the nation. Imperceptibly, chemicals had become an invisible aspect of Michigan's landscape. The state's motto, *Si quaeris peninsulam amoenam circumspice* ("If you seek a pleasant peninsula, look around you"), was an invitation to enjoy the countryside—and, indeed, visitors were still attracted by vistas of lakes and lagoons, left by the ebbing of prehistoric seas. Residents, however, were obliged to live with what man had made of mineral deposits from the same ancient origins.

The PBB crisis created an awareness of these hazards. Before it happened, state authorities had only a limited realization of mercury and PCB contamination of Great Lakes fish. By the end of the 1970s Michigan's Department of Natural Resources had discovered fifty thousand sources of potential pollution, threatening some of the state's underground water supplies. People in some Michigan communities were warned that they might have to drink bottled water for years. Dioxins —the most toxic class of man-made chemicals—were found in river fish near the Midland plant of Dow Chemical; state investigators discovered more than a hundred chemical compounds in the vicinity of the Hooker Chemical Company plant near Montague; the small town of Adrian was seriously contaminated by Curene 442, a chemical made in a local plant, known to cause cancer in animals.

Michigan formed its own Toxic Substances Control Commission with Rick Halbert as its first chairman. His was an obvious, yet ironic, appointment. Here he was, charged with responding to other toxic crises in the state, when his own was still plaguing him: PBB, he knew, was so bonded to the soil of his farmland that he might never be rid of it. One of his commission's tasks was to alert the appropriate state agencies to chemical contaminations so that the delays of the PBB incident would not be repeated. But the commission had only limited authority and was resented by state officials in other departments. Rick became impatient with some of the staff at the Department of Public Health who, in his opinion, were still too slow to recognize the health effects of environmental pollution. "They have grown up in a world of infectious diseases and unless they can see and measure the cause in a petri dish, they still conclude that the illness must be in a person's mind," he complained.

Little more could be done about the PBB contamination, yet the episode might leave its mark indefinitely. After Lois Zuiderveen's funeral, some of the mourners gathered at Doc Clark's house, and when the conversation inevitably turned to PBB Doc passed around a recent snapshot of a calf born in the neighborhood. "That animal is supposed to be black and white," he remarked, quietly. "Instead it's gray, with hyperkeratotic skin and swollen joints and contracted tendons. It's a second generation PBB animal, and its mother was normal. I believe it is a mutation—that there has been a change in the genes which skipped a generation. I can't prove this, but if I'm right it's mind-boggling. And the most mind-boggling thing of all is that this kind of thing could happen with any of the other chemical contaminations we are hearing about."

In Lansing, Francis Spaniola—whose legislation for a lowered PBB tolerance level would have only a limited life before testing ceased—expressed a similar sentiment.

"I doubt if we can ever deal with the PBB problem properly," he said. "I have a notion that as time goes on Michigan people may suffer from a lot of different diseases, probably due to PBB, and that the bureaucrats who failed us will attribute these illnesses to every other circumstance they can imagine. They will evade the issue, and there never will be a way for any of us to prove that the poisoning of Michigan was the cause."

REFERENCES

Chapter I

Firemaster BP-6—A New Flame Retardant Additive from Michigan Chemical Corp. Technical Bulletin dated May 6, 1970. *Firemaster BP-6—Product Information Bulletin.* Michigan Chemical Corp., Chicago.

Chapter II

Deposition of Frederic L. Halbert of March 7, 1975, in the case of Gene Goering, his wife, Janet Goering, and their children, Michelle, Gene, Jr., Jana, Kristi Goering, *vs.* Farm Bureau Services et al.; also Carl and Julia Wynkoop *vs.* Farm Bureau Services et al., Circuit Court of Hillsdale, Michigan.

Testimony of Frederic L. Halbert before the Sub-Committee on Science, Technology & Space, U.S. Senate Committee on Commerce, March 28, 1977.

"Is PBB Poisoning Partly X-Disease?" Article by Dick Lehnert in *Michigan Farmer.* May 15, 1976.

"A Toxic Syndrome Associated with the Feeding of Polybrominated Biphenyl-Contaminated Protein Concentrate to Dairy Cattle" by Ted F. Jackson, DVM, and Frederic L. Halbert, MS, *Journal of the American Veterinary Medical Association.* September 1, 1974.

Statement on History of PBB Accident—the Role of the Federal Government in Chemical Contaminations by Donald R. Armstrong, executive vice president, Farm Bureau Services Inc., before the Sub-Committee on Science, Technology & Space, U.S. Senate Committee on Commerce. March 31, 1977.

Correspondence of the late Dr. Ted F. Jackson, 1973–74, supplied by Lois Jackson.

Chapter III

Correspondence of the late Dr. Ted F. Jackson, 1973–74, supplied by Lois Jackson.

Chapter IV

Farm Bureau Services: inter-office memorandum from Dr. James D. McKean to Donald Armstrong re Halbert Dairy Problem, dated February 22, 1974.

Farm Bureau Services: Minutes of Staff Meetings of October 8, 15, 29, 1973, and January 2 & 7, 1974.

Farm Bureau Services: inter-office memorandum from Dr. James D. McKean to Donald Armstrong re telephone conversations with Tony Grusczynski and Rick Halbert, dated January 25, 1974.

"Experimentele leverporfyrie bij vogels," with summary in English, thesis of J. J. T. W. A. Strik. Utrecht. September 19, 1973.

Letter from Dr. Joel Bitman of the U.S. Dept. of Agriculture Research Station, Beltsville, Maryland, to the Editor of the *Chemical and Engineering News.* October 12, 1971.

Chapter VI

Statement of Dr. Francis J. Mulhern, administrator, Animal & Plant Health Inspection Service, U. S. Dept. of Agriculture before Sub-Committee on Science, Technology & Space, U.S. Senate Committee on Commerce. March 31, 1977.

Farm Bureau Services: inter-office memorandum from Dr. James D. McKean to Donald Armstrong re report of phone conversation with Agway Research Farm. March 12, 1974.

Chapter VII

Farm Bureau Services: inter-office memorandum from Dr. James D. McKean to Donald Armstrong re Yale Visitation. April 10, 1974.

Chapter VIII

The written opinion of Circuit Court Judge William R. Peterson in the case of Roy M. Tacoma and Marilyn K. Tacoma *vs.* Michigan Chemical Corp. and Farm Bureau Services et al., Circuit Court for the County of Wexford, Michigan. October 26, 1978.

Statement of Dr. Albert C. Kolbye, Jr., associate director for Sciences, Bureau of Foods, Food & Drug Administration, before the U. S. Senate Committee on Commerce. March 31, 1977.

Chapter IX

Cook. Charles W., *The Brine and Salt Deposits of Michigan: Their Origin, Distribution and Exploitation.* Wynkoop Hallenbeck Crawford Co. of Lansing, Michigan, 1914

Guide to Fabric Flammability. Revised June 1976. U. S. Consumer Product Safety Commission.

Statistics on fire-related deaths and injuries from the President's Commission on Fire Protection and Control.

"Children Absorb Tris-BP Flame Retardant from Sleepwear: Urine Contains the Mutagenic Metabolite 2,3-Dibromopropanol" by Arlene Blum, Marian Deborah Gold, Bruce N. Ames, Christine Kenyon, Frank R. Jones, Eva A. Hett, Ralph C. Dougherty, Evan C. Horning, Ismet Dzidic, David I. Carroll, Richard N. Stillwell, Jean-Paul Thenot, published in *Science,* Vol. 201. September 15, 1978.

"Estimating the Cancer Hazard to Children from Tris-treated Sleepwear" by Robert H. Harris, Ph.D., Environmental Defence Fund: A Memorandum to the Consumer Product Safety Commission in Support of E.D.F's petition to Ban the Sale of Tris-Treated Wearing Apparel. March 8, 1977.

"Acute Toxicity and Irritation Studies on Firemaster BP-6 Hexabromobiphenyl" for Michigan Chemical Corp: Report U-169 by Hill Top Research Inc., Miamiville, Ohio. May 22, 1970.

Firemaster BP-6, Health & Safety Warning: Michigan Chemical Corp., Chicago. December 28, 1971.

Firemaster BP-6, A New Flame Retardent Additive from Michigan Chemical Corp. Second Revision April 26, 1971. Brochure from Michigan Chemical Corp., Chicago.

Statement of Dr. Paul Hoffman, chairman of Velsicol Chemical Corp., before Sub-Committee on Science, Technology & Space of the U. S. Senate Committee on Commerce. March 31, 1976.

Grievance report, Local 7-224 of the Oil, Chemical and Atomic Workers' International Union, February 21, 1974, with the management's written reply.

Written statement from Local 7-224 of the Oil, Chemical & Atomic Workers' International Union, presented to the Environmental Protection Agency. December 1, 1977.

"The Toxicology of Brominated Biphenyls—Haskell Laboratory for Toxicology & Industrial Medicine," E. I. du Pont de Nemours & Company Inc., of Wilmington, Delaware. Paper presented at the Society of Toxicology Meetings at Williamsburg, Virginia. March 8, 1972.

Statement by Dr. Perry Gehring, assistant director of toxicology at Dow Chemical U.S.A., presented to the Sub-Committee on Science, Technology and Space of the U.S. Senate Committee on Commerce. March 30, 1977.

Carson, Rachel. *Silent Spring.* Houghton Mifflin Co. 1962.

"Water Pollution Aspects of Polybrominated Biphenyl Production: Results of Surveys in the Pine River in the Vicinity of St. Louis, Michigan." Report by John L. Hesse, presented at the Second National Conference on Complete Water Re-use, Chicago. May 4–8, 1975.

Press release from Michigan Dept. of Public Health on PBB contamination of water and fish in the Pine River. November 21, 1974.

Testimony of members of Local 7-224 of the Oil, Chemical and Atomic Workers' International Union, before the Michigan House of Representatives Special Committee to Study Effects of Polybrominated Biphenyls. January 18, 1978.

Testimony of Steve Wodka of the Oil, Chemical & Atomic Workers' International Union before the Environmental Protection Agency hearing at Lansing, Michigan. December 1, 1977.

Testimony in the suit by Roy M. Tacoma and Marilyn K. Tacoma *vs.* Michigan Chemical Corp., Farm Bureau Services et al, Circuit Court for Wexford County,

Michigan. Testimonies by management officers of Michigan Chemical Corp.: Dr. A. Fred Kerst, vice president of Research & Development, Chicago; Howard A. Washer, product manager for Magnesia Chemicals, Chicago; Lynn Hahn, manager of the Analytical Group, Research & Development Dept., Chicago; Charles L. Touzeau, plant manager, St. Louis; William Thorne, manager of operations, St. Louis; Richard Jeffries, asst. to the manager, St. Louis.

The written opinion of Circuit Court Judge William R. Peterson in the above case. October 26, 1978.

Chapter X

"Contamination of Some Dairy Cattle Feed is Indirectly Linked to '73 Paper Shortage." *Wall Street Journal.* May 8, 1974.

Depositions of Paul Mullineaux, Charles Szeluga, Ronald Lee Jex (all of Farm Bureau Services) filed in Missaukee County Clerk's office, Michigan, in the suit of Farm Bureau Services et al *vs.* Northwest Industries, Michigan Chemical Corporation and Michigan Salt Company. Depositions taken between March and July, 1975.

Testimony in the suit by Roy M. Tacoma and Marilyn K. Tacoma *vs.* Michigan Chemical Corp., Farm Bureau Services et al, Circuit Court for Wexford County, Michigan. Testimonies by officers of Farm Bureau Services: Donald Shepard, sales manager of the Feed Department; Donald R. Armstrong, general manager and executive vice president.

Chapter XI

Depositions of Paul Mullineaux, Jack Galbreath, Ronald Lee Jex, Robert Wonegeshik and Wayne Edwards (all of Farm Bureau Services) filed in Missaukee County Clerk's office, Michigan, in the suit of Farm Bureau Services et al. *vs.* Northwest Industries, Michigan Chemical Corp. and Michigan Salt Company. Depositions taken between March and July 1975.

The written opinion of Circuit Court Judge William R. Peterson in the case of Roy M. Tacoma and Marilyn K. Tacoma *vs.* Michigan Chemical Corp. and Farm Bureau Services et al., Circuit Court for the County of Wexford, Michigan. October 26, 1978.

"Poison Residue Traces Shut Three Grain Elevators". News report in the *Detroit News.* March 21, 1975.

United Press International report on dust contamination at Stanwood elevator, and dismissal of Michael Creighton. April 17, 1977.

"Witness Says Co-op Close Order Not Carried Out". News report in the *Cadillac Evening News.* March 16, 1977

"PBB Recycled for Pets, Livestock." Article by Roy Howard Beck and Lois Servaas, *Grand Rapids Press.* March 20, 1975.

Michigan Department of Agriculture news release. May 13, 1974.

Memorandum from Farm Bureau Services addressed to all Farm Bureau Services feed dealers. May 14, 1974.

Memorandum from the McBain-Falmouth Co-operative Company of McBain, Michigan, headed "Attention All Concerned Dairy Farmers." May 17, 1974.

Letter from B. Dale Ball, director of Michigan Department of Agriculture, to the Hon. William G. Milliken, governor of Michigan. May 13, 1974.

Memorandum from Dr. George L. Whitehead, deputy director of Michigan Department of Agriculture, to B. Dale Ball, director, Michigan Department of Agriculture, headed "PBB Investigation—Dairy Herds." May 9, 1974.

Testimony of B. Dale Ball before Sub-Committee on Science, Technology and Space of the U. S. Senate Committee on Commerce. March 31, 1977.

Chapter XII

Michigan Department of Agriculture: Agenda and minutes of PBB Feed Contamination Conference at W. C. Geagley Laboratory, East Lansing, May 17, 1974.

Transcript of tape recording of the above conference.

Notes of a participant at the above conference.

"Epidemiologic Study on *Yusho,* a Poisoning Caused by Ingestion of Rice Oil Contaminated with a Commercial Brand of Polychlorinated Biphenyls." Report by Masanori Kuratsune, Takesumi Yoshimura, Junichi Matsuzaka, and Atsuko Yamaguchi. *Environmental Health Perspectives.* April 1972.

Huddle, Norie and Reich, Michael. *Island of Dreams: Environmental Crisis in Japan.* Autumn Press.

Testimony by Dr. A. Fred Kerst of Michigan Chemical Corp., and by Donald R. Armstrong and Donald Shepard, both of Farm Bureau Services, in the trial of Roy M. Tacoma and Marilyn K. Tacoma *vs.* Michigan Chemical Corp. and Farm Bureau Services et al., Circuit Court for the County of Wexford, Michigan, 1977–78.

Testimony of Dr. I. Eugene Wallen, deputy director, Office of Toxic Substances, Environmental Protection Agency, before the Sub-Committee on Oversight and Investigations of the U. S. House Committee on Interstate and Foreign Commerce. August 3, 1977.

"Federal Efforts to Protect Consumers from Polybrominated Biphenyl Contaminated Food Products." Report of the Comptroller General of the United States. June 8, 1977.

Michigan Department of Agriculture press release dated May 17, 1974.

Chapter XIII

Draft of the Albosta Committee report for Michigan State Legislature (Undated. No final report).

Statements of B. Dale Ball, director of Michigan Department of Agriculture, and Dr. George L. Whitehead, deputy director, before Sub-Committee on Science, Technology and Space of the U. S. Senate Committee on Commerce. March 31, 1977.

Text of House Bill no. 6115 before the 77th Legislature, State of Michigan, regular session of 1974.

Letter from Frank J. Kelly, Attorney General, State of Michigan, to B. Dale Ball, director of Michigan Department of Agriculture. June 10, 1974.

Letter from Elton R. Smith, president of Michigan Farm Bureau, to B. Dale Ball. June 17, 1974.

Letter from B. Dale Ball to Elton R. Smith. June 28, 1974.

"The Contamination Crisis in Michigan: Polybrominated Biphenyls." Report from the Michigan State Senate Investigating Committee on Polybrominated Biphenyls. July 1975.

Michigan Water Resources Commission, Bureau of Water Management, Department of Natural Resources—staff report. June-August, 1974.

Associated Press report, datelined Detroit and published August 23, 1974, quoting Harry Iwasko, assistant attorney general, state of Michigan.

Memorandum from B. Dale Ball to County Commissioners headed "Information on Cattle Contamination." August 9, 1974.

Memorandum from the McBain-Falmouth Co-operative Company of McBain, Michigan, marked "Attention: All Concerned Dairy Farmers," May 17, 1974.

Chapter XIV

Testimony of Dr. George Fries before the Public Health Committee of Michigan House of Representatives. March 7, 1977.

Statements and testimony by Dr. George L. Whitehead, deputy director, Michigan Department of Agriculture; Dr. Maurice S. Reizen, director, Michigan Department of Public Health; Dr. Albert C. Kolbye, associate director for Sciences, Bureau of Foods, the Food and Drug Administration, before the Sub-Committee on Science, Technology and Space of the U. S. Senate Committee on Commerce. March 31, 1977.

Statement and testimony of Dr. Albert C. Kolbye at Michigan Department of Agriculture's public hearing on PBB standards for food, Lansing, Michigan. May 29, 1975.

Letter from Dr. Maurice S. Reizen, director of Michigan Department of Public Health, to Bobby D. Crim, Speaker of the Michigan House of Representatives. February 24, 1977.

"Annals of Chemistry: A Compelling Intuition" by Paul Brodeur. The *New Yorker.* November 24, 1975.

Corbett, M.D. Thomas H. *Cancer and Chemicals.* Nelson-Hall, Chicago, 1977.

Memorandum of telephone conversation between Dr. Donald Islieb, Michigan Department of Agriculture, and Sam D. Fine, associate commissioner for Compliance, Food and Drug Administration. November 1, 1974.

Chapter XV

Statement and testimony of Dr. Howard A. Tanner, director of Michigan Department of Natural Resources, before the Sub-Committee on Science, Technology and Space of the U. S. Senate Committee on Commerce. March 30, 1977.

"A Toxic Syndrome Associated with the Feeding of Polybrominated Biphenyl-Contaminated Protein Concentrate to Dairy Cattle" by Ted F. Jackson, DVM, and Frederic L. Halbert, MS, *Journal of the American Veterinary Medical Association.* September 1, 1974.

Michigan Department of Agriculture press release dated July 18, 1974.

Chapter XVI

Letter from the McBain-Falmouth Co-operative Company addressed to all customers. October 31, 1974.

Letter from Dr. Richard A. Copeland of the Environmental Research Group Inc. of Ann Arbor, Michigan, to Dr. James McKean of Farm Bureau Services, enclosing graphs of a series of tests of the PBB content in Jim Cronin's No. 6 and No. 12 cows.

Testimony of Roy M. Tacoma in the trial of Roy M. Tacoma and Marilyn K. Tacoma *vs.* Michigan Chemical Corp. and Farm Bureau Services et al., Circuit Court for the County of Wexford, Michigan. March 11, 1977.

Chapter XVII

Testimony of Yvonne Yarnell and Ronald Creighton before Sub-Committee on Science, Technology and Space of the U. S. Senate Committee on Commerce, March 28, 1977. (Mrs. Yarnell's first name is incorrectly stated in the printed record as Bonnie.)

Testimony of Dr. Walter Meester, director and toxicologist, Western Michigan Poison Center and Dept. of Clinical Toxicology, Blodgett Memorial Medical Center, before the above committee. March 29, 1977.

Testimony of Dr. Walter Meester at Michigan Department of Agriculture's public hearing on PBB standards for food, Lansing, Michigan. May 29, 1975.

"Human Toxicology of Polybrominated Biphenyls" by Walter D. Meester, MD, PhD, and Daniel J. McCoy, Sr., PhD, of Blodgett Memorial Medical Center, Grand Rapids. Presented in the symposium of environmental toxicology at the joint meeting of the American Academy of Clinical Toxicology, the American Association of Poison Control Centers, and the Canadian Academy of Clinical Toxicology, in Seattle, Washington. August 4, 1976.

Critique on the Michigan Department of Public Health Study on the Short Term Effects of PBB on Health. Walter D. Meester, MD, PhD. June 6, 1975.

"The Short Term Effects of PBB on Health." Study by Michigan Department of Public Health. May 1, 1975.

Letter from Dr. David T. Salvati to Michigan State Representative Donald Albosta re PBB and tolerance level. March 4, 1975.

"Farmer to Sell Sheep He Says Are Tainted." News report in the *Detroit News.* March 4, 1975.

Chapter XIX

"PBB: the Poison Puzzle" by Roy Howard Beck and Lois Servaas. Thirteen-part series published in the *Grand Rapids Press* during March 1975.

Reply to State Task Force charges, *Grand Rapids Press.* April 2, 1975.

Three part series of articles by Ellen Grzech and Kathy Warbelow. *Detroit Free Press.* March 13, 14, and 15, 1977.

"PBB: Answers Taking Shape" by Paul Courter and Dick Lehnert. *Michigan Farmer.* November 1, 1975.

"Nightmare on Michigan Farms" by Dan Morgan. *Washington Post.* March 22, 1976.

"Cheap Chemicals and Dumb Luck" by Curtis K. Stadtfeld. *Audubon* magazine. January 1976.

"T.V. Drama on PBB Story Rescheduled." Associated Press report, datelined Detroit. October 1978.

Transcript of Public Hearing on Polybrominated Biphenyls—Standards for Food, held by Michigan Department of Agriculture. Lansing, Michigan. May 29, 1975.

Chapter XX

Biologic Response of Cattle to Low Levels of Firemaster FF-1. F. R. Robinson, DVM, PhD, professor, Toxicology-Pathology, School of Veterinary Medicine, Purdue University, West Lafayette, Indiana.

Chapter XXI

"Another Halogenated Hydrocarbon" by Larry W. Robertson and David P. Chynoweth. *Environment.* September 1975.

Industrial hygiene walk-through survey report of the Cincinnati Chemical Processing Company, Batavia, Ohio, by the Public Health Service of the Department of Health, Education and Welfare. August 5, 1977.

"PBB Testing May Lead to Vet's Arrest" by Jane Haradine. News report in the *Grand Rapids Press.* February 26, 1976.

"No Charges Pressed Against Clark." News report in the *Cadillac Evening News,* Michigan. May 19, 1976.

Memorandum from Kathy Stariha to Pat Babcock (both members of Governor William Milliken's staff). Subject: PBB—Confiscation of files from Purdue University. March 15, 1976.

Berger, Samuel. *Dollar Harvest: the Story of the Farm Bureau.* Heath Lexington Books, 1971.

Statement of Robert E. Braden, administrative director, Michigan Farm Bureau to the Sub-Committee on Science, Technology and Space of the U. S. Senate Committee on Commerce. March 31, 1977.

Investigation of Problem Dairy Herds. Donald Hillman, PhD, David Bolenbaugh, BS, and Edward M. Convey, PhD, Department of Dairy Science, Michigan State University. March 12, 1976.

Memorandum by Donald R. Armstrong of Farm Bureau Services of his telephone conversation with B. Dale Ball, director of Michigan Department of Agriculture, on May 28, 1974.

Testimony of Donald R. Armstrong in the trial of Roy M. Tacoma and Marilyn K. Tacoma vs. Michigan Chemical Corp. and Farm Bureau Services et al., Circuit Court for the County of Wexford, Michigan. June 9, 1977.

Memorandum from Ken Jones, risk manager, to Paul D. Mullineaux, plant manager, both of Farm Bureau Services. Subject: dates when the Battle Creek feed plant was cleaned. January 1, 1975.

"Formation of Brominated Dibenzofurans from the Pyrolysis of the Polybrominated Biphenyl Fire Retardant, Firemaster FF-1." Patrick W. O'Keefe, Department of Biochemistry and Molecular Biology, Harvard University.

"Persistent Liver Lesions in Rats after a Single Oral Dose of Polybrominated Biphenyls (Firemaster FF-1) and Concomitant PBB Tissue Levels." Renate B. Kimbrough, Virlyn W. Burse and John A. Liddle. Center for Disease Control, Atlanta, Georgia.

"Firemaster BP-6: Fractionation, Metabolic and Enzyme Induction Studies." S. Safe, J. Kohli and A. Crawford, Department of Chemistry, University of Guelph, Ontario, Canada.

"Toxicity of PBBs with special reference of porphyrinogenic action and spectral interaction with hepatic cytochrome P-450." J. J. T. W. A. Strik, Department of Toxicology, Agricultural University, Wageningen, The Netherlands.

"Responses of Nonhuman Primates to a Polybrominated Biphenyl Mixture." L. K. Lambrecht, D. A. Barsotti and J. R. Allen, University of Wisconsin Medical School Department of Pathology and Regional Primate Research Center.

"Some General Effects of Polybrominated Biphenyls on Nonhuman Primates." J. R. Allen, L.K. Lambrecht and D. A. Barsotti, University of Wisconsin Medical School Department of Pathology and Regional Primate Research Center.

Chapter XXII

"The Contamination Crisis in Michigan: Polybrominated Biphenyls." Report from the Michigan State Senate Special Investigating Committee. July 1975.

Text of Enrolled House Bill no. 5033 (an act to provide loans to persons having livestock or poultry destroyed as a result of polybrominated biphenyl poisoning), State of Michigan, 78th Legislature.

Communication from Governor William G. Milliken to members of the House of Representatives (veto of House Bill no. 5033). September 2, 1975.

"U.S. Sells PBB-Laced Herd for Beef" by Jane Haradine. News report in the *Grand Rapids Press.* November 19, 1975.

Transcript of Michigan House of Representatives Special Committee to Study PBB hearings, February and March 1976.

Chapter XXIII

"Report on the Former Richard Eddington [sic] Cattle Herd" by Michigan Department of Agriculture. April 15, 1976.

Michigan Department of Agriculture press release (about the Edington herd). April 23, 1976.

Chapter XXIV

Testimony of Dr. Isadore Bernstein, professor of Biological Chemistry, University of Michigan, before the Sub-Committee on Science, Technology and Space of the U. S. Senate Committee on Commerce. March 28, 1977.

Testimony of Dr. Walter D. Meester, director and clinical toxicologist, Western Michigan Poison Center and Department of Clinical Toxicology, Blodgett Memorial Medical Center, Grand Rapids before the above committee. March 29, 1977.

PBB Scientific Advisory Panel Report to William G. Milliken. governor, State of Michigan, on Polybrominated Biphenyls. May 24, 1976.

Press release from the Executive Office, Lansing, reporting the governor's response to taped questions submitted by citizens in various parts of Michigan. Dated April 29, 1976, released May 2, 1976.

"Panel Urges Near-Ban on PBB" by David Johnston. News report in the *Detroit Free Press.* May 25, 1976.

Transcript of public hearing by Michigan Department of Agriculture concerning a tolerance level for PBB. Lansing. June 10, 1976.

Transcript of Governor William Milliken's news conference on PBB. June 22, 1976.

Memorandum from Elton R. Smith, president of Michigan Farm Bureau, addressed to County Farm Bureau presidents, County Women's chairmen, County Young Farm chairmen, Michigan Farm Bureau directors, State Women's Committee, State Young Farmer Committee. June 14, 1976.

Chapter XXV

Michigan Department of Public Health press release on levels of PBB in the milk of Michigan nursing mothers. October 15, 1976.

Statements by Dr. Albert C. Kolbye Jr., Alan L. Hoeting, and Raymond K. Hedblad of the Food and Drug Administration before the Sub-Committee on Science, Technology and Space of the U. S. Senate Committee on Commerce. March 29, 1977.

Memorandum from Raymond K. Hedblad to Alan L. Hoeting. November 9, 1976.

Transcript of joint news conference with Governor William Milliken, House Speaker Bobby Crim, Dr. Maurice Reizen and Dr. Irving Selikoff. Lansing. January 4, 1977.

Reports on the progress of Michigan House Bill 4109 ("the Spaniola Bill") and House Bill 4115 ("the Porter Bill") from *Michigan Farmer* of March 5, March 19, April 2 and May 21, 1977.

"Finally, PBB Action." Editorial in the *Grand Rapids Press.* February 17, 1977.

"Firm PBB Action Can Wait No Longer." Joe H. Stroud, Editorial comment in the *Detroit Free Press.* February 27, 1977.

Client Matters of Williams, Connolly and Califano relating to the Department of Health, Education and Welfare—written statement issued on behalf of Joseph A. Califano to the Sub-Committee on Science, Technology and Space of the U. S. Senate Committee on Commerce, March 29, 1977.

Letter from Dr. Peter Greenwald, director of the Cancer Control Bureau, New York State Department of Health, to Michigan State Senator Richard J. Allen. May 17, 1977.

Memorandum from Elton R. Smith, president of Michigan Farm Bureau, to County Farm Bureau Presidents, County Farm Bureau board members and selected Farm Bureau leaders. March 9, 1977.

Memorandum of meeting between Concerned Dairy Farmers of Michigan and Alan L. Hoeting, district director (Detroit) of the Food and Drug Administration, signed by Alan L. Hoeting. February 23, 1977.

Statement of Larry K. Crandall before the Sub-Committee on Science, Technology and Space of the U. S. Senate Committee on Commerce. March 31, 1977.

Chapter XXVI

Transcripts of testimony in the trial of Roy M. Tacoma and Marilyn K. Tacoma *vs.* Michigan Chemical Corporation and Michigan Farm Bureau Services et al., Circuit Court for the County of Wexford, Michigan. February 1976–May 1977.

Opinion of the Hon. William R. Peterson, Circuit Court Judge, in the above case, dated October 26, 1978.

Chapter XXVII

"Symptoms and Clinical Abnormalities Observed Following Ingestion of Polybrominated Biphenyl Contaminated Food Products." H. A. Anderson, M. S. Wolff, R. Lilas, E. C. Holstein, J. A. Valciukas, M. Petrocci, and I. J. Selikoff (all of the Environmental Sciences Laboratory, Mount Sinai School of Medicine of the City University of New York); K. Anderson of the Rockefeller University, New York City; and L. Sarkozi of the Department of Pathology, Mount Sinai School of Medicine, New York City. Presented at the International Conference on Health Effects of Halogenated Aromatic Carbons, the New York Academy of Sciences. June 25, 1978.

Health Effects of Polybrominated Biphenyl Exposure in the General Population of Michigan: interim report, February 1, 1978–September 15, 1978, submitted to the Michigan Department of Public Health by Irving J. Selikoff and Henry M. Anderson of the Environmental Sciences Laboratory, and Harry Smith and Ford Calhoun, Department of Biostatics, Mount Sinai School of Medicine, New York City.

Testimony of Irving J. Selikoff, MD, Mount Sinai School of Medicine, New York City, and George Bekesi MD, professor, Department of Neoplastic Diseases, Mount Sinai School of Medicine, before the Sub-Committee on Oversight and Investigations of the U. S. House of Representatives' Committee on Interstate and Foreign Commerce. August 2, 1977.

Testimony of Thomas Ostrander before Special Committee of Michigan State Legislature to Study Effects of Polybrominated Biphenyls. St. Louis, Michigan. January 18, 1978.

Opinion of the Hon. William R. Peterson, Circuit Court Judge in the suit of Roy M. Tacoma and Marilyn K. Tacoma *vs.* Michigan Chemical Corp. and Michigan Farm Bureau Services et al, Circuit Court for the County of Wexford, Michigan. October 26, 1978.

"Hope, or No Hope?" by Dick Lehnert. *Michigan Farmer.* October 20, 1979.

Employer's Basic Report for Injury for Thomas Ostrander on form supplied by Bureau of Workers' Disability Compensation, Lansing, Michigan and submitted by Velsicol Chemical Corp. for injury sustained by August 26, 1977.

Epilogue

"Toxic Effects of Dietary Polybrominated Biphenyls on Mink." Richard J. Aulerich and Robert K. Ringler. *Archives of Environmental Contamination and Toxicology,* 1979, pp. 487–98.

"Small Chemical Firm Has Massive Problems With Toxic Products" by Frederick C. Klein, *Wall Street Journal.* February 13, 1978.

Testimony of Dr. Howard A. Tanner, director of Michigan Department of Natural Resources, before the Sub-Committee on Science, Technology and Space of the U. S. Senate Committee on Commerce. March 30, 1977.

Letter from Dr. Howard A. Tanner to Frank J. Kelly, attorney general, State of Michigan, describing disposition of PBB wastes in various Michigan landfills. Sept. 26, 1978.

Testimony of Dr. I. Eugene Wallen, deputy director, Office of Toxic Substances, Environmental Protection Agency, before the Sub-Committee on Oversight and Investigations of the U. S. House of Representatives' Committee on Interstate and Foreign Commerce. August 3, 1977.

Food Contamination Problems: Research at Michigan State University, supported by a grant from Michigan Department of Agriculture. Principal investigator, Steven D. Aust, professor, Department of Biochemistry. Final report 1977–78.

Medical report (with appendices), health hazard evaluation of workers at Velsicol (formerly Michigan Chemical), St. Louis, Michigan. Published by the National Institute of Occupational Safety and Health. August 7, 1979.

Environmental Protection Agency press release: "EPA Reports Finding of PBB Residues in Northern New Jersey, Staten Island." June 17, 1977.

Legislative History of the Toxic Substances Control Act, printed for use of the U. S. House Committee on Interstate and Foreign Commerce (U. S. Government Printing Office). December 1976.

Environmental Contaminants in Food—report by the Office of Technology Assessment, Congress of the United States. December 1979.

INDEX